Cooperation and Conflict
in General Evolutionary Processes

Cooperation and Conflict in General Evolutionary Processes

Edited by
JOHN L. CASTI
ANDERS KARLQVIST

A Wiley-Interscience Publication
JOHN WILEY & SONS, INC.
New York / Chichester / Brisbane / Toronto / Singapore

Library of Congress Cataloging in Publication Data:
Cooperation and conflict in general evolutionary processes / edited by
  John L. Casti and Anders Karlqvist.
     p.   cm.
  "A Wiley-Interscience publication."
  Includes bibliographical references and index.
  ISBN 0-471-59487-3 (cloth)
    1. Evolution (Biology)—Mathematical models. 2. Evolution—
  Mathematical models. 3. Cooperation—Mathematical models.
  I. Casti, J. L. II. Karlqvist. Anders.
  QH366.2.C655   1994
  575—dc20   94-14384

Contents

Chapter 12 LANGUAGE, EVOLUTION, AND THE THEORY OF GAMES **405**

Karl Wärneyrd

Contributors

W. Brian Arthur—Food Research Institute, Stanford University, Stanford, CA 94305, USA

Gerold Baier—Physics Laboratory III, The Technical University of Denmark, DK–2800 Lyngby, Denmark

Clas Blomberg—Department of Theoretical Physics, Royal Institute of Technology, S–100 44 Stockholm, Sweden

John Casti—Santa Fe Institute, 1660 Old Pecos Trail, Santa Fe, NM 87501, USA

Gregory J. Chaitin—IBM Research Division, Box 704, Yorktown Heights, NY 10598, USA

Michael Cronhjort—Department of Theoretical Physics, Royal Institute of Technology, S–100 44 Stockholm, Sweden

N. Katherine Hayles—Department of English, University of California, Los Angeles, CA 90024, USA

Anastassis A. Katsikas—Department of Mathematics, University of Patras, Patras, Greece

Philip Lieberman—Department of Cognitive and Linguistic Sciences, Brown University, Providence, RI 02912, USA

Erik Mosekilde—Physics Laboratory III, The Technical University of Denmark, DK–2800 Lyngby, Denmark

John S. Nicolis—Department of Electrical Engineering, University of Patras, Patras, Greece

Robert Rosen—Department of Physiology and Biophysics, Dalhousie University, Halifax, Nova Scotia B3H 4H7, Canada

Peter Schuster—Institute for Molecular Biology, Beutenbergstraße 11, D–07708 Jena, Germany

Erik Skarman—Saab Missiles, AB, S–581 88 Linköping, Sweden

Heidi Stranddorf—Physics Laboratory III, The Technical University of Denmark, DK–2800, Lyngby, Denmark

Jesper Skovhus Thomsen—Physics Laboratory III, The Technical University of Denmark, DK–2800, Lyngby, Denmark

Karl Wärneryd—Stockholm School of Economics, Box 6501, S–113 83 Stockholm, Sweden

Introduction

Beginning in 1983, the Swedish Council for Planning and Coordination of Research has organized an annual workshop devoted to some aspect of the behavior and modeling of complex systems. These workshops are held in the scientific research station of the Royal Swedish Academy of Sciences in Abisko, a rather remote location far above the Arctic Circle in northern Sweden. In May 1992, during the period of the midnight sun, this exotic venue served as the gathering place for a small group of researchers from across a wide disciplinary spectrum to examine the twin problems of cooperation and conflict in evolutionary processes.

In light of the extremely stimulating presentations and discussions at the meeting itself, each participant was asked to prepare a formal written version of his view of the meeting's theme. The book you now hold contains those views, and can thus be seen as the distilled essence of the meeting itself. Regrettably, one of the meeting participants, John Maynard Smith, was unable to prepare a written contribution of his very provocative views, due to the pressure of other commitments. Nevertheless, we feel that the twelve chapters presented here, spanning as they do fields as diverse as philosophy, physics, biology, and economics, give an excellent overview of how successful evolutionary adaptations rely on a judicious combination of self-interest and altruism.

Knee-jerk knowledge about evolution suggests that it's a jungle out there, and that the role of evolution is indeed to enforce the "survival of the fittest." But knee-jerk knowledge and folk wisdom are wrong. The more thoughtful among us discovered long ago that the essence of evolution is a finely balanced highwire act between individual self-interest and collective group interests. These complementary effects can be seen as the point and counterpoint between conflict and cooperation that makes up the dynamics of evolution.

A good recent illustration of this interplay between cooperation and conflict arises in the determination of optimal playing strategies for the simple Prisoner's Dilemma game, in which the two players can act either selfishly by "selling out" their playing partner or altruistically by trusting their opponent. Numerous experimental and computer studies have shown clearly that the best way to play is to judiciously mix cooperative and selfish decisions.

The contributions to this volume provide ample evidence of the importance of cooperation in the working out of nature's formula, which we can write compactly as adaptation = heredity + variation + selection.

Ranging from one side of the intellectual landscape to the other, the papers presented here make a compelling case for the crucial role that cooperation plays not only in biology, but also in economics, urban structure, linguistics, the chemistry of molecular formation in the primeval environment, microbiology, and computer science.

It is a pleasure for us to acknowledge the generous support, both intellectual and financial, from the Swedish Council for Planning and Coordination of Research (FRN), and its director, Dr. Hans Landberg, for their willingness to entertain even the idea of such a multidisciplinary gathering, let alone provide the facilities and resources to make it happen. Special thanks are also due to Mats-Olof Olsson of the Center for Regional Science Research (CERUM) at the University of Umeå for his skills in attending to the myriad administrative and organizational details that such meetings necessarily generate.

John Casti, Santa Fe
Anders Karlqvist, Stockholm

CHAPTER 1

Urban Systems and Evolution*

W. Brian Arthur

Food Research Institute
Stanford University
Stanford, CA 94305, USA

1. Introduction

Is the pattern of cities we have inherited inevitable and foreordained, or has it evolved in a way that reflects the fortunes and vagaries of history? If small events in history had been different, would there be a different formation of urban centers than the one that exists today?

The German Industry Location School debated this question in the earlier part of this century, but it was never settled conclusively. Von Thünen (1826), the early Weber (1909), Predöhl (1925), Christaller (1933), and Lösch (1941) all tended to see the spatial ordering of cities (viewed as clusters of industry) as preordained—by geographical endowments, shipment possibilities, firms' needs, and the spatial distribution of rents and prices that these induced. In their view, history did not matter: the observed spatial pattern of industry was a unique "solution" to a well-defined spatial economic problem. Therefore, early events in the configuration of an industry could not affect the result. But others, the later Weber, Engländer (1926), Ritschl (1927), and Palander (1935), tended to see industry location as process-dependent,

*This paper appeared in slightly different form as "Urban Systems and Historical Path-Dependence," in *Cities and Their Vital Systems*, R. Herman and J. Ausubel (eds.), National Academy of Engineering, Washington, D.C., 1988.

Cooperation and Conflict in General Evolutionary Processes, edited by John L. Casti and Anders Karlqvist.

ISBN 0-471-59487-3 ©1994 John Wiley & Sons, Inc.

almost geologically stratified, with new industry laid down layer by layer upon inherited, previous locational formations. Again geographical differences and transport possibilities were important, but here the main driving forces were agglomeration economies—the benefits of being close to other firms or to concentrations of industry. In the simplest formulation of this viewpoint (Maruyama, 1963), an industry starts off on a uniform, featureless plain; early firms put down by "historical accident" in one or two locations; others are attracted by their presence, and others in turn by their presence. The industry ends up clustered in the early-chosen places. But this spatial ordering is not unique: a different set of early events could have steered the locational pattern into quite a different outcome, so that settlement history would be crucial.

These two viewpoints—determinism versus history-dependence, or "necessity" versus "chance"—are echoed in current discussions of how modern industrial clusters have come about. The determinism school, for example, would tend to see the electronics industry in the United States as spread over the country, with a substantial part of it in Santa Clara County in California (Silicon Valley) because that location is close to Pacific sources of supplies, and because it has better access to airports, to skilled labor, and to advances in academic engineering research than elsewhere. Any "small events" are overridden by the "necessity" inherent in the equilibration of spatial economic forces, and Silicon Valley is part of an inevitable result. Historical dependence on the other hand, would see Silicon Valley and similar concentrations as largely the outcome of "chance." Certain key persons—the Packards, the Varians, the Shockleys of the industry—happened to set up near Stanford University in the 1940s and 1950s, and the local labor expertise and interfirm markets they helped to create in Santa Clara County made subsequent location there extremely advantageous for the thousand or so firms that followed them. If these early entrepreneurs had other predilections, Silicon Valley might well have been somewhere else. In this argument "historical chance" is magnified and preserved in the locational structure that results.

While the historical dependence-agglomeration argument is appealing, it has remained problematical. If history can indeed steer the spatial system down different paths, there are multiple "solutions" to the industry location problem. *Which* of these comes about is indeterminate. In the 1920s, analysts could not cope with this difficulty, and the argument did not gain enough rigor to become completely respectable.

In this chapter I investigate the importance of "chance" (as represented by small events in history) and "necessity" (as represented by determinate economic forces) in determining the pattern of industry location. I contrast three highly stylized locational models in which small events and economic forces are both present and allowed to interact. In each model an industry is allowed to form, firm by firm, and build up into a locational pattern. In each model I examine whether historical chance can indeed alter the location pattern that emerges. I use the insights gained from the three models to derive some general conditions under which long-run locational patterns may be affected by small historical events. And I discuss the role of cooperation and competition in the formation of spatial patterns.

2. The Evolution of Locational Patterns: Three Models

Model 1: Pure Necessity: Locational Under Independent Preferences

Let us begin with a very simple model indeed of the emergence of an industry location pattern. Starting from zero firms, we allow an industry to form firm by firm, with each new firm that enters deciding "at birth" which of N possible regions (or sites) it will locate in. Once located, each firm will stay put. Firms in this industry are not all alike; there are of I different types. The net present value or payoff to a firm of type i for locating in region j is π_j^i, and each firm choosing selects the location with the highest return for its type. In this model firms are independent: the presence or absence of other firms does not affect what they can earn in each region.

We now inject a small element of "chance" by assuming that the particular historical circumstances that lead to the next firm's being of a particular type are unknown. We do know, however, that a firm of type i will occur next with probability p_i. The question is: What pattern of industrial settlement will emerge in this model, and can it be affected by a different sequence of historical events in the formation of the industry?

It is not difficult to work out the probability that at any time of choice region j will be chosen. This is simply the probability q_j that the newest firm is of a type which has its highest payoff in region j, which is given by $q_j = \sum_{k \in K} p_k$, where K is the set of firm types that prefer j. Repeating this calculation for each of the N regions, we have a set of probabilities of choice $q = (q_1, q_2, \ldots, q_N)$ that are constant no matter what the current pattern of location is. Starting from zero firms in any region, concentrations of the industry in the various regions will fluctuate, considerably at first. But the strong law of large numbers tells us that as the industry grows, the proportions of it in the N regions must settle down to the expectation of an addition being made to each region. That is, regions' shares of the industry must converge to the constant vector q. In this simple model, then, even though well-defined "chance historical events" are present, a unique, predetermined locational pattern emerges and persists.

Figure 1 shows a simple three-region simulation of this process, with three possible firm types that prefer (clockwise from the top) region 1, region 2, and region 3, respectively, with probabilities of occurrence 0.5, 0.25, and 0.25. After 16 firms have located, the regions' shares of the industry are 0.75, 0.125, 0.125, respectively—not yet close to the long-run predicted pattern. After 197 firms have located, however, the shares are 0.528, 0.221, 0.251—much closer to the predetermined theoretical long-run shares.

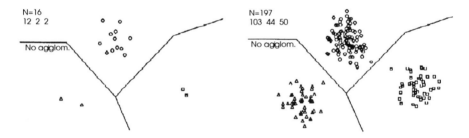

Figure 1. A three-region example of the independent preferences location model.

In this model, chance events, represented as randomness in the sequence of firm types that enter the industry, are important early on. But they are progressively averaged away to become dominated by the economic forces represented by firms' payoffs in each region. Different sequences of firm types caused by different historical events would, with probability 1, steer the system into the same locational pattern. Here historical chance cannot affect the outcome. Necessity dominates.

Model 2: Pure Chance: Location by Spin-off

We now assume a quite different mechanism driving the regional formation of an industry—one in which chance events become all-important.

Once again the industry builds up firm by firm, starting with some set of initial firms, one per region, say. This time new firms are added by "spinning off" from parent firms one at a time. [David Cohen (1984) has shown that such spin-offs have been the dominant "birth mechanism" in the U.S. electronics industry.] We assume that each new firm stays in its parent location, and that any existing firm is as likely to spin off a new firm as any other. With this mechanism we have a different source of "chance historical events": the sequence in which firms spin off daughter firms.

It is easy to see that in this case firms are added incrementally to regions with probabilities exactly equal to the proportions of firms in each region at that time. This random process, in which unit increments are added one at a time to one of N categories with probabilities equal to current proportions in each category, is known in probability theory as a *Polya process.* We can use this fact to examine the long-term locational patterns that might emerge. From Polya theory we know that once again the industry will settle into a locational pattern (with probability 1) that has unchanging proportions of the industry in each region. But although this vector of proportions settles down and becomes constant, surprisingly it settles to a constant vector that is *selected randomly* from a uniform distribution over all possible shares that sum to 1. This means that each time this spin-off locational process is "rerun" under different historical events (in this case a different sequence of firms spinning-off) it will in all likelihood settle into a different pattern. We could generate a representative outcome by placing $N-1$ points on the unit interval at random, and cutting at these points to obtain N "shares" of the unit interval.

Figure 2 shows four realizations of this location-by-spin-off mechanism starting from the same three original firms in a three-region case. Each of the four "reruns" has settled into a pattern that will change but little in regional shares with the addition of further firms. But each pattern is different from the others. In this model industry, location is highly path-dependent. Although we *can* predict that the locational pattern of industry will indeed settle down to constant proportions, we cannot predict what proportions it will settle into. Any given outcome—any vector of proportions that sum to 1—is as likely as any other. "History" in the shape of the early random sequence of spin-offs becomes the sole determining factor of the regional pattern of industry. In this model "chance" dominates completely.

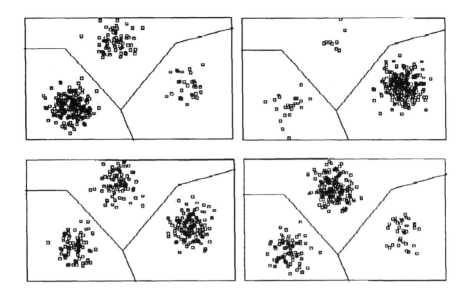

Figure 2. Four realizations of location by spin-off.

Model 3: Chance and Necessity: Location Under Agglomeration Economies

Firms that are not tied to raw-material localities and that do not compete for local customers are often attracted by the presence of other firms in a region. More densely settled regions offer better infrastructure, deeper labor markets (David 1984), more specialized legal and financial services, better local availability of inventory and parts, and more opportunity to do business face-to-face. For our third model we go back to model 1 and extend it by supposing that new firms gain additional benefits from local agglomerations of firms.

Suppose now that the net present value or payoff to a firm of type i for locating in region j is $\pi_j^i + g(y_j)$, where the "geographical benefits" π_j^i are enhanced by additional "agglomeration benefits" $g(y_j)$ from the presence of y_j firms already located in that region. We can recalculate the probability that region j is chosen next, given that $y_1, \ldots, y_N$ firms are currently in regions 1 through N, once again as $q_j = \sum_{k \in K} p_k$, where K is now the set of firm types for which $\pi_j^i + g(y_j) > \pi_m^i + g(y_m)$ for all regions $m \neq j$. Notice that in this case the probability that region j is chosen is a function of the number of firms in each region at the time of choice.

Starting from zero firms in the regions, once again we can allow the industry to grow firm by firm, with the appearance of firm types subject to known probabilities as in model 1. Again the pattern of location of the industry will fluctuate somewhat, but in this model if a region by a combination of luck and geographical attractiveness gets ahead in numbers of firms, its position is enhanced. We can show (see Arthur, 1986, for proof) that if agglomeration benefits increase without ceiling as firms are added to a region (that is, if the function g is monotonically increasing without upper bound), then eventually (with probability 1) one of the regions will gain enough firms to offer sufficient locational advantages to shut the other regions out in all subsequent locational choices[1]. From then on, each entering firm in the industry will choose this region, and this region's share of the industry will tend to 100 percent, with the others' shares tending to 0 percent.

Figure 3 shows two realizations of a three-region example with agglomeration economies. The first three panels (read left to right, top to bottom) show the buildup of firms, with geographical preferences dominating in panel 1, but with region three in panel 2 by good fortune in the sequence of arrival of firm types just gaining enough firms to cause another firm type to favor it instead of its pure geographical preference. In panel 3 region three has come to dominate the entire industry in a Silicon Valley-like cluster. Panel 4 show the outcome of an alternative run. Here the industry is locked in to region two.

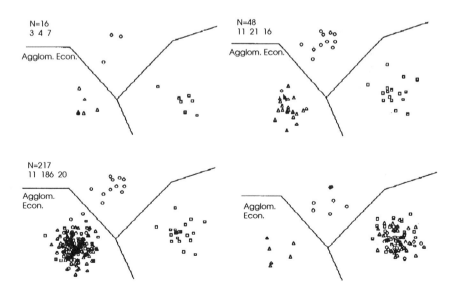

Figure 3. Two realizations of a locational process with agglomeration economies.

In this unbounded agglomeration economies model, monopoly of the industry by a single region must occur. But which region achieves this "Silicon Valley" locational monopoly is subject to historical luck in the sequence of firm types choosing. Chance, of course, is not the only factor here. Regions that are geographically attractive to many firm types—regions that offer high economic benefits—will have a higher probability of being selected early on. And this will make them more likely to become the single region that dominates the industry. To use an analogy borrowed from genetics, chance events act to "select" the pattern that becomes "fixed"; but regions that are economically attractive enjoy "selectional advantage," with correspondingly higher probabilities of gaining dominance. In this third model the long-run locational pattern is due both to chance and necessity.

3. Path Dependence and Convexity

Each of our three stylized industry location models includes both determinate economic forces and some source of chance events. But each behaves differently. Determinate forces, or historical chance, or a mixture of the two, are in turn responsible for the long-run pattern of industry settlement that emerges.

To explain these results, and to provide some precise conditions under which historical chance can be important, it is useful to introduce a general framework that encompasses all three models (as well as many others). In this general framework, suppose there are N regions, and that industry locates, one firm at a time, starting from a given number of firms in each region. Different economic forces, different sources of chance events, and different mechanisms of locational choice would be possible within this framework, but we do not need to know these. What we do need to know are the probabilities that region 1, region 2, ..., region N will be chosen next, as a *function* of current regional shares of the industry $x_1, x_2, \ldots, x_N$.

Plotting this function (as in Figure 4 for the two-region case), we might expect that where the probability of a region's receiving the next firm exceeds its current proportion of the industry, it would tend to increase in proportion; and where the probability is less than current proportion, it would tend to decrease in proportion. Moreover, as firms are added, each new addition changes proportions or shares by an ever smaller magnitude. Therefore, proportions should settle down, and fluctuations in proportions should die away. In the long run, then, we might expect that regions' proportions (the industry's location pattern) ought to converge to a point—to a vector of locational shares—where

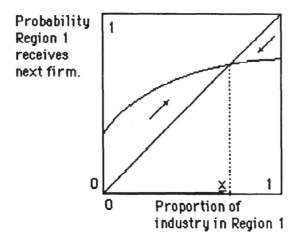

Figure 4. Proportion-to-probability mapping (arrows indicate expected motions).

proportions equal probabilities, and to one that expected motions lead toward (point x in Figure 4). That is, it ought to end up at a stable fixed point of this proportions-to-probabilities function. It takes powerful theoretical machinery to prove this conjecture, but it turns out to hold under unrestrictive technical conditions (see Hill, Lane, and Sudderth, 1980, and Arthur, Ermoliev, and Kaniovski 1983, 1986a, 1986b).[2] Further, and significantly for us, where there are multiple stable fixed points, each of these would be candidates for the long-run locational pattern, with different sequences of chance events steering the process toward one of the multiple candidates.

We can now see what happened in our three locational models (Figure 5). The first model, "independent preferences," has constant probabilities of choice, thus a single fixed point. Therefore, it has a unique, predetermined outcome. The second model, "spin-off," with probabilities equal to proportions, has *every point* a fixed point, so "chance" can drive this locational process to any outcome. The third model, "agglomeration economies," has 0 and 1 as candidate stable fixed points. Thus the outcome is not fully predetermined and one of the candidate solutions is "chosen" by accumulation of chance events.

When does history count in the determination of industry location patterns? We can now answer this question, at least for the broad class of models that fit our general framework. "History"—the small elements outside our economic model, that we must treat as random—

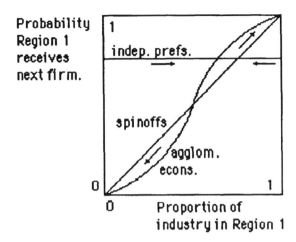

Figure 5. Probability mappings for the three models.

becomes the determining factor when there are multiple solutions or multiple fixed points in the proportions-to-probabilities mapping. More intuitively, history counts when expected motions of regions' shares do not always lead the locational process toward the same share. It is useful to associate with each probability function a *potential function V* whose downhill gradient equals the expected motion of regions' shares[3]. (See Figure 6.) Intuitively, we can think of the process as behaving like a particle attracted by gravity to the lowest points on the potential, subject to random fluctuations that die away. If this potential function is convex (looking upward at it), it has a unique minimum, and therefore the locational process that corresponds to it has a unique determinate outcome which expected motions lead toward and which historical chance cannot influence. If, on the other hand, it is nonconvex, it must have two or more minima, with a corresponding split in expected motions, and with "historical chance" determining which of these is ultimately selected.

To establish nonconvexity, all we need is the existence of at least one unstable point, a "watershed" share of the industry, above which the region with this share exerts enough attraction to increase its share, below which it tends to lose its share. But in a way, this is another definition of the presence of agglomeration economies: If above a certain density of settlement, a region tends to attract further density, and if below it tends to lose density, there must be some agglomeration mechanism present. The underlying system will then be nonconvex, and history will count.

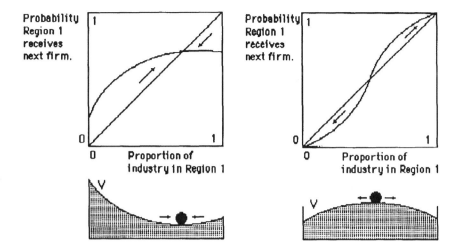

Figure 6. Convex and nonconvex potential functions.

4. Cooperation and Competition

The last model discussed above, of chance and necessity, has several properties reminiscent of ones that occur in evolutionary systems. This might not seem obvious at first. But where agglomeration forces are present, there is a multiplicity of possible long-run spatial patterns, and we can think of these as in *competition* with each other. In the long-run, of course, only one of the potential patterns can survive; so that something like a *Gause exclusion principle*—that in the long run only one species can occupy one niche—operates. Some locational patterns may be more likely than others—these will have *selection* advantage. And a pattern that manages to establish itself early on can become locked in; so that the equivalent of a *founder effect* may be in operation.

If we pressed the evolutionary metaphor, we could also apply it to the first model. But here the outcome that emerges is preordained and inevitable[4]. There is little in the way of competition, because one asymptotic pattern enjoys complete selectional advantage. In fact, it is the existence of cooperative effects—the presence of agglomeration economies—that gives rise to a multiplicity of potential patterns, and hence to entities that we might think of as being in evolutionary competition. We see this repeatedly in systems with cooperative effects or positive feedbacks. These amplify asymmetries in systems they occur in, thereby inducing multiple possible outcomes. The result of such cooperative effects is the creation of different "solutions" or long-run patterns in competition with one another—and these in turn induce properties that are strongly evolutionary.

5. Concluding Remarks

We can conclude from this essay that whether small events in history matter in determining the pattern of spatial or regional settlement in the economy reduces, strangely enough, to a question of topology. It reduces to whether the underlying structure of locational forces guiding the locational pattern as it forms is convex or nonconvex. And for this structure to be nonconvex, so that history will matter, there must be some mechanism of agglomeration present.

Our models were highly stylized. They considered populations of firms, not people; they assumed that firms lived forever and never moved; and they dealt with the formation of one industry only over time, not several. Nevertheless, even if the mechanisms creating urban systems in the past and present are a great deal more complex, it is still likely that a *mixture* of economic determinism and historical chance—not either alone—has formed the spatial patterns we observe. Certain firms, like steel manufacturers, do need to be near sources of raw materials, and for them spatial economic necessity dominates historical chance. Certain firms, such as gasoline distributors, do need to be separated from their competitors in the same industry, and for them the necessity to spread apart again dominates historical chance. But most firms need to be near other firms, if not in their own industry then in other industries that act as their suppliers of parts, machinery, and services, or as consumers of their products and services. And for this reason they are attracted to existing and growing agglomerations. After all, it is this need of firms to be near other firms that causes cities—agglomerative clusters—to exist at all.

Thus it is highly likely that the system of cities we have inherited is only partly the result of industries' geographical needs, raw material locations, the presence of natural harbors, and transportation costs. It is also the result of where immigrants with certain skills landed, where early settlers met to market goods, where wagon trains stopped for the night, where banking services happened to set up, and where politics dictated that canals and roads and railroads be built.

We cannot explain the observed pattern of cities by economic determinism alone, without reference to chance events, coincidences, and circumstances in the past. And without knowledge of chance events, coincidences, and circumstances yet to come, we cannot predict with accuracy the shape of urban systems in the future.

Notes

[1] In the case where g is bounded, several locations can share the industry in the long run. But again typically there are multiple possible outcomes, so that chance events matter here too (see Arthur 1986, 1985, 1990).

[2] The set of fixed points needs to have a finite number of components. Where the proportions-to-probabilities function itself changes with the number of firms located, as in the agglomeration case, the theorem applies to the limiting function of these changing functions, provided that it exists. (See Arthur, Ermoliev, and Kaniovski, 1986a, 1986b.)

[3] For dimension $N > 2$, a potential function may not exist. This would be the case if there were cycles or more exotic attractors than the single-point ones that are considered here.

[4] The second model suffers from the opposite extreme: an infinity of potential outcomes. We can also think of these in competition, but again due to the fact that *any* outcome is equally possible here, the evolutionary metaphor is less useful in this case.

References

Arthur, W. B. (1985), "Competing Technologies and Lock-In by Historical Small Events: The Dynamics of Choice Under Increasing Returns," C.E.P.R. Paper 43. Stanford.

Arthur, W. B. (1986), "Industry Location Patterns and the Importance of History," C.E.P.R. Paper 84. Stanford.

Arthur, W. B. (1990), "'Silicon Valley' Locational Clusters: When Do Increasing Returns Imply Monopoly?" *Mathematical Social Sciences*, 19, 235–251.

Arthur, W. B., Ermoliev, Yu. M. and Kaniovski, Yu. M. (1983), "A Generalized Urn Problem and Its Applications," *Cybernetics*, 19, 61–71.

Arthur, W.B., Ermoliev, Yu. M., and Kaniovski, Yu. M. (1986a), "Strong Laws for a Class of Path-Dependent Urn Processes," in *Proc. International Conf. on Stochastic Optimization*, Kiev, 1984, A. Arkin, A. Shiryayev, and R. Wets (eds.), Springer: Lecture Notes in Control and Info. Sciences, vol. 81.

Arthur, W. B., Ermoliev, Yu. M., and Kaniovski, Yu. M. (1986b), "Path-Dependent Processes and the Emergence of Macro-Structure", forthcoming, *European Journal of Operations Research*.

Christaller, W. (1933), *Central Places in Southern Germany.* Prentice Hall, 1966.

Cohen, D. L. (1984), "Locational Patterns in the Electronics Industry: A Survey," Mimeo. Stanford.

David, P. (1984), "High Technology Centers and the Economics of Locational Tournaments", Mimeo. Stanford.

Engländer, O. (1926), "Kritisches and Positives zu einer allgemeinen reinen Lehre vom Standort," *Zeitschrift für Volkswirtschaft und Sozialpolitik,* Neue Folge, 5.

Hill, B., Lane, D., and Sudderth W. (1980), "Strong Convergence for a Class of Urn Schemes," *Annals of Probability,* 8, 214-226.

Lösch, A. (1941), *The Economics of Location.* Yale University Press, 1954.

Maruyama, M. (1963), "The Second Cybernetics: Deviation Amplifying Mutual Causal Processes," *American Scientist,* 51, 164-179.

Palander, T. (1935), *Beiträge zur Standortstheorie.* Almqvist and Wicksell.

Predöhl, A. (1925), "Das Standortsproblem in der Wirtschaftslehre," *Weltwirtschaftliches Archiv,* 21, 294–331.

Ritschl, H. (1927), "Reine und historische Dynamik des Standortes der Erzeugungszweige," *Schmollers Jahrbuch,* 51, 813–870.

Thünen, J. H. von, (1826), *Der Isolierte Staat in Beziehung auf Landwirtschaft und Nationalkonomie.* Hamburg.

Weber, A. (1909), *Theory of the Location of Industries.* Univ. of Chicago Press, 1929.

CHAPTER 2

Modeling Errors and Parasites in the Evolution of Primitive Life: Possibilities of Spatial Self-Structuring

CLAS BLOMBERG AND MIKAEL CRONHJORT

Department of Theoretical Physics
Royal Institute of Technology
S–100 44 Stockholm, Sweden

1. Introduction

The mathematical modeling of the evolution of primitive life is an important and relevant task, as there exist a manifold of ideas about possible scenarios, but what can be learned from experiments is quite limited. The aim of the models is not to give a realistic account of the evolutionary pathways, but rather to point out qualitative features and to try to answer questions such as what kind of scenarios are possible, how could they have evolved, and what are the conditions for stability? Questions about stability in this context are crucial—but often neglected. It is important to know more about how stability could have arisen, how instability effects could be avoided, and what kind of restrictions the possibilities for instability impose on the evolutionary pathways. A crucial stability issue is the accuracy of molecule reproduction, and it has been pointed out repeatedly that the level of accuracy puts severe restrictions on how complexity could have evolved.

In this chapter we will do two things: First, survey accuracy effects at various stages in the evolution of primitive life, emphasizing the destructive effects of errors. For this, we use models starting

Cooperation and Conflict in General Evolutionary Processes, edited by John L. Casti and Anders Karlqvist.
ISBN 0–471–59487–3 ©1994 John Wiley & Sons, Inc.

from kinetic, nonlinear ordinary differential equations describing self-reproducing systems. And second, we will include diffusion effects and other possibilities for obtaining spatial distributions. Such spatial patterns are suggested as a possible way of coping with destructive error effects, in particular, parasites.

In view of the interdisciplinary goal of this volume, we want to point out that the "error propagation effects" that we discuss may be important not only in these biological contexts, but also for other evolving dynamic systems. They may well occur in many complex systems in which significant cooperative effects enter. However, we will not develop that point further here.

In Chapter 10 in this volume, Peter Schuster describes some aspects of an RNA world in an early stage of the origin of life. This is a scenario built up by RNA molecules, which serve perfectly as carriers of genetic information. But these molecules can also catalyze and to some extent control the reproduction processes. Thus, the same macromolecules can both act as catalysts ("ribozymes") and carriers of genetic information. The models of Eigen and Schuster and their co-workers (Eigen, 1971; Eigen and Schuster, 1978; Hofbauer and Sigmund, 1988) are an important starting-point for any modeling efforts in this context. These models also emphasize the significance of accuracy, and study restrictions that accuracy of reproduction generate. However, past discussions of error effects have been confusing, as they occur in quite different ways at various stages of the early evolution of life. Concepts like the "error catastrophe" have been used for various situations without any cross-reference. A primary purpose for us here is to give a unified picture of the error effects, (see Blomberg, 1993).

In the papers by Eigen, Schuster, and their co-workers (see Eigen and Schuster, 1978), an error catastrophe occurs as a breakdown of accuracy in a noncooperative system. For the most part, this puts a restriction on the polymer length and on the complexity that can be generated. On the other hand, the very same term is found in a different context, one originally discussed by Orgel (1963, 1970) and studied in models by Hoffman and Kirkwood and co-workers (Hoffman, 1974; Kirkwood and Holliday, 1975). In that case, the term refers to the possibility that errors in a cooperative control system can be propagated by a feedback effect, destroying a seemingly stable situation. (This is considered in more detail below.) The occurrence of parasites in cooperative systems has similar effects and can be quite destructive (Maynard Smith, 1979; Bresch, Niesert, and Harnasch, 1980; Niesert,

Harnasch, and Bresch, 1981). It is important to study such effects more fully, and to try to understand their implications for primitive life. It has been proposed that they can be managed in a spatially varying situation, and a spiral mechanism has recently been suggested as a way of saving the system from parasites (Boerlijst and Hogeweg, 1991). It is also clear that these error effects are less destructive in a cellular organization. However, that requires a complex control machinery, and it is difficult to see how it could have occurred at an early stage in the development of life (e.g., see Eigen, Gardiner, and Schuster, 1980). Other suggestions of cell-like structures include coacervate droplets (Feistel, Romanovskii and Vasil'ev, 1980).

2. Stages of the Origin of Life

Of course, details about the first steps toward life are not known. Still, certain stages should have occurred, although we do not know the order in which the important features appeared. In this section we give a general account of a possible scenario, emphasizing the bad effects of errors in replication.

A crucial property on the path to life is the ability of an object to make a copy of itself, i.e., *self-reproduction*. This is an essential aspect of all scenarios of the origin of life. Certainly, the nucleic acids of present-day cells with their ability to bind bases in a complementary way are ideally suited for self-reproduction (see Chapter 10 in this volume). We will not go in any details in this chapter about the molecular biology, but rather, accept the possibility that molecules like nucleic acids can build copies of themselves. At some primitive stage, molecules should have occurred having this ability. We assume that these were macromolecules (polymers), which as with present-day molecules, were built up from monomer sequences, where the monomers are selected to make a proper copy. Perhaps there was some external catalyst (such as a metal compound) for these polymerization processes. These details are not needed for our development here. In our present discussion, we will not take into account the fact that a molecule such as RNA first makes a "negative copy" with complementary bases, which after a further replication makes a strict copy of the original molecule. In our models, we merely assume that a molecule can be replicated in a single step. This simplifies the models, yet contains all the relevant features. The basic model we employ is illustrated in Figure 1.

It is also important that there be an energy source that could have driven the polymerization. For this, it is reasonable to assume that the monomers constituting the copy were in an "activated form." The

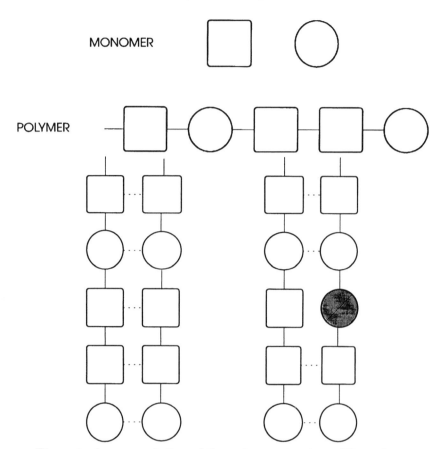

Figure 1. A representation of the polymer structure. The polymers are assumed to be composed of two different types of monomers, which provide a certain polymer sequence in a chain-like structure. The polymers may be further stabilized by forming bonds between the building blocks. The gene polymers are reproduced by a template action: monomers are coupled to those of the same kind along the polymer chain, and are then coupled together to form a new polymer. Sometimes this selection goes wrong, and an erroneous monomer is incorporated in the sequence as in the figure on the right.

activation should make use of some external energy source, such as sunlight and/or catalyzed chemical reactions.

There were probably various polymers that used the same monomers. These could have been produced by errors or "mutations" in the reproduction of the existing polymers, or they could have occurred spontaneously, as must have been the case with the first polymer. There was no cooperation between the polymers at this stage. They reproduced essentially independently of each other, but competed for the same monomer molecules. What all models of this situation show is that there could not be any appreciable coexistence of different polymers. The competition for monomer resources always leads to a situation in which one polymer survives, that one being the polymer that could use the monomers most efficiently. There could have been some coexistence, for instance, if the polymers used the monomers in a different way or if some special mechanism were involved (see Blomberg, von Heijne and Leimar, 1981). For a simple, independent reproduction, a clear "fitness value" can be defined as the quotient between the growth and the decay rate of the polymer. There is a strict *irreversibility* at this stage, as polymers with larger values of this ratio always triumph over those with lower values. This also implies decreasing monomer concentrations (the polymer reproduction is making a more efficient use of the monomers).

There is always some coexistence as various polymers are produced by errors (mutations). Eigen (1971) introduced the concept of a "quasispecies" to represent a kind of established equilibrium, usually between a dominating species and mutations. For many aspects, the quasispecies occur equally as pure polymer species and obey similar kinetic equations. When we discuss the competition between polymer species, it will refer to competition between quasispecies. Quasispecies can be defined by the eigenvectors of a matrix whose entries are the probabilities that different species are produced by each other (as errors or mutations). We will consider such matrices in later sections.

Another point of interest is that the external conditions can change. The scenario expressed above is valid for a constant external influence. In particular, it holds when the external monomer production remains constant. But, for instance, a drastic decrease in the formation rate of monomers (the energy influx) or a change in the reproduction accuracy can lead to new stability requirements and may cause a successful polymer state to decay. However, unless there are such external changes, the irreversible evolution always leads to greater stability. At this stage, *a system can decay only by external changes, not by the appearance of polymers having destructive properties.*

What has been noted for this stage (Eigen, 1971; Swetina and Schuster, 1982) is that there is a severe restriction on the possible lengths of the polymers. If the length is too large, the probability of making exact copies, as well as the production rate of exact copies, goes down. Long polymers cannot occur because their production is too slow, mostly because too many erroneous polymers are made. Of course, this stage is limited in scope since there is no cooperation. So it's necessary to propose a different situation, in which complexity and larger systems can arise.

One possibility is a *cooperative system,* in which polymers act as information carriers (genes) **and** catalysts (enzymes). Polymers *within* the system have the ability to catalyze the reproduction of the system, with the information of the system also being carried by the polymers. The information carriers and the catalysts may be different kinds of polymers (nucleic acids and proteins), or they may be of the same type. The latter situation would be the case in the RNA world discussed in Chapter 10 in this volume. As already noted, RNA molecules are perfect information carriers, as by their structure they can be templates for direct reproduction. They can also have catalytic activity, and can catalyze the reproduction reactions. The simplest possibility is when there is only one type of RNA molecule that can catalyze the reproduction of itself. This double effect of the macromolecules is the main feature of what Eigen and Schuster call "hypercycles" (Eigen and Schuster, 1978). These can contain one molecule type or a set of molecules that are perhaps ordered sequentially, so that each polymer catalyzes the reproduction of the next polymer in the sequence. We will consider this type of hypercycle for the formation of spatial structures in the last part of this chapter.

The above type of cooperative system does not have any simple evolutionary features (cf. Maynard Smith, 1979). Different hypercycles can compete, but the outcome is not always clear, and it usually depends on the initial distribution of quasispecies (Eigen, 1971). Further, the enzyme activity should not be restricted to one particular information carrier. For catalyzing a certain polymer reproduction, the enzyme should recognize some part of the polymer. Polymers having this part in common should also be catalytically reproduced by the same enzyme. The important step in evolution is that new, possibly more efficient, species occur through mutations (errors in reproduction). In order that new molecules be able to grow and compete with existing ones, their reproduction in their initial stage must be catalyzed by existing catalytic polymers. *What is important is that the growth of a new species depends primarily on how efficiently it can use existing enzymes.* This does not necessarily provide a more "efficient" polymer, as it does not depend on the catalytic activity of the new enzyme. In fact, successful parasites without catalytic activity may grow faster than the existing polymers, making the existing polymers decay. This eventually leads to a total extinction.

This possibility of extinction is a very great danger at this stage. It is this possibility that may be coped with in a spatial organization, and which will be the focus of the last part of this chapter (see Boerlijst and Hogeweg, 1991). Certainly this can be achieved in a cellular organization with only one information carrier (RNA/DNA) per cell. Of course, cell division and the first production of macromolecules in a new cell both depend on a previous generation of enzymes. However, these will decay, and later the cell efficiency depends on the enzymatic activity of molecules belonging to the cell's own genome. Cells compete, and those with no catalytic activity cannot be reproduced and so will simply die without causing any further harm. In spatial patterns, mutations can be established *locally*. As stated above, these mutations rely on efficient use of the existing enzymes rather than establishing an efficient independent entity. In different parts of the space, different mutations may take over. These may then interact *globally*, which may lead to new rules. Then the catalytic efficiency of the overall hypercycle is important. Up to now, this has not been investigated in any detail, but it is an important possibility that should be pursued further.

Still further dangers due to errors may appear in other stages of evolution. Consider a cooperative stage with different kinds of molecules as information carriers (nucleic acids) and catalysts (proteins). This is the situation we have with the organisms alive today. In such a system the catalytic molecules catalyze their own reproduction from coded information and they control the accuracy of the reproduction. There is a bad feedforward effect here, as erroneous catalysts control the accuracy less efficiently. Thus errors may lead to more errors, which propagate and can lead to what's termed an "error catastrophe." This kind of error catastrophe was proposed by Orgel (1963, 1970) and later described in models by Hoffman and Kirkwood and co-workers (Hoffman, 1974; Kirkwood and Holliday, 1975). Their models have mostly emphasized the possibilities for obtaining stabilized situations, as well as possible connections to "aging," in which the accumulation of erroneous products can lead to the crossing of an error threshold, whereupon the cells decay. Certainly, these features are very relevant in the early stages in the origin of life, where this leads to a much more drastic situation than that of the simple self-reproducing system introduced by Eigen (1971).

The error situation is enhanced by the existence of an adaptor, the transfer molecule that couples the monomers of the catalyst (amino acids) to the code of the information carrier. This is the task of present-day tRNA. In a primitive situation there could have been some type of precursor. There are proteins that recognize the monomers and couple them to the adaptor. The accuracy of this step is crucial, and if erroneous recognizing proteins have low accuracy and relatively high activity, then this step can easily develop into an error catastrophe.

In more complex machinery approaching that of present-day organisms, this error propagation can afflict many stages. The production of activated monomers may also be overseen by catalysts. If these are erroneous, the activation can decline, which is a possible source of instability. So we see that it is always the active, erroneous catalysts that provide the danger. The situation can be improved by enzymes (scavengers) that can distinguish and break down the erroneous proteins more efficiently than the correct ones. For instance, they may recognize certain structural features. This reduces the erroneous proteins, thereby increasing the stability. Still, it couples to the catastrophe effect: When the accuracy goes down, the efficiency of scavenger proteins also goes down. More wrong proteins are produced, which enhances the appearance of erroneous catalysts. These factors may affect a fully developed cellular organization, which must have evolved a complex control mechanism for this purpose.

The errors can also be propagated to the gene. Wrong enzymes controlling the reproduction of the information carriers (genes) (polymerases) may introduce errors in the genes. From the erroneous (mutated) genes, more erroneous proteins are then produced. In present-day cells, this is not a problem, as cells with errors introduced at the gene level produce deficient enzymes and are not able to survive; these cells just decay. At a primitive stage in the evolution of life, however, this kind of error may be coupled to the parasite difficulty mentioned earlier. Existing "good" enzymes reproduce the "bad" genes. Other efficient enzymes produce more deficient proteins from these genes, which enhances the error catastrophe effect.

As stated, these bad effects are generally lessened in a cellular organization. Cells compete by the efficiency of the enzymes coded by their own genome. In that way, there is a kind of "group selection" by the cooperative units. This has led to the proposal that a cellular organization of some kind should have arisen at a very early stage in the development of life. However, a complete cellular organization depends on a number of subtle control mechanisms, which are susceptible to the error propagation effects. It's not easy to see how these control mechanisms could have developed at an early stage. Certainly, some primitive organizations can easily be established. There might have been droplets (coacervate) of mixtures of the essential cooperative molecules (see Feistel, Romanovskii, and Vasil'ev, 1980). Also, as mentioned above, spatial patterns can be formed by the nonlinearity of the kinetic equations, which may be less sensitive to the error effects. It is possible that different regions in space may compete by some type of "group selection rules" similar to those of the cells.

In the following sections we discuss the models of the basic self-reproducing system and the simplest cooperative hypercycle systems. A fuller account of the error features will be given in a subsequent paper (Blomberg, 1993).

3. The Basic Kinetic Model

Here we present the basic model for examining the situation described in the preceding section. As usual in kinetic models, notations such as X or M represent both the molecule species and their respective concentrations. Throughout the discussion, we shall adopt a formalism that explicitly considers the role of the monomer concentration in the kinetic equations. This provides the essential limitation to growth in our systems. We will use a formalism developed by Blomberg, von Heijne, and Leimar in 1981 (see also Blomberg, 1991, 1993). It is always necessary to have some constraint relation like this that limits the polymer growth. For this, Eigen and Schuster (1978) use mostly an assumption of "constant organization," which means that the sum of all concentrations remains constant. Many qualitative features are independent of how growth limitation is imposed. We believe that our approach is physically appealing, and also has some implications that are lacking under the constant organization assumption. In particular, it provides a way for systems to decay completely. As we are interested in stability conditions, it's important to include the possibility for the system to die out completely. Nevertheless, our assumption is still rather simplified.

In this chapter we will not go into any of the details of the underlying molecular biology. Some of this is treated in Chapter 10 in this volume. The main objects of our interests are assumed to be large, linear molecules, called *polymers* or *macromolecules,* which are taken here to be synonymous concepts. The polymers are composed of a number of similar units, the *monomers,* that are selected from a small number of variants. Normally, we will assume that there are two alternatives for each unit (see Figure 1, top). In today's macromolecules there are more than two variant monomers, but this fact does not influence the qualitative features that interest us here. Often there is only one main alternative that is difficult to distinguish from the correct monomer. The sequence of monomer units in the composition of the polymers determines the efficiency of the polymer in its possible biological function. In the living organisms of today, the important biological macromolecules are the nucleic acids, which contain the genetic information, and the proteins, which as enzymes catalyze, direct and control all relevant reactions, including reproduction and synthesis. It's likely that some complex organic polymers had similar roles in the early stages in the origin of life. At present, we cannot really tell whether these were of the same type as today's molecules or if they were of a simpler kind. It is rather probable that the latter is the case. In this chapter we try to be noncommittal about the character of the polymers, and may use terms such as *information carriers* or *genomes* for those polymers that carry information, and *catalysts* or *enzymes* (and sometimes *proteins,* especially in situations where real proteins should have occurred).

If the polymers X are composed of N monomers M, we have the following simplified kinetic equation for their self-reproduction:

$$\frac{dX}{dt} = kMX - qX, \tag{1}$$

where k is a polymerization rate and q is a decay rate. In principle, k and q depend on the polymer length. For instance, k may be inversely proportional to the length, while q is directly proportional to it. The monomers M are assumed to be available and used in an activated form, which is created by an available energy source (e.g., sunlight). It is necessary to include a step that can drive the processes, but we need not specify this point further. For the monomers, we have the following equation:

$$\frac{dM}{dt} = a - bM - kNMX. \tag{2}$$

The activated monomers are created by a constant rate a (which is a measure of the energy influx) and decay by a rate b. The monomer concentration in the absence of a polymerization process is then equal to a/b, which can be regarded as a measure of the turnover of external energy.

Consider now the possible stationary states of Eqs. (1) and (2) when the time derivatives are set equal to zero. There is always the possibility of a "dead" state, in which $X = 0$. We are of course mainly interested in other possible stationary states, and then (1) gives a value of $M = q/k$. If the concentration of monomers is larger than this, X will grow. But if M is smaller than q/k, X decays. From (2), the stationary value of X is $X = (a - bq/k)/(Nq)$. There is a requirement for this solution to make sense:

$$\frac{a}{b} > \frac{q}{k}. \tag{3}$$

This has an appealing interpretation. As noted above, a value of M larger than q/k is required in order that the polymer X shall be able to grow. In the absence of polymerization, the value of M is a/b, and the condition (3) means that this must be large enough to allow X to grow. In other words, the energy turnover must be large enough for X to develop.

Competition with another species, X_2, is easily described in this framework. For this, there should be a kinetic equation of a type similar to (1) for X_2:

$$\frac{dX_2}{dt} = k_2 M X_2 - q_2 X_2. \tag{4}$$

This relation should then be considered together with (1). It is easily seen that these two equations cannot lead to a stationary state in which both X and X_2 are greater than zero (unless q/k is exactly equal to q_2/k_2). Consider a situation where X, described by (1), has already arisen. Then X_2 may also come into existence and grow if q_2/k_2 is less than q/k. The monomer concentration M then decreases. When the latter becomes less than q/k, X decays. But X_2 can still grow, and eventually it will displace X.

This is the normal scenario for the evolution of these systems. Here we can use the Eigen concept of a "quasispecies" (1971), which may mean a dominating species with a "tail" of erroneous polymers that are the inevitable by-products of the reproduction process. In our scheme, X may represent a quasispecies.

Next, consider errors. Assume there are n vital units in the sequence of X that must be reproduced correctly if the molecule is to be able to reproduce itself. If there is a probability p at each step that a correct unit is chosen, the probability that a correct sequence is reproduced is then p^n. The growth rate in Eq. (1) should then be reduced by this amount, i.e., k should be replaced by kp^n. With this replacement, the requirement (3) becomes

$$\frac{a}{b} > \frac{q}{kp^n} \quad \text{or} \quad p^n > \frac{bq}{ak} \quad \text{or} \quad n(1-p) < \log\left(\frac{ak}{bq}\right). \tag{5}$$

For the last inequality, we use the relation $n \log(1/p) > n(1 - p)$. The condition (5) is a typical relation for the restriction due to accuracy. In the last form, it is a restriction on the sequence length n. Normally, the logarithm is of order 1, which allows a large variation of the rate constants. This means that n cannot be larger than order $1/(1 - p)$. A species violating condition (5) can never appear in the evolution, as it could never grow and contribute to a stationary situation.

4. Matrix Formalism: The Swetina-Schuster Scenario

Allowing several possible polymer species X_i and the possibility of errors (mutations) in which wrong monomers may be selected and inserted in the reproduction of a template, one gets a more general equation in place of (1):

$$\frac{dX_i}{dt} = M \sum_j P_{ij} k_j X_j - q_i X_i, \tag{6}$$

where P_{ij} is the probability that the reproduction of species X_j yields species X_i. Of course, this is a correct reproduction if $i = j$, but erroneous if $i \neq j$. The quantities k_i and q_i are the growth and decay rates of species i, respectively. This relation is of the same type as that employed by Eigen (1971) for describing molecular evolution through mutations. But here it is modified to our monomer formalism. As throughout this entire paper, we look for a stationary solution when the time derivatives are zero. This can be obtained by diagonalizing a matrix whose elements are

$$P_{ij} k_j / q_i. \tag{7}$$

An eigenvector of this matrix represents a distribution of the X_i as a stationary solution of (6). The monomer concentration M equals the inverse of the eigenvalue. There is also an equation for the monomers M that is the direct generalization of (2), and which is needed to completely determine the X_i. The eigenvector provides their distribution and their values up to a constant multiple.

Of course, no elements of this matrix are negative. If it is also assumed that no probability is identically zero, we have a matrix with strictly positive elements. For this, some results that are relevant for our development arise out of Frobenius's Theorem (Karlin and Taylor, 1975; Hofbauer and Sigmund, 1988). Such a matrix has a largest eigenvalue, which must be positive and simple (i.e., of multiplicity 1).

This corresponds to an eigenvector with positive elements, which is the only eigenvector having this property. *The conclusion is that only the X_i corresponding to the eigenvector of the largest eigenvalue of (7) are meaningful for a stationary solution, representing the final survivors in the scheme (6).*

This requirement for the largest eigenvalue is the analog of the condition on the largest value of k/q in the earlier discussion. The distribution for X_i obtained in this way corresponds to the surviving "quasispecies." [The probabilities in (6) may be very small, and there may be subspaces in the full distribution space that are very weakly coupled to each other. In that case, there are metastable distributions with long lifetimes, which for practical purposes appear stationary. The number of such states can be large. This leads to a "spin-glass" or "rugged landscape" model (Kauffman, 1987). We will not consider this possibility further here.]

Swetina and Schuster (1982) have studied a particular model for errors in this context. We will use their model here as a prototype for further applications. In this model, one "good" polymer is compared with a large group of less efficient, possibly destructive polymers. The model assumes that all polymers in the deficient group are equivalent. In particular, the kinetics and the accuracy are determined by exactly the same parameters. These may represent average parameters for a more realistic situation. The model is very instructive and we will consider it in some detail here, suitably modified for our purposes.

Among the monomers of the polymer sequence, some are essential for the polymer's functioning, while some are important for its structure and stability. In the model, it is assumed that there is a sequence of m units that is crucial for its reproduction. This is not necessarily the entire polymer sequence. Of all possible polymers, there is one species X_0 having a particular sequence that is reproduced with the largest value of k/q. Polymers differing from X_0 by one or more monomer units in this main sequence reproduce themselves with growth rates k_1 and decay rates q_1. In the model, these rates are assumed to be the same for all polymers of the deficient group. The assumption that X_0 has the largest value of the growth-to-decay ratio means that $k_1/q_1 < k_0/q_0$. As stated above, we consider two alternatives for each monomer unit in the polymer sequence that can be mistaken for each other. Thus, the total number of polymers of arbitrary composition in the main sequence of m units is 2^m. For this model and for the further extensions of the following sections, it is not necessary to specify the sequences other than the main one. Some part of the polymer sequences may be required for any kind of action, and that should be the same in X_0 and in the large group. Other parts may be less relevant, and need not be specified either for X_0 or for the other polymers.

Further assumptions: Let the probability that a unit of X_0 is correctly selected be p_0, while the corresponding probability for the other polymers is p_1. Then the probability of getting a perfect copy of the main sequence in the two cases is p_0^m and p_1^m, respectively. The quantity p_0 should be close to 1. In the original Swetina-Schuster model, p_1 is assumed to be $1/2$, which means that the deficient polymers lack any kind of accuracy. But that is not very important, since the results are pretty much independent of p_1 (which may well be equal to p_0). We refer to X_0 as the "best" polymer, calling the others "the large group of deficient polymers."

The matrix to be diagonalized is given by (7). For this, we need an expression for P_{ij}, which is the probability that the reproduction of species j leads to species i. The probability that any polymer sequence is correctly reproduced is p_i^m, where p_i is p_0 or p_1, depending on whether the polymer is the "good" one, X_0, or belongs to the deficient group. In general, P_{ij} depends on how the monomer sequences of species i and j differ from each other. We introduce here the notation that X_i represents polymers differing from X_0 by i monomers in the relevant sequence. The probability that X_0 gives rise to X_i equals $p_0^{m-i}(1-p_0)^i$. The expression for P_{ij} is more complicated, as it depends on how the i units of X_i and the j units of X_j that differ from X_0 differ from each other. This expression equals

$$P_{ij} = \sum_{k=\max(0,\, j-i)}^{\min(j,\, m-1)} \binom{i}{j-k}\binom{m-i}{k} p_j^{m+j-i-2k}(1-p_j)^{i+2k-j}. \quad (8)$$

This formula can be understood more easily if one notes that we have divided the m sites into two groups: the i sites by which X_i differs from X_0 and the remaining $m-i$ sites that these have in common. The quantity k is the number of units in the latter group that differ in X_j from both X_0 and X_i. Then there are $j-k$ elements in the first group of i units that are the same in X_j and X_i but which differ from X_0. See also Figure 2.

Now consider the eigenvalue relations of the matrix (7), with P_{ij} given by (8). For X_0, we have the equation

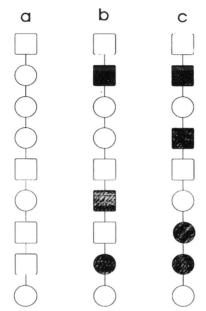

Figure 2. In the model of the text we consider all types of transitions between all possible monomer sequences. Here a is the "correct" polymer, b is a representative of X_3 having three errors, and c is a representative of X_4 with four errors, each as compared to the polymer to the left. The general kinetic scheme includes reproductions by error from any of these polymers to the others. A reproduction from b to c involves three specific errors. This makes a contribution to P_{43} in (8) when $k = 2$, i.e., there are two monomers in c that differ from the other two polymers.

$$\lambda X_0 = \frac{k_0}{q_0} p_0^m X_0 + \frac{k_1}{q_0} \sum_i \binom{m}{i} p_1^{m-i} (1-p)^i X_i. \tag{9}$$

Since we are interested primarily in a solution for which X_0 dominates, we can neglect the contribution of the higher X_i. This is an appropriate approximation for the situation in which we are interested, and it can easily be checked by direct calculation of the full matrix. It means that we neglect mutations from the X_i to X_0. Using this approximation, we find an eigenvalue approximately equal to $k_0 p_0^m / q_0$.

The equations for other X_i are of the form $\lambda X_i = k_0/q_1 P_{i0} + (k_1/q_1) \sum P_{ij} X_j$. We can sum these relations multiplied with binomial factors $\binom{m}{i}$. This yields λX_T for the left-hand side of the summed equation, where X_T is the total number of incorrect polymers. To this we add Eq. (9), multiplied by q_0/q_1 (in order to get the same q-denominators). The probabilities then add up to 1. We get the (exact) relation

$$\lambda \left(X_0 \frac{q_0}{q_1} + X_T \right) = X_0 \frac{k_0}{q_1} + X_T \frac{k_1}{q_1}. \tag{10}$$

Using the approximate eigenvalue above, this yields

$$\frac{X_T}{X_0} = \frac{k_0 q_0 (1 - p_0^m)}{k_0 q_1 p_0^m - k_1 q_0}. \tag{11}$$

For this quantity to be finite and positive, it is necessary that

$$\frac{k_0}{q_0} p_0^m > \frac{k_1}{q_1}. \tag{12}$$

The quantity (11) diverges when these expressions are equal. The left-hand side of (12) is the eigenvalue above. This is a condition that the eigenvalue be the largest one of our main matrix. There is also another type of eigenvector in which all the deficient polymers of the large group have approximately the same concentrations. As they constitute a large group, they dominate the system. The specificity of the X_0 polymer is then lost. That eigenvalue is obtained from (10) if the X_0 terms are neglected. It equals k_1/q_1. The inequality (12) is of the same form as (5), which was derived for a different situation but which gives similar conclusions. Only if that condition is fulfilled do we obtain a meaningful situation with a self-reproducing polymer and a high information content in its monomer sequence. Both these expressions for the eigenvalues are good approximations to the exact values for this model as m becomes large. Figure 3 shows the two largest exact eigenvalues and their approximate expressions.

If the inequality (12) is not fulfilled, we obtain the other type of eigenvector that is dominated by the polymers of the large deficient group, which arbitrarily reproduce each other. The large group is a self-reproducing entity containing no information. We note that this state need not be unstable. It could well fulfill a condition that is the generalization of (3), which in this case would be $a/b > q_1/k_1$. Unlike Eq. (5), it does not contain the probability and the length. This state can be highly viable; it need not be a "dead end," and may well be a part of an evolutionary pathway that could later provide more efficient polymers with more information content. Note that "fitness measures" as the stability criteria do not involve "information measures." Information may evolve if it is coupled to the efficiency of the single polymers, but it has no primary survival value.

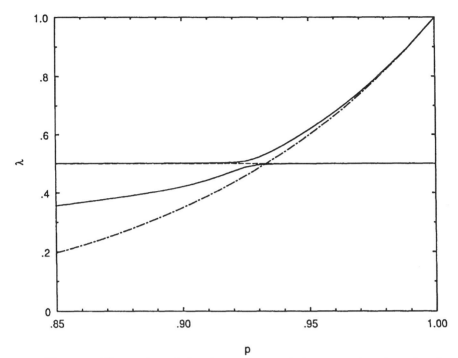

Figure 3. Illustration of the eigenvalues and the approximations for the case $n = 10$, $k_0/q_0 = 1$, and $k_1/q_1 = 0.5$. The solid curves are the two largest eigenvalues of the model matrix, while the dashed and dash-dotted curves are the two approximations which intersect each other. The true eigenvalues never intersect but get close to each other. The largest eigenvalue moves from one of the approximate expressions to the other. The value $n = 10$ is not particularly large, yet the approximations describe the largest eigenvalue rather well except close to the intersection point.

5. Cooperative Systems and Parasites

In the self-reproducing system treated above, the basic reaction was

$$NM + X \rightarrow X + X. \tag{13}$$

A more developed stage of evolution involves a cooperative system, where the molecules act as templates and catalysts. This situation is found in the kind of "RNA world" discussed in a previous section. It means that there are reproductive reactions of the form

$$NM + X_C + X_T \rightarrow X_C + X_T + X_T. \tag{14}$$

In Eq.(14), the subscripts C and T denote the catalytic and template molecules, respectively. As mentioned earlier, these can be the same molecule. Cooperative systems of this type are what Eigen and Schuster refer to as *hypercycles.* The catalytic molecules also serve as templates and are reproduced by the same reactions. Here we will mostly neglect reactions of type (13) in the cooperative models, and consider only reactions of type (14). This enhances some of the effects that we're interested in, making these effects more explicit.

In many models of this kind, one includes both schemes (13) and (14). This can make it possible for a species to grow from low concentrations, which is not possible in the scheme (14). In the latter case, catalysts must be present in a significant amount for reproduction to be possible. In the models of this section, we consider a cooperative system with a single component. This means that the catalytic and template molecules are the same species. As noted above, the scheme (14) requires a significant presence of catalysts for molecules to grow. New template molecules may grow and outnumber existing templates *if the value of the quotient between the growth rate as catalyzed by an existing catalytic molecule and the decay rate is greater than that of the old polymers.* This holds *irrespective* of the catalytic activity of the new polymers, which may be completely lacking. In that case the new templates are *parasites* that can be reproduced but that cannot catalyze further reproductive reactions. It is also possible that the new molecules have some limited catalytic activity, which may not be enough to sustain itself. The existing system is driven completely to extinction—even if the existing cooperative system is quite efficient. This is the essence of the parasite problem.

There is no reason why parasites should have lower reproductive rates or higher decay rates than the active polymers. Indeed, if the parasites involve a reduction of length, they will grow faster than the active polymers and could be more stable. The parasite problem also creates a difficulty in reaching a high complexity and length. An erroneous reduction of length can have fatal consequences.

There are many possibilities for making variants through errors (mutations) in the replication process, so it's reasonable to assume that there will be many possible parasites in the system. Here we consider a possible group of parasites, obtaining a formalism quite similar to the one used in the simple self-reproduction case considered above, and described by formula (6).

We assume that there is a set of polymers X_i, of which X_0 is the only catalyst, which can catalyze the reproduction of all the others. With the same notations as employed previously, and taking P_{ij} as the probability that the reproduction of polymer j leads to polymer i, we get the following set of equations:

$$\frac{dX_i}{dt} = MX_0 \sum_j k_j P_{ij} X_j - q_i X_i. \tag{15}$$

This is similar to (6), differing only by the factor X_0. The stationary solution can be analyzed by the same method: we look for the eigenvalues and eigenvectors of the matrix whose elements are $k_j P_{ij}/q_i$, which is exactly the same matrix as that encountered in the preceding section. The eigenvector associated with the largest eigenvalue is the only relevant one for the stationary solution, and provides the X_i-values up to a constant multiplier. The product MX_0 is equal to the reciprocal of the largest eigenvalue.

We can consider this more explicitly by analogy with the Swetina-Schuster model of the preceding section. For this we make the following assumptions: There is a sequence of m units of the polymers that is crucial for the catalytic activity but not for recognition by the catalyzing enzymes. The quantity X_0 has a correct version of this sequence and is the only one with catalytic activity. There are other polymers that differ from X_0 by one or several bases in this main sequence. They can be reproduced by catalytic assistance from X_0, but do not have any catalytic activity themselves. We assume that all polymers are reproduced by the same growth rate k, and that the probability of getting a correct monomer unit is p for all units. The polymers differ in the decay rates. All parasites have the same decay rate $q_1 > q_0$. This leads to the same matrix diagonalization as in the preceding section, and the model itself is the same as that of the previous section. So its solution is given by the preceding expressions, in particular (10). They are the same type of eigenvectors and eigenvalues, but we do not distinguish here the growth rates k and the probabilities p. As stated above, the largest eigenvalue of the matrix equals the inverse of the product MX_0, and the values of M and X_0 are given from the M-equation, which now takes the form

$$\frac{dM}{dt} = 0 \Rightarrow a = bM + NMX_0 k(X_0 + X_T). \tag{16}$$

As before, X_T is the total number of incorrect polymers, while N is the length of the polymers. The condition that the X_0-dominated solution is the correct stationary one is the same as above, and is given by formula (12). Now (16) cannot offer a stable possibility under realistic conditions if (12) is not satisfied. This can be seen rather directly.

If (12) does not hold, the eigenvalue is k/q_1, which is equal to the reciprocal of the product $X_0 M$. As the erroneous polymers dominate, X_T/X_0 is very large, of the order 2^m. When X_0 varies and the product $X_0 M$ is constant, the right-hand side of (16) is always large, having a minimum on the order of $2^{m/2}$. Then X_0 is of order $2^{-m/2}$, and M is large, of order $2^{m/2}$. So the equation cannot be satisfied if m is large.

If any of the parasites has a decay rate lower than q_0, then these parasites always grow and engulf the entire system. For this to be lethal, the difference in decay rates needs only to be greater than the order of the probability that parasites are erroneously formed from X_0.

6. Error Propagation in Protein Synthesis

In the systems described up to now, there is only one type of molecule, which naturally restricts the possibility for developing any real complexity. In a more evolved system, there are at least two kinds of molecules: those that carry the genetic information (nucleic acids) and those that perform the reactions, the catalysts (proteins, enzymes). It certainly is the case that the possibility of using proteins for efficient catalysis was a big step in the evolutionary process. But this also means a more complex system, in which disturbances due to errors are even more difficult to overcome.

In the protein-nucleic acid world there are three obvious stages in synthesis: the reproduction of nucleic acids catalyzed by proteins, the synthesis of proteins, which is controlled and catalyzed by proteins, and the transfer of information from the nucleic acids to the protein synthesis machinery. These processes are coupled into a complex network. And as we shall discuss later, the situation may be still more complex in more developed networks.

First, we consider the protein synthesis step. Assume that there is only one type of control protein (e.g., a ribosome enzyme) that is coded by a certain gene D_0. The protein X_0 is a "correct copy" of the gene (but not necessarily the best one). A number of variants X_i occur through errors. The protein X_0 and perhaps some of the variants work well with a high production rate and a high accuracy, while others have a lower activity and less accuracy. Moreover, most of the variants probably do not have any activity.

There is a kinetic equation for the protein synthesis of the same form as with the previous cases:

$$\frac{dX_i}{dt} = \sum_{k=0}^{n} K_k P_{ik} X_k A D_0 - q_i X_i. \qquad (17)$$

The notation is similar to that used in previous formulas, such as (6): P_{ik} is the probability that the product will be the polymer X_i produced by the enzyme X_k with the production rate K_k. The quantity A corresponds to the monomer and is assumed to be an adaptor (tRNA) that carries the amino acid monomer and recognizes the genetic code instruction. We will discuss the adaptor features further in the next section.

There is a fundamental difference with the previous cases, for example Eq. (6). Here the genetic message is given, and the correct copy made by any of the enzymes is X_0. Equation (17) does not describe any evolving genetic information, but rather the evolution of errors in proteins. Thus, P_{0k} is the probability that the protein X_k makes a perfect copy of the code. As before, the physically meaningful stationary solution is the one corresponding to the largest eigenvalue of the matrix $K_k P_{ik}/q_i$. For stability, this eigenvalue must be larger than a certain value.

Here again, we have a model that is analogous to the previous ones. This model also has essential features in common with a model used by Kirkwood and collaborators (Kirkwood and Holliday, 1975) for describing error propagation. As before, we consider a "best" protein X_0, along with a group of erroneous ones differing from X_0 by one or more erroneous monomers in an essential sequence of m units. The protein X_0 has a production rate K_0, a decay rate q_0, and a probability p_0 of selecting the correct adaptor in the protein synthesis. As in the previous models, we assume that there are two choices at each site, which means that the "erroneous" group consists of $2^m - 1$ possible proteins. It is further assumed that all proteins in that group act in the same way: they provide the same production rates K_1, the same probability p_1 of selecting a correct unit, and have the same decay rates q_1. Thus, the probability that a protein in the erroneous group gives rise to a correct protein is $P_{0k} = p_1^m$.

As stated, this model is related to that of Kirkwood and his collaborators, which, however, is formulated in another way. Their formulation uses relations between features of successive generations rather than in terms of kinetic differential equations. They normally assume that there are more than two possible choices for the monomer units, and also that the erroneous proteins do not have any accuracy, which in our case means that $p_1 = 1/2$. The latter assumption is unnecessary for our model, and these differences are not important for the qualitative features of the model.

As in the previous cases, we treat a matrix with elements A_{ij}, which refer to the situation in which a protein "i is produced by the enzyme j." As before, X_i denotes a protein that differs from X_0 at i sites. We have

$$A_{ij} = K_1 q_1 \binom{m}{j} p_1^{n-i}(1 - p_1)^i, \qquad i, j > 0. \tag{18}$$

The binomial factor is the ratio of possible X_j proteins.

Most of the eigenvalues of this matrix are zero, and it's possible to obtain exact expressions for the (two) nonzero eigenvalues. These can be further simplified. The accuracy p_1 of the erroneous proteins is probably not very close to 1, so we may neglect terms which contain p_1^m, as m is fairly large (at least 10–20). Under that assumption, one obtains the same eigenvalues and similar expressions for the eigenvectors as in the previous models. Here and in the following sections it's convenient to introduce some quotients as the most relevant parameters, and also to employ a unified scheme of notation.

We write:

1. k_x for the quotient K_1/K_0 of the production rates of the deficient and correct proteins, respectively. It is reasonable to assume that $k_x < 1$.

2. q_x for the quotient q_1/q_0 of the decay rates of the deficient and correct proteins. One may assume that this quotient is greater than 1, but this is not absolutely necessary, as the deficient molecules may have a more stable structure.

3. x for the quotient $\sum X_i/X_0 = X_T/X_0$ of the total number of deficient proteins and the correct ones. Clearly, x is a measure of the accuracy of the protein synthesis. If $x = 1$, then there are as many deficient proteins as correct ones.

The results are now (a more detailed account is found in Blomberg, 1993):

$$
\begin{aligned}
&\text{If } \ p_0^m > \frac{k_x}{q_x} \ \text{ then } \ \lambda \approx \frac{K_0}{q_0} p_0^m \ \text{ and } \ x \approx \frac{1 - p_0^m}{q_x p_0^m - k_x}. \\
&\text{If } \ p_0^m < \frac{k_x}{q_x} \ \text{ then } \ \lambda \approx \frac{K_1}{q_1} \ \text{ and } \ x \approx \frac{k_x - q_x p_0^m}{k_x q_x p_1^m}.
\end{aligned}
\tag{19}
$$

Although the situation is now different and the matrix structure quite different, the results are very much the same as with the previous models. As in those cases, X_0 is much larger than any of the erroneous components in the first situation of (19), which is the reasonable case of interest. In the second situation, the total number of erroneous proteins dominates over the correct one, since p_1^m is small.

The condition for a stable situation again comes from the monomer equation. Here this will be similar to relation (3), and as in the model of Section 4, the second situation of (19) may well be stable. As in Section 4, the stability requirement involves only the rate coefficients, not the polymer length, together with the large value of the quotient x. This kind of system may not die out even if the erroneous proteins dominate.

Nevertheless, the condition for the first and most meaningful situation in (19) is again primarily a restriction on the accuracy and the length of the polymers. The erroneous proteins must not be too active, and the quotient k_x/q_x must be substantially smaller than 1. The effect is qualitatively enhanced if we include a further monomer recognition step, which introduces a feedback effect.

7. Adaptors

The genetic information for building up proteins needs an adaptor system. This means that the monomers of the proteins (amino acids) are coupled to an adaptor molecule (tRNA) able to recognize the genetic code. This recognition of the monomers and the coupling to the adaptor is made by synthetase enzymes, which we denote here by S. As in the other cases, there may be "good" synthetases having high accuracy, and "bad" ones with less accuracy. The probability that a synthetase S_i will make a correct adaptor will be denoted P_i^A, and its production rate of adaptors by K_i^A. As with other proteins, the synthetases are made by the same general protein synthesis machinery described by Eq. (17) of Section 6. The production of S-proteins, adaptors A, and correct adaptors A_C can then be expressed by the equations

$$\frac{dS_i}{dt} = \sum_j K_j^S P_{ij}^S X_j A - q_i^S S_i,$$

$$\frac{dA_C}{dt} = \sum_i K_i^A P_i^A S_i M - q^A A_C, \tag{20}$$

$$\frac{dA}{dt} = \sum_i K_i^A S_i M - q^A A.$$

In general, the rate and probabilities may differ for the S-proteins from those for the X-proteins in Eq. (17). Here, however, we will assume that the parameters with an S-index are the same as those of (17). (Or, at least, the relevant quotients are the same.) The relevant parameter here is *the ratio of correct adaptors,* which we write as y. The wrong adaptors introduce possible errors at all sites of the polymer, and to provide an acceptable accuracy y must be close to 1. The stationary state arises from the last two equations in (20):

$$y = \frac{A_C}{A} = \frac{\sum K_i^A P_i^A S_i}{\sum K_i^A S_i}. \tag{21}$$

This quantity is crucial for error propagation. The adaptor ratio in (21) must be included in the previous kinetic equations for protein synthesis, as those equations represent an independent error source: Adaptors that recognize the code at the protein synthesis step are accepted, but may contain the wrong monomer. In expressions such as (18) and (19), the probabilities p_0 and p_1 must be multiplied by y.

We consider a model of the same type as in the previous sections. Thus, the S-proteins are of two types: the "correct" S_0 and a group of similar, but erroneous proteins S_i. We will continue to use the quotient notation of the preceding section, writing

1. k_A for the quotient K_1^A/K_0^A of the adaptor synthesis rate and the deficient and correct proteins, respectively. As before, it's reasonable to assume that this number is less than 1;

2. s for the quotient $\sum S_i/S_0$ of the total number of deficient S-proteins and the correct ones. The decay rates of correct and incorrect adaptors are assumed to be equal.

For our subsequent development, we need a relation between the quotients s and x. These are certainly related, but as the proteins may be of different length, we cannot generally expect them to be the same. It is possible to write down complete relations, where a length difference is taken into account (see Blomberg, 1993). But for the description presented here, we simply make the assumption that they are equal. That is, $s = x$, which is the case if the proteins are of the same length. Then we substitute the x-value of the first relation in (19) into (20), which leads to the following expression for y:

$$
\begin{aligned}
y &= \frac{P_0^A + P_1^A k_A x}{1 + k_A x}, \\
&= \frac{P_0^A(q_x(yp_0)^m - k_x) + P_1^A k_A(1 - (yp_0)^m)}{q_x(yp_0)^m - k_x + k_A(1 - (yp_0)^m)} = f(y).
\end{aligned}
\tag{22}
$$

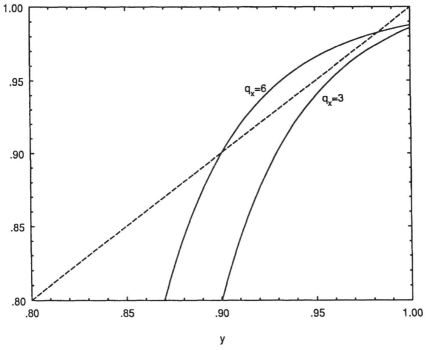

Figure 4. Curves, illustrating formula (22), and the graphical way of obtaining y. The full curves show the right-hand side of (22) with $q_x = 6$ and $q_x = 3$. For $q_x = 6$, there are two intersections with the line $y = f(y)$, while the value $q_x = 3$ leads to no intersection and thus does not offer any high-accuracy possibilities. The other parameters values are $k_x = k_A = 0.3, p_{0x} = P_0^A = 0.99, P_1^A = 0.8$.

A necessary condition for this expression to hold is that the first relation of (19) holds, which means that $p_0^m > k_x/q_x$. Otherwise, s becomes much larger than 1. Equation (22) implicitly determines y, s, and further quantities. It can be displayed graphically by plotting $f(y)$ against y. This is shown in Figure 4. Possible solutions are obtained at the intersections of the line y with $f(y)$. The curves in Figure 4 show one example in which there are solutions of (22) (for the case $q_x = 6$), and one example where there is no solution (when $q_x = 3$). If there are two intersections, the one with largest value of y is a stable solution, while the other intersection is an unstable solution. The latter represents a threshold situation: initial states with y lower than the threshold value never stabilize, while all states with y-values above this value eventually lead to the stable solution. There is always a solution with a low-accuracy branch, represented by the second part of formula (19). This leads to a y-value close to P_1^A. This generates very large values of s (no accuracy), and it is reasonable to assume that this represents an unstable situation that we need not consider further. The general features are similar to those found in previous models (Kirkwood and Holiday, 1975; Blomberg, 1990), although the formalism is quite different. Indeed, the $f(y)$ curve here proceeds more drastically from a high-accuracy branch described by (22) to the low-accuracy, unstable branch. The condition for a solution requires that the quotient k_x should be small and q_x should be large. Figure 4 shows what that may mean, and fuller representations of this condition are shown in a later section (Figure 7).

8. Coupling to the Gene

The feedback loops are more complete if we include the coupling to the gene. There are three steps, of which the first two have been described in the previous sections:

1. Protein reproduction, governed by equation (17);

2. Protein monomer selection and adaptor synthesis, described by (20).

To these we add

3. Reproduction of the genome D, catalyzed by a polymerase protein R.

A schematic representation of this network is represented in Figure 5. The genome D contains the genes of the three proteins that are translated in step 1. As in the previous framework, we consider variants D_i with different monomer composition. We distinguish one variant D_0, which is the "best" genome, and is also the most stable with lowest decay rate (this is a necessary but far from sufficient condition). There are also variants of the proteins, in particular the polymerase enzymes, where the "best" one is denoted R_0. The genome reproduction is described by kinetic equations similar to those of Section 5.

Compared to formula (15), here there are separate genes and enzymes, as well as active erroneous enzymes (which are deficient to the cognate R_0), not merely the "parasites" of Section 5. We have

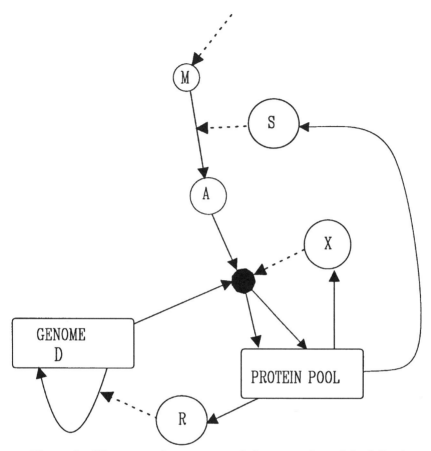

Figure 5. The network structure of the general model of Section 8. The solid lines describe the reproduction paths from M to A, from A and the genome to the proteins, and from the genome to itself. The protein production is represented by the filled circle, which gives the different kinds of proteins: X, S, and R. The dashed curves show the catalytic control of the enzymes on the production steps.

$$\frac{dD_i}{dt} = \sum_j \sum_k BK_k^D P_{ij,k}^D R_k D_j - q_i^D D_i.$$ (23)

Here B represents the monomers of the genome, which are analogous to M in previous formulas. The quantity K_k^D is the reproduction rate by the polymerase R_k. The probability that the polymerase R_k makes the genome D_i in the reproduction of D_j is $P_{ij,k}^D$. (This is similar to the earlier notations. The reproduction is correct if $i = j$; otherwise, erroneous.) We use R_k/R_0 as a parameter, obtaining the stationary features of (23) from the largest eigenvalue and the corresponding eigenvector of the matrix whose elements are

$$\sum_k \frac{K_k^D P_{ij,k}^D R_k}{q_i^D R_0}. \tag{24}$$

The quantity MR_0 then is the reciprocal of the eigenvalue. The R_k are given by the kinetic equations of the protein production. As before, we write X_i for a general protein. For simplicity, in the explicit formulas we assume that only X_0 is active at the protein synthesis stage. This means that $k_x = 0$. More general formulas are found in Blomberg (1993), and in the figures we will show some results based on these with nonzero k_x. As will be seen, this does not change any qualitative features of the model. The kinetic equations are direct generalizations of (17), but now include gene variants:

$$\frac{dX_0}{dt} = AX_0 K^X P_{00}^X D_0^X + AX_0 K^X \sum_k P_{0k}^X D_k^X - q_0^X X_0,$$

$$\frac{dX_i}{dt} = AX_0 K^X \sum_k P_{ik}^X D_k^X - q_i^X X_i. \tag{25}$$

Here D^X stands for the X-gene, and the indices refer to the "correctness" in an obvious way. The quantity D_0^X is the gene that correctly corresponds to X_0, while D_k^X is the gene of X_k. The equations also describe the synthesis of the R- and S-proteins.

We use assumptions for the genome reproduction that are similar to those above, and introduce several new quantities:

1. $k_D = K_1^D/K_0^D$, the quotient of the reproduction rate of the genome by deficient enzymes and the correct ones. This is analogous to the previous quantities k_x and k_A.

2. $q_D = q_1^D/q_0^D$, the ratio of the decay rates of the erroneous genomes and the correct one. This is a crucial parameter which must be larger than 1 (because of the parasite effect of Section 5). There is no obvious reason that it must be large.

3. $d = (D_0 + \sum D_i)/D_0$, the quotient of all genomes and the correct one. Of course, this quantity is greater than 1.

We also write the new probabilities: p_0^X is the probability of the selection of a cognate adaptor by X_0, while p_{D0} and p_{D1} are the probabilities that the correct R_0 and the deficient R_i select a correct gene monomer at the gene replication.

We now want relations for the stationary values of the different variables. As before, the decay rates of the erroneous X_k are assumed to be equal. One relation for the X is obtained by summing over all variants in (25), whereby the probabilities sum up to 1. Another comes from neglecting all contributions in the first relation from genes other than D_0^X. We get

$$
\begin{aligned}
AK^X (yp_0^X)^{n_x} X_0 D_0^X &= q_0^X X_0, \\
AK^X X_0 D_{\text{tot}} &= q_0^X X_0 + q_1^X X_T.
\end{aligned}
\tag{26}
$$

The quantity P_{00}^X of (25) is equal to $(yp_0^X)^n$, and includes the adaptor accuracy y. The term D_{tot} is the sum of all genomes, while D_0^X signifies those genomes that have the X-gene correct (but may have errors in other parts of the genome). As before, X_T represents the erroneous X-proteins.

This leads to a further quotient

$$
d_x = D_{\text{tot}}/D_0^X.
$$

From this, we obtain the following relation from (26):

$$
d_x(yp_{X0})^{-n_x} = 1 + xq_x \Rightarrow x = \frac{1}{q_x}\left[d_x(yp_{X0})^{-n_x} - 1\right].
\tag{27}
$$

The treatment of the matrix (24) follows the same lines as in previous sections. We get the d-quotient from a generalization of (10):

$$
d = \frac{(p_{D0}^N + k_D p_{D1}^N r)(q_D - 1)}{q_D(p_{D0}^N + k_D p_{D1}^N r) - 1 - k_D r}.
\tag{28}
$$

Here r is the quotient of the total number of deficient reproduction enzymes and the correct ones. We also need the somewhat more complex concept of "genes," which are parts of the complete genome. The "correct" X-gene of (25) includes the correct genome D_0 and those that have the X-gene correct but may contain errors in other parts:

$$
D_0^X = D_0 + \sum_{k=1}^{N-n_X} \binom{N - n_x}{k} D_k.
\tag{29}
$$

We calculate this from the equations for the D_k by summing over all equations to get λD_0^X in a manner similar to Eq. (10). We then keep only the terms that contribute to D_0^X, neglecting certain low-order terms in the inverse lengths. The result is

$$\lambda D_0^X = \lambda \left(1 - \frac{1}{q_D}\right) D_0 + \frac{1}{q_D} \left(p_{D0}^{n_X} + r k_D p_{D1}^{n_X}\right) D_0^X,$$

$$\frac{D_0^X}{d_0} = \frac{d}{d_x} = \frac{\lambda(q_X - 1)}{\lambda q_D - \left(p_{D0}^{n_X} + r k_D p_{D1}^{n_X}\right)}. \tag{30}$$

The notation here is as before, where λ is the eigenvalue of the matrix (25) multiplied by q_0^D / K_0^D. It is approximately equal to $(p_{D0})^N$, where N is the total length of the relevant genome sequences. This length is at least equal to the sum of the corresponding lengths n_X of the protein genes.

We now have all relations for the coupled system, at least for the simplified situation in which the proteins are of equal length, which means that their quotients are the same:

$$x = \frac{\sum_{i=1}^{n_X} \binom{n_X}{i} X_i}{X_0} = r = \frac{\sum_{i=1}^{n_R} \binom{n_X}{i} R_i}{R_0} = s = \frac{\sum_{i=1}^{n_S} \binom{n_X}{i} S_i}{S_0}. \tag{31}$$

Equation (27) yields x as a function of y and d_x. The latter is given in (30) as a function of the r, now assumed to be equal to x. The relation for y is that of the preceding section:

$$y = \frac{A_C}{A} = \frac{K_0^A p_0^A + K_1^A p_1^A s}{K_0^A + K_1^A s}.$$

We can write this as two relations for the gene quotient in (30), obtaining one implicit relation for x expressed in the rate and probability parameters. For this we neglect high powers of the probabilities p_{D1}.

$$d_x = \left[\frac{p_0^A + p_1^A k_A x}{1 + k_A x}\right]^{n_x} p_{x0}^{n_x}(1 + x q_X) = \frac{q_D p_{D0}^N - p_{D0}^{n_x}}{q_D p_{D0}^N - (1 + x k_D)}. \tag{32}$$

9. Further Error-Feedback Effects

The effects we have described up to now are the most relevant for error propagation and possible error catastrophes. There are, however, further effects that are connected to these and that are also important for controlling the accuracy. These are all enzyme-controlled processes, connected to the protein synthesis kinetics as described above.

The disintegration process of the proteins is an important tool for controlling the reaction schemes. This is carried out by "scavenging" enzymes that make it possible to distinguish between "correct" and "incorrect" macromolecules. This leads to a stabilizing effect. Further, as in the other cases, it is natural to assume that the distinguishing capability worsens if the scavenging enzymes are erroneous. This can easily be treated along the same lines as the previous effects, with the result that these enzymes can stabilize a high-accuracy situation, but they also lead to a more drastic increase in errors, errors that may increase and contribute to an "error catastrophe."

Another feedback effect relevant for the stability and a possible source of a drastic extinction is coupling to the monomer activation. In most of this work, we have simply assumed that activated monomers are created at a constant rate. In a more complex system, it would be natural to suppose that the main activation is catalyzed by some enzymes that use an external energy source. In a pure case, this would mean that the synthesis rate a introduced in (2) might be replaced by a term

$$a \rightarrow a_0 + a_1 E_M.$$

Here E_M is a metabolic enzyme. Of course, this couples to the other kinetic expressions. With an increasing number of errors the number of correct enzymes goes down, and so does the activation rate. This leads to a more rapid extinction as the activated monomers decline. This also strengthens the catastrophic features of error propagation. We will not go into this any further.

10. Results of the Error Propagation Models

Here we present some results based mostly on formulas (32) and (22), as the relation (32) contains most of the other formulas so far. An appropriate way to represent solutions is to look at the protein accuracy quotient x as a function of some rate quotient. The general appearance is the same for all the various types of rate quotients, so in Figure 6 we use q_x. A direct way to treat (32) is to calculate this quotient as a function of x when other parameters are fixed at some "standard values." The typical solution has a branch, which for sufficiently large values of the decay quotient provides stable, low values of x (i.e., few erroneous proteins). The quantity x increases with decreasing values of q_x. This curve has a turning point: At a certain point, the branch disappears. This means that there is no solution below a certain value of the decay quotient. The turning point represents a stability limit, yielding a maximum value of x that is not particularly large—normally around 0.5 and rarely much greater than 1. (That is, there are fewer erroneous proteins than correct ones.)

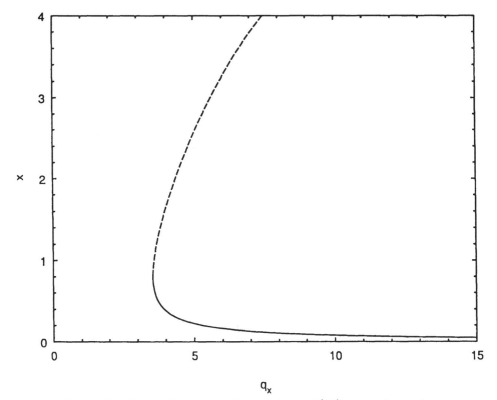

Figure 6a. Typical behavior of a solution of (32) as a relation between x, the quotient of deficient proteins and correct ones, and the protein decay parameter q_x defined in the text. Typical behavior of a solution of (32) as described in the text. Other parameters are as in Figure 4, with the additions $k_D = 0.3, q_D = 6, p_{D0} = 0.99$. Here $k_x = 0$ as was assumed in Section 8.

The stable branch turns into a new branch where x increases with increasing values of the decay quotient. This is an unstable branch, which represents a threshold situation: For initial values below that branch, the system stabilizes and goes to the stable branch, while no stability is attained from initial values above it.

The limit of stability can be represented as a relation between a decay quotient, q_x or q_D, and a function of a quotient of the growth rates, i.e., any of the k. This is shown in Figure 7, where the different growth quotients are shown as functions of q_x. A striking feature of

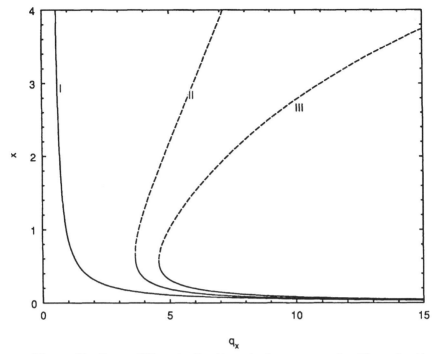

Figure 6b. Some different situations to be compared with each other and with Figure 6a. In these cases, $k_x = 0.3$. Curve I is that obtained for the pure protein synthesis model in Section 6 and formula (19). Curve II results from (22), i.e., for the adaptor model, but where the error propagation over the gene is not included. Points with $q_x = 6$ correspond to the two intersections in Figure 4. The limit of stability occurs for a value of q_x slightly smaller than 4. Finally, curve III represents a more general situation with $k_x = 0.3$. We note that curve II is very close to that in Figure 6a, and that these and curve III are qualitatively similar. Other parameters are as in Figure 4, with the additions $k_D = 0.3, q_D = 6, p_{D0} = 0.99$.

both Figures 6 and 7 is that the adaptor step of Section 7 is always the one providing the widest variation and the largest contribution. It can certainly be regarded as the most sensitive step. This comes from the fact that its parameters contribute to the y-parameter, which is raised to a high power in the equations, for instance in the first factor in (32).

The error propagation over the gene is less important than that of the adaptor synthesis. In fact, the leading possible failure of the gene step is still the same as that treated in Section 5, the parasite problem. Genes with deficient enzymes occur like the parasites and

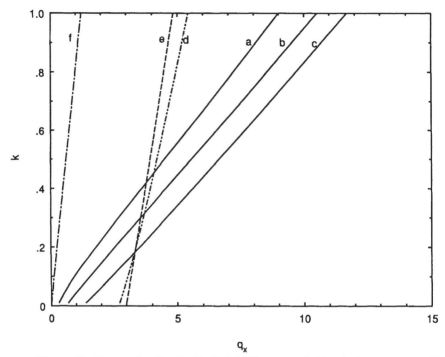

Figure 7. Curves for the limit of stability as a relation between a growth quotient k_x, k_A, or k_D and the decay quotient q_x. The curves a, b, and c show the adaptor growth quotient k_A. The quantity a is obtained from the pure adaptor model of Section 7, formula (22) with $k_x = 0$. The quantity b comes from Section 8, formula (32), while c is a more general situation in which $k_x = 0.3$. Thus, a shows the pure adaptor influence, the difference between a and b displays the influence of the propagation over genes, while the difference between b and c is the influence of the protein production of Section 6. The adaptor influence is the most important. Curve d shows k_x in the adaptor model of Section 7, equation (22). Curve e involves the gene parameter k_D from formula (32) with $k_x = 0$. Finally, f is given by relation (19) in Section 6. Most of the curves are straight lines, which can be demonstrated analytically.

provide similar constraints, which are now modified and enhanced due to the error propagation. For the rate coefficients, the protein synthesis constraints may be more restrictive than those of the gene reproduction.

   In real situations, the problem may be more complex than this. It is not unreasonable to suppose that a deficient protein is less stable than the "best" one. However, there is no reason that a gene corresponding to defective or nonfunctioning proteins should be less stable than one that leads to efficient proteins. It is important to consider these problems, as well as to account for the possibilities for a system to stabilize. We should then try to figure out possible paths that could have been followed for a primitive system to avoid some of the worst consequences seen here.

As previously noted, the parasite difficulty is a problem for the early stages of a cooperative system. The error propagation is worse in a more evolved system. It seems that the full complexities and handling of different genes and proteins with an adaptor mechanism require rather advanced methods of error control.

11. Many-Component Hypercycles

Hypercycle systems are often regarded as consisting of many components coupled together in a network. This can mean proteins and nucleic acids, where the proteins catalyze all synthesis processes—the reproduction of the nucleic acids, as well as the synthesis of the proteins from their codes in the nucleic acids. In an "RNA world," all molecules are of the same type and act both as information carriers and catalysts. The polymers catalyze the reproduction of other polymers in the network. Disregarding mutations, this can be represented by a coupled set of kinetic equations:

$$\frac{dX_i}{dt} = \sum_j M K_{ij} X_j X_i - q_i X_i. \tag{33}$$

Here K_{ij} is the rate at which the polymer X_i is reproduced by the catalytic action of the polymer X_j. Other notations are as in previous sections.

A simple model that has been much studied involves a circular sequence of polymers $X_1, X_2, \ldots, X_n$, where X_i catalyzes the production of X_{i+1} and eventually X_n catalyzes the production of X_1. We regard X_0 as being identical to X_n, and get the following set of equations when the production and decay rates are the same for all polymers:

$$\frac{dX_i}{dt} = M K X_{i-1} X_i - q X_i, \qquad i = 1, 2, \ldots, n; \quad X_0 \equiv X_n,$$
$$\frac{dM}{dt} = a - bM - NMK \sum_{i=1}^{n} X_{i-1} X_i. \tag{34}$$

In this expression, N is the number of monomers in the polymers. It is assumed to be the same for all species.

With more than four components, this system oscillates. For a large number of components n, the oscillations are very drastic since the component levels go down to very low values. At all times, there are only two components with appreciable concentrations. In our scheme, the concentrations decrease exponentially after each maximum. For instance, when $n = 9$ the concentrations go down by a factor 10^8 from their maximum values, and this factor increases by about two orders of magnitude when we add just one more component.

Of course, the regular, circular pattern of this model makes it rather special. It may contain features that are not found in more general cases. Nevertheless, limit cycles and chaotic behavior may well be encountered in schemes of the general form (33). The simplicity of this model makes it very suitable for study, although the conclusions may not be generally valid for schemes of the form (33).

12. Schemes for Spatial Variation

If the reactions (33) are combined with diffusion to provide a reaction-diffusion equation, spatial patterns may arise. In particular, spatial patterns from reactions of the type (34) have attracted a lot of attention recently due to their ability to resist parasite attacks (Boerlijst and Hogeweg, 1991). The rest of this chapter is devoted to such patterns.

Combining (34) with diffusion, we get the reaction-diffusion equations

$$
\frac{\partial X_i}{\partial t} = KMX_{i-1}X_i - qX_i + D_X\nabla^2 X_i,
$$
$$
\frac{\partial M}{\partial t} = a - bM - NMK\sum X_{i-1}X_i + D_M\nabla^2 M.
$$
(35)

In this model, the X-components are assumed to have identical kinetic parameters. In particular, they have the same diffusion coefficients. The latter are normally different for the polymers D_X and the monomers D_M. Practical calculations are usually limited to two-dimensional patterns. Nonenzymatic reproduction as in (13) is not included here. If it is included, a system with a low concentration may possibly grow, having a stable final state with only one of the components in (32) or (33). This is impossible in the scheme considered here.

Numerical treatments of equation (35) rewrite it in discrete form as a set of equations on a (two-dimensional) lattice:

$$
X_i(r, s, t + 1) = X_i(r, s, t) + K'M(r, s, t)X_{i-1}(r, s, t)X_i(r, s, t)
$$
$$
- q'X_i(r, s, t) + D_X''[X_i(r - 1, s, t) + X_i(r + 1, s, t)
$$
$$
+ X_i(r, s + 1, t) + X_i(r, s - 1, t) - 4X_i(r, s, t)].
$$
(36)

The primed kinetic parameters include time steps, and D'' also incorporates spatial steps. For this system, we have examined two different kinds of boundary conditions: *periodic,* which means that for all variables

$$X_i(r, s, t) = X_i(r + R, s, t) = X_i(r, s + S, t) \tag{37}$$

on an $R \times S$ lattice, and the assumption of zero flux:

$$
\begin{aligned}
X_i(0, s, t) &= X_i(1, s, t); & X_i(r, 0, t) &= X_i(r, 1, t), \\
X_i(R + 1, s, t) &= X_i(R, s, t); & X_i(r, S + 1, t) &= X_i(r, S, t).
\end{aligned} \tag{38}
$$

Besides partial differential equations, one can use cellular automata (CA) models for producing spatial patterns. In this case, there is a two-dimensional lattice, each point of which can be in one of $n + 1$ states: "0" means that the point is empty, while "i" means that the point contains one polymer of the type X_i. The states of the lattice points are changed at discrete time steps according to the following rules:

1. A polymer can decay: any state "i" can become empty "0."

2. A new polymer can be produced: state "0" becomes state "i" if some neighboring point contains a state "i" (self-reproduction) and, in particular, if there is also a state "$i - 1$" nearby (catalyzed production).

The rules are **random rules,** which means that an event is determined by choosing a random number: States decay or are reproduced with certain probabilities. Random rules are necessary for diffusion to occur.

There are two possibilities for carrying out the CA calculations:

I. There can be a **synchronous updating,** in which all points in the lattice are updated at each step and the states are changed simultaneously according to the rules. (This is similar to what is done for the PDE.)

II. There is **asynchronous updating,** in which one point is chosen at random at each step. The state of this point is then changed according to the rules. This is essentially a kind of simulation calculation.

The decay and production rules are the same in the two cases, but diffusion is treated quite differently. For synchronous updating, a diffusion algorithm is used in which the lattice is divided into 2×2

blocks that are rotated in random directions. There are two sets of such a division, and one goes through both these sets at each diffusion step. Diffusion cannot be varied arbitrarily, and the states change sites and may pass through each other. For asynchronous updating, diffusion can be introduced in a more natural manner and simply means that a polymer state moves to an *empty* neighboring site with some probability. The diffusion can be varied arbitrarily, and polymers do not move through each other. This represents a *cage effect* with reduced motion in a region of high density. Such effects are less direct in the synchronous updating and in the PDE. It can be introduced as an extra constraint. It is, of course, possible to have an algorithm for asynchronous updating in which occupied states may change sites.

Another factor that can be of importance is growth limitation. There must be some mechanism that prevents growth at high densities. In our previous description, this is accomplished by the couplings to the monomers. In most of the works by Eigen, Schuster, and co-workers, a limitation is introduced by assuming a "constant organization," which means that the sum of all concentrations shall be constant. This is more difficult to justify and use in a spatially varying situation. For the CA, there is a natural density limitation: Growth is possible only at empty states, and is slow in regions where the density of empty states is low. Thus, an explicit growth limitation is unnecessary. In the Boerlijst-Hogeweg calculations (Boerlijst and Hogeweg, 1991) there are no other rules.

Of course, other rules can be used in the CA. We have made CA calculations with a coupling to monomers, in which the monomer concentration varies in time but remains the same over the entire lattice (which corresponds to a high diffusion rate for the monomers). For this kind of CA, one needs a continuous monomer description coupled to the discrete formulation. For the PDE, we use the monomer coupling represented in (35). Density limitations can be introduced in the PDE by extra growth factors of the form $c - \sum X_i$.

There are some important conceptual differences between cellular automata models and the partial differential equation framework. In the PDE, each species is present at all lattice points in some concentration, which may be very small. A drawback of the continuous description is that small "unphysical" concentrations may occur and sometimes be multiplied by an exponential growth factor. In this way, a very improbable event (represented by a very small concentration) can eventually become physically relevant. This should not really occur if the low concentration means less than one molecule in a certain

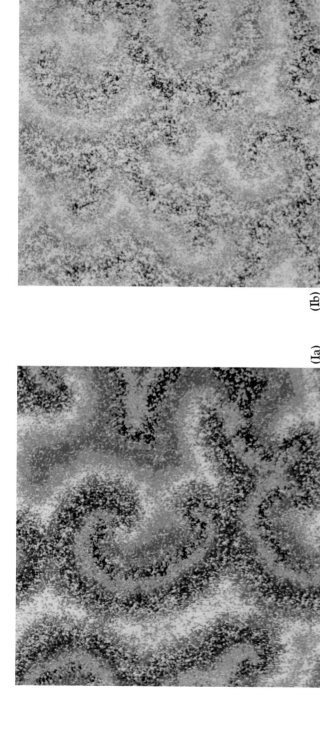

(Ia)　　　(Ib)

PLATE I. Typical patterns of the CA models with synchronous updating. Each point represents a lattice point in the CA, and the colors represent the different molecule species. The patterns rotate but remain essentially stationary during long time periods.

PLATE Ia. Pattern from the Boerlijst–Hogeweg algorithm with no restriction on monomers. Six molecule species are considered on a 400×400 lattice. The probability of enzymatic reproduction at each step is equal to the number of empty sites around a molecule with a catalysing species next to it. We also include non-enzymatic reproduction with a probability of 0.01 at each step. The death probability at each step is 0.05.

PLATE Ib. A pattern with monomer turnover included and seven species on a 600×600 lattice. The reproduction probabilities are multiplied by (monomer concentration)/100. The steady-state monomer concentration without polymer reproduction is 100. At a reproduction, the monomer concentration is decreased by 10, and at each time step restored by the relation $M(t + 1) = 0.999M(t) + 0.1$.

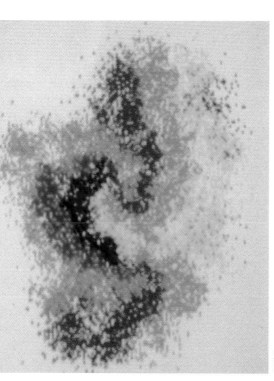

(IIa)

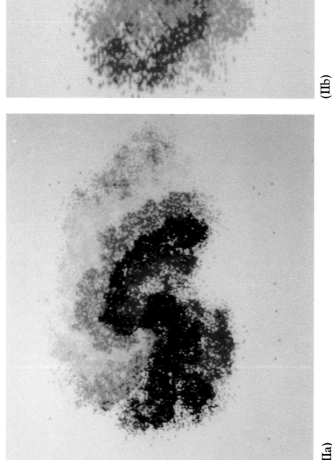

(IIb)

PLATE II. Cluster pattern for the CA with asynchronous updating. The probabilities for growth are the same as in Plate I.

PLATE IIa. A cluster with seven molecule species on a 400 × 400 lattice which rotates counter-clockwise around a single center.

PLATE IIb. A cluster with two centers on a 200 × 200 lattice. The geometry is here just being cut off by the growing light-green species, for which the yellow one is a catalyst. The region of the yellow species is here just being cut off by the growing light-green species, for which the yellow one is a catalyst. The outer yellow species have very little catalytic support at this stage (only from the decaying red points), and will decay, while the confined, inner yellow region grows.

cell volume. This can and should be modified by some type of cutoff that eliminates very low concentration values at any point. In the CA, this poses no problem as the concentration is "quantized." There is a low concentration cutoff and cage effects occur (it is this aspect that sometimes leads to statements claiming that the CA may be closer to reality than the PDE).

As already noted, the Boerlijst and Hogeweg model involves CA calculations with a synchronously updated model and with only the "natural density" growth limitation in a CA description. We have made calculations with the PDE as shown above, as well as for various CA models with synchronous and asynchronous updating, with and without a coupling to monomers. All give rise to spatially oscillating patterns. But these can be quite different. In all models, one can study parasites, the reproduction of which is catalyzed by some species in the hypercycle sequence. It should have at least some rate constant that differs from the hypercycle species.

13. Spatial Pattern Results

We now discuss the patterns arising from these approaches, restricting ourselves to situations with five or more components in which there are clear oscillating structures. With certain initial conditions, the PDE approach provides a homogeneous, oscillating structure with some stability. Other initial distributions lead to spatially varying spiral patterns, which are studied in another paper (Cronhjort and Blomberg, 1993). The patterns depend on the initial conditions: With zero flux as in (38), single spirals may form, while for periodic boundary conditions, as in (37), spirals always occur in pairs. Typical patterns are depicted in Figure 8, and have rotating arms in which one component is dominant. The motion of an arm with a particular dominating species is naturally directed toward a region with the component that catalyzes its reproduction.

If the monomers diffuse much faster than the polymers or the replication of polymers consumes a substantial part of the monomers (i.e., the polymers are long), we get a cluster distribution with clusters of homogeneous internal structures oscillating in time. In that case, neither the homogeneous state nor the spiral patterns are stable.

The Boerlijst-Hogeweg synchronously updated CA (Boerlijst and Hogeweg, 1991) yields a spiral pattern very reminiscent of our PDE pattern. The patterns fill the entire lattice, which contains rather well-delimited regions with only a single polymer component. These move toward the region of the component that catalyzes its reproduction.

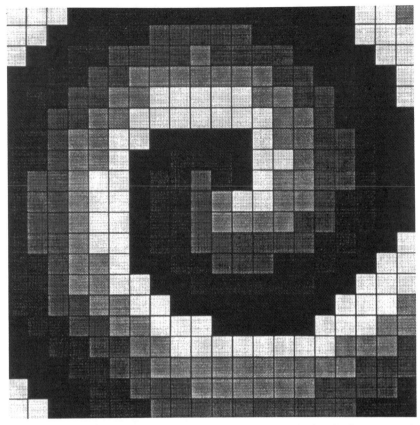

Figure 8a. This figure and Figure 8b show typical spiral patterns in a PDE model. Every lattice point is represented by a square, the gray scale of which indicates which polymer species is dominant at that lattice point. This figure gives no information about the concentration of the molecular species that are not dominant. But one or more of them may be almost as strong as the dominant species. A single spiral is obtained using zero-flux boundary conditions on a 20×20 lattice. The initial state is divided into sectors, in each of which one polymer species X_i is given a higher concentration than the others. A steady state is established at $T = 1,000$ time units. The rotation time of the spiral is slightly less than 100 time units. The values of the parameters are the same for all polymer species: $K = 1.0, q = 0.1, D_X = 0.01$, and $N = 10$. For the monomers, $a = 1.0, b = 0.1$, and $D_M = 0.1$.

The motion is outward from spiral centers, where the essential production takes place. The arms eventually move to regions with no catalysts, where they decay. Spiral centers always occur in pairs.

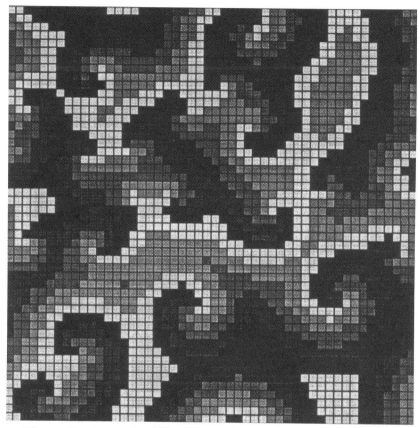

Figure 8b. A complex 50×50 spiral system obtained by starting from random concentrations of X_i on all lattice points and using periodic boundary conditions. A steady state is established at time $T = 5,000$. Parameter values for all polymer species are: $K = 1.0, q = 0.1, D_X = 0.01$, and $N = 100$. For the monomers, $a = 5.0, b = 0.1$, and $D_M = 1.0$.

This type of CA, as well as other CA models that use only the natural growth limitation, normally leads to a very high polymer density (the number of empty sites is low). We have studied this algorithm and included monomers (which then are not spatially varying). If the monomer uptake per polymer is not too large (the polymers are not too long), we obtain a pattern with the same general appearance as that without monomers, but which has a lower density. There seems to be a general tendency toward such patterns. When we increase the monomer uptake or vary parameters to provide lower densities, the patterns do not persist, but fall apart. No stable configuration is obtained

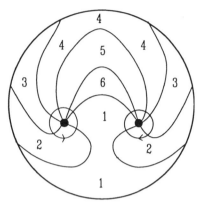

Figure 9. Schematic picture of a cluster with a double spiral structure in a cellular automaton model with six molecular species. The regions of the respective species are numbered. Each species grows toward its catalytic support, i.e., to the region with one unit lower (and "1" to "6"). Thus, the cluster rotates as shown in the figure, and the smaller regions close to the centers grow outward. At a later stage, outer parts are being cut off, e.g., region 1 will soon be cut off by region 2. Molecules in the outer regions have no catalytic support and will decay. Compare this with the cluster of Plate IIb.

in such a case. It seems as if a certain minimum density is required for stability. Pictures of typical structures for this kind of CA are shown in Plate I (see the color plate). The structures rotate around centers at which all species meet. They are essentially stable, remaining the same for a long time. We have not been able to get such a sparse pattern without monomers by varying relevant parameters.

We have also considered CA with asynchronous updating and a restriction of diffusion to empty sites. Without monomers, we obtain dense structures similar to the previous cases, in which movements are highly restricted. If growth and decay parameters are varied in order to decrease the concentration, the system decays, and it is only possible to get configurations with a high density. However, by introducing monomers, other structures are obtained. We then get confined, rotating clusters. In what seems to be a stable configuration, clusters rotate around one center by a kind of wheel formation, or they can have rotations around two rotation centers with a kind of "double-spiral" structure. The density in the cluster is high, and the size (not density) of the clusters is related to the monomer uptake by the polymers. Typical clusters are shown in Plate II (see the color plate). Features of a two-center cluster are shown in Figure 9. Note that single centers are possible only in confined regions, either clusters or in limited spaces with zero flux.

At present, we do not know which features determine this model dependence of the rotating spatial structures. Work is in progress to investigate this point.

14. The Response to Parasites

As described in the previous sections, parasites with a growth-to-decay ratio greater than that of the original polymers will destroy a homogeneous system. They will grow faster than the other polymers, reducing the monomer concentration (or take up most of the space if that is growth limiting) and then cause the other polymers to decay. Eventually everything decays. We now investigate their appearance in spatially varying situations.

The Boerlijst-Hogeweg CA (1991) with its typical spiral pattern shows resistance to parasites. Parasites can destroy a spiral if they occur in its center, but they cannot move against the outward motion of a spiral. Thus, these spirals are resistant to parasite attacks from the outside. In some cases with parasite infection, clusters are found.

For the spiral pattern of our PDE with a monomer coupling (Cronhjort and Blomberg, 1993), we do not get this behavior. In our case, parasites can always move into a spiral and destroy it. To investigate if this is due to the fact that the PDE allows unphysically low concentrations of parasites, which can move in all directions, we introduce a cutoff factor: If the concentration of any polymer species at any point is lower than a fixed limit, it is set equal to zero. A high cutoff limit prevents parasite attacks, but not as in the manner of Boerlijst and Hogeweg. In our work, the cutoff restricts the spread of the polymer species in all directions—toward the spiral center as well as outward.

The consequences of parasites are more pronounced if one allows different diffusion constants for them and for the other species. We then find coexisting states, in which the hypercycle and the parasites survive, although the latter have the largest growth-to-decay ratio. If the parasites diffuse faster than other species, one first sees the typical features when parasites grow. They seem to take over and the species decay. However, at that stage the faster motion of the parasites leads to a more rapid decline of parasite concentration. At this stage, the monomer concentration grows and becomes fairly high as the polymer production is low. The hypercycle can then be restored locally. Later, such a restored region can be infected with parasites, and the same process repeats itself. In this case, we get an irregular pattern and highly irregular, chaos-like fluctuations of the concentrations. If the parasite diffusion constant is not large enough, the system eventually dies out, seemingly by fluctuations after a long "transient period." For larger diffusion constants, the system seems to survive for a long time. The diffusion constants needed for this depend on the relation between the growth-to-decay rates for the parasite and the hypercycle components. We show the general features of this situation in Figure 10.

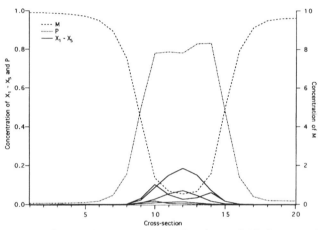

Figure 10a. Coexisting states may be formed if the parasites diffuse faster than the other species. A typical cross section showing the concentration of all species. The hypercycle is in the form of a cluster, enclosed by the parasite.

The hypercycle can also survive if the diffusion of the parasites is significantly slower than that of the other components. In that case, when the system begins to decay the hypercycle components can move to regions where the parasite concentration is low and where there is a high concentration of monomers. In these regions the polymers are restored. Later, these regions will be invaded by the parasites, and the process repeats itself. The hypercycle components move in a kind of irregular, traveling wave fashion and seem to be able to survive for long periods. This behavior is illustrated in Figure 11.

Note that the monomers play an important role in these pictures of hypercycle survival: Polymers are able to be restored as the monomers grow rapidly in regions where the species have previously declined. It is difficult at present to judge the possibilities of hypercycle survival

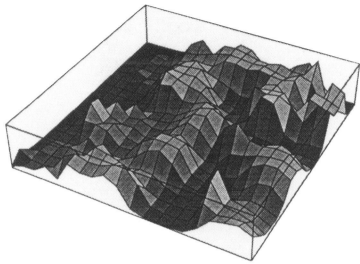

Figure 10b. The irregular spatial structure of this coexisting state is illustrated by the concentration of the parasite.

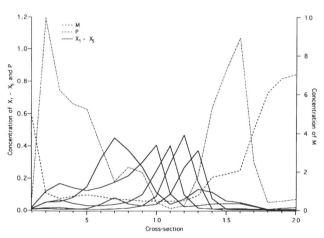

Figure 11a. When the parasites diffuse more slowly than the polymers of the hypercycle, a different coexisting state may be formed. A cross section showing the concentrations of all species. The hypercycle is in the form of traveling waves. The parasite is slowly advancing into the area of the hypercycle waves, but when the parasite has killed the hypercycle and declined itself, new hypercycle waves soon move into the area.

of "parasite infection." We have made estimates of how the growth and decay parameters, as well as the diffusion constants, depend on the lengths of the polymers (see Cronhjort and Blomberg, 1993). A re-

Figure 11b. The irregular spatial structure is illustrated by a matrix showing the dominant species. Black represents the parasite, while the different gray shades represent the species of the hypercycle, the lightest being X_5 and the darkest X_1. No information is given in the figure about the concentration of species that are not dominant.

duction in polymer length could lead to fatal parasites. Unless specific molecular factors play a role, a smaller polymer would be reproduced faster and decay slower than a longer one. Thus, the growth-to-decay ratio can easily be larger for a small parasite than for the catalytic larger hypercycle components. The smaller molecule will have a larger diffusion constant, but a general estimate shows that the expected increase is less than what is required for our results. It is questionable if this can save the hypercycle.

References

Blomberg, C. (1990). Modeling efficiency, error propagation and the effect of error-enhancing drugs in protein synthesis. *Biomed. Biochem. Acta,* 49, 879-889.

Blomberg, C. (1991). Considerations of stability in models of primitive life. Effect of errors and error propagation. In *Complexity, Chaos and Biological Evolution,* E. Mosekilde and L. Mosekilde, eds. New York: Plenum Press.

Blomberg, C. (1993). The effect of errors and error propagation in simple models of self-reproducing systems and the origin of life. Manuscript, submitted.

Blomberg, C., von Heijne, G., and Leimar, O. (1981). Competition, coexistence and irreversibility in models of early evolution. In *Origin of Life,* E. Wollman, ed. Dordrecht, The Netherlands: D. Reidel Publ. Company, pp. 385–392.

Boerlijst, M. C. and Hogeweg, P. (1991). Spiral wave structure in pre-biotic evolution. Hypercycles stable against parasites. *Physica D,* 48, 17–28.

Bresch, C., Niesert, U., and Harnasch, D. (1980). Hypercycles, parasites and packages. *J. Theor. Biol.,* 85, 399–405.

Cronhjort, M. and Blomberg, C. (1993). Hypercycles versus parasites in a two-dimensional partial differential equations model. Submitted.

Eigen, M. (1971). Self organization of matter and the evolution of biological macromolecules. *Die Naturwissenschaften,* 65, 7–41.

Eigen, M. and Schuster, P. (1978). The hypercycle. A principle of natural self-organization. Part B: The abstract hypercycle. *Die Naturwissenschaften,* 65, 47–61.

Eigen, M., Gardiner, W. G., and Schuster, P. (1980). Hypercycles and compartments. *J. Theor. Biol.,* 85, 407–411.

Eigen, M. and Schuster, P. (1982). Stages of emerging life: five principles of early organization. *J. Mol. Evol.,* 19, 47–61.

Feistel, R., Romanovskii, Yu. M., and Vasil'ev, V. A. (1980). Evolution of Eigen hypercycles occurring in coacervates. *Biofizika,* 25, 882–887.

Garcia-Tejedor, A., Moran, F., and Montero, F. (1987). Influence of the hypercyclic organization on the error threshold. *J. Theor. Biol.,* 127, 393–402.

Hofbauer, J. and Sigmund, K. (1988). *The Theory of Evolution and Dynamical Systems.* Cambridge: Cambridge University Press.

Hoffman, G. W. (1974). On the origin of the genetic code and the stability of the translational apparatus. *J. Mol. Biol.,* 86, 349–362.

Karlin, S. and Taylor, H. M. (1975). *A First Course in Stochastic Processes.* New York: Academic Press.

Kauffman, S., (1987). Towards a general theory of adaptive walks on rugged landscapes. *J. Theor. Biol.,* 128, 11–45.

Kirkwood, T. B. and Holliday, R. (1975). The stability of the translational accuracy. *J. Mol. Biol.,* 97, 257–265.

Maynard Smith, J. (1979). Hypercycles and the origin of life. *Nature,* 280, 445–446.

Niesert, U., Harnasch, D., and Bresch, C. (1981). Origin of life between Scylla and Carybdis. *J. Mol. Evol.,* 17, 348–353.

Orgel, L. E. (1963). The maintenance of the accuracy of protein synthesis and its relevance to aging. *Proc. Nat. Acad. Sci. USA,* 49, 517–521.

Orgel, L. E. (1970). The maintenance of the accuracy of protein synthesis and its relevance to aging: a correction. *Proc. Nat. Acad. Sci. USA,* 67, 1476.

Swetina, J. and Schuster, P. (1982). Self-replication with errors. A model for polynucleotide replication. *Biophys. Chem.,* 16, 329–345.

CHAPTER 3

Cooperation: The Ghost in the Machinery of Evolution

JOHN L. CASTI

Santa Fe Institute
Santa Fe, NM 87501, USA

1. Neo-Darwinism and the Problem of Competition

Neo-Darwinian evolution can be described compactly by the formula

$$\text{adaptation} = \text{variation} + \text{heredity} + \text{selection}.$$

Here "variation" refers to the fact that individual organisms each have a different genetic endowment bequeathed to it by its parents, while "heredity" means that these genetic characteristics can be passed on to an organism's own offspring. Since the pioneering work by Mendel on genetic inheritance, there has been little controversy in the evolutionary biology community about these two processes. But not so for the mysterious process of "selection," whereby nature bestows its favors on some and ignores others.

The traditional view of evolution handed down to us by Darwin can be encapsulated in two pithy phrases: "Nature, red in tooth and claw," and "survival of the fittest." These epigrams paint a picture of a cruel, hostile world indeed, one in which each organism has to battle literally to the death with every other organism in a kind of "all-against-all" war of attrition. Of course, if we open our eyes and just look around, the kind of world we see does not match this picture, at

Cooperation and Conflict in General Evolutionary Processes, edited by John L. Casti and Anders Karlqvist.
ISBN 0-471-59487-3 ©1994 John Wiley & Sons, Inc.

all. Instead, what we see is a world in which there is at least as much cooperation as there is competition, and where living entities organize themselves into social groups and communities in order to bolster their individual chances for survival.

By the tenets of a strictly Darwinian view of the world, the emergence of cooperative behavior poses a real problem: If "survival of the fittest" is indeed nature's golden rule, how can it be that an individual's fitness is increased by a cooperative act, when by definition "cooperation" means foregoing benefits that you might have taken so that those benefits can go to another? We might term this the "Problem of Competition."

Modern sociobiologists have put together a variety of explanations for how such "altruistic" cooperative acts might arise in the natural course of evolutionary events. Basically, these mechanisms can be divided into four categories:

• *Group selection*— In his work on animal behavior, Konrad Lorenz noted that animals hardly ever engaged in behavior dangerous to their own species. Lorenz's explanation of why potentially harmful aggression in animals appeared to be confined to interspecies competition is that an individual within a group would be willing to suffer a personal loss in fitness only if that loss was more than compensated for by an increase in overall group fitness. But as the result of theoretical models, as well as ingenious alternative explanations, there is more or less universal agreement today that group selection is a pretty rare phenomenon, taking place only under very special circumstances.

• *Kin selection*—It's often observed that close relatives tend to look after each other more than they look after strangers, and the closer the relationship (e.g., identical twins vs. distant cousins), the greater the willingness to sacrifice. Experiments in the animal world involving social insects such as wasps and bees demonstrate how this kind of mechanism works, and it doesn't take too much exercise of the imagination to see how it would work for humans, as well.

• *Parental manipulation*—This is a type of enforced altruism in which a parent coerces a child to give help to another for the parent's benefit. A typical situation of this sort might arise, for instance, if a mother cat has a litter of, say, five kittens but can raise only three of them to maturity using her own resources. Then it would pay her (genetically speaking) to employ her position of authority to force some of her older offspring to devote a part of their resources to helping her raise the litter. She can do this in many ways, perhaps the most common being a threat to withhold some of her attention from certain offspring if they refuse to help out. In nature the strategy of parental manipulation often takes the form of cannibalism, in which the weaker members of the litter are sacrificed for the benefit of the stronger. Of course it might be argued that putting yourself on your brother's dinner plate hardly constitutes an "altruistic" act, in the sense that the term is normally used in polite conversation. But in nature "altruism" means only an act that decreases your own fitness in order to enhance the fitness of another. So such an act of sacrifice is indeed altruistic, at least in Nature's dictionary.

At first glance it may appear that there is no real difference between parental manipulation and kin selection—they both involve the sacrifice of an individual for the benefit of another. However, there is one critical difference: In kin selection, one individual helps another because they share some genes; in parental manipulation, one person helps another for the benefit of a third party (the parent). So in parental manipulation, the fact that the two parties might share genes is incidental, although it often happens that they do. In practice, however, it may not be easy to distinguish between the two forms of altruism, and any given situation may involve both. In fact, it has been suggested that the main causal factor at work in the development of sterile castes in the social insects is parental manipulation and not kin selection. This is because when the queen sets up the nest, she chooses to make workers rather than reproductives by virtue of what she feeds her initial offspring. But this is still a matter of some controversy and the jury is out as to which of the two altruistic mechanisms is really at work in that situation.

• *Reciprocal altruism*—By far the largest share of altruistic acts, at least among humans, involve parties who are not related at all. Robert Trivers introduced the idea of reciprocal altruism to account for these sorts of sacrificial acts. In essence, the principle governing reciprocal altruism is "If you'll scratch my back, I'll scratch yours." Briefly, the claim is that individuals engage in altruistic acts because they expect that by doing so they will benefit by someone else's altruism toward them at some time in the future. Note the very great difference here between an act of reciprocal altruism and an act of kin selection altruism. In the reciprocal case, the giver expects to see a direct return from a sacrifice; in the latter situation, the giver sees no direct reward but only the satisfaction of seeing his or her genes being given a better chance to make it into future generations.

The most convincing example of reciprocal altruism in nature seems to be the case of the "cleaner fish." Certain species of fish clean parasites off fish of a different species. This is a situation in which both parties gain: The cleaners get a hearty meal, while the fish being cleaned avoid the sores and diseases that would otherwise result from the parasites. The most remarkable aspect of this situation is that the cleaner fish are never eaten by those they're cleaning, even though this could easily happen. Furthermore, it's often the case that other types of fish try to imitate the cleaners, rushing in to bite big chunks off the fish being cleaned. In these cases, the big fish happily gobble up the pretenders despite the fact that the pretenders have developed high-level camouflage techniques to fool them. Since the cleaners and the cleaned have no genetic relationship at all, Trivers argues persuasively that this situation can be explained only as a case of reciprocal altruism. We'll return to a deeper consideration of reciprocal altruism later on when we consider the evolution of cooperative behavior.

In order to study how these various forms of cooperative behavior might emerge naturally in a population, theoreticians have noted that many of the encounters organisms have in their environment take the form of a game, in the sense in which mathematicians use that term. We devote the next section to an account of how one makes the transition from behaviors in nature to games.

2. Games, Strategies, and the Prisoner's Dilemma

The mathematical formulation of a game of strategy involves two or more players, each of whom has at his or her disposal a set of actions. Each play of the game consists of the players choosing an action, with the combined choice of actions leading to a payoff to each player. A strategy for each player is simply a rule by which the player selects an action for each play of the game.

The simplest type of game is the two-person game. In this situation, we have two players, call them Player I and Player II, who choose actions i and j, respectively. This results in a payoff a_{ij} to Player I and a payoff b_{ij} to Player II. If the game's payoff structure is such that

$$a_{ij} + b_{ij} = 0,$$

we have what is termed a *two-person, zero-sum game.* As it turns out, this is really the only broad class of games for which there is a satisfactory mathematical theory for how rational players will act so as to optimize their expected returns from playing the game. Here's a sketch of how this theory goes.

Suppose the components of the vector $p = (p_1, p_2, \ldots, p_m)$ represent the probability that Player I chooses action $i, i = 1, 2, \ldots, m$. We call p the *strategy vector* for Player I. Similarly, the vector $q = (q_1, q_2, \ldots, q_n)$ is the strategy vector for Player II. Further, let r_{ij} be the payoff to Player I when he selects action i and Player II selects action j. Suppose the game is zero-sum. Then the payoff to Player II is simply $-r_{ij}$, and the *expected* return to Player I is just the sum

$$\sum_{i=1}^{m} \sum_{j=1}^{n} p_i q_j r_{ij}.$$

Player I clearly wants to choose his strategy vector p so as to maximize this quantity. Similarly, Player II tries to select the strategy vector q to minimize this sum. The question of the moment is to ask if there is a choice of strategy vectors for the two players that will result in

$$\max_{p} \min_{q} \sum_{i=1}^{m} \sum_{j=1}^{n} p_i q_j r_{ij} = \min_{q} \max_{p} \sum_{i=1}^{m} \sum_{j=1}^{n} p_i q_j r_{ij}. \qquad (*)$$

In 1928, following up on work by Emile Borel, who analyzed this question in the special case when $m, n \leq 4$, and Ernst Zermelo, who conjectured that equality should hold in $(*)$, John von Neumann proved the famous

MINIMAX THEOREM. *For a two-person, zero-sum game with payoff matrix $R = [r_{ij}]$, there exists a unique number V such that*

$$\max_{p} \min_{q} \sum_{i=1}^{m} \sum_{j=1}^{n} p_i q_j r_{ij} = V = \min_{q} \max_{p} \sum_{i=1}^{m} \sum_{j=1}^{n} p_i q_j r_{ij},$$

and strategy vectors p^* and q^* such that

$$\min_{q} \sum_{i=1}^{m} p_i^* q_j r_{ij} = V = \max_{p} \sum_{j=1}^{n} p_i q_j^* r_{ij}.$$

Here the quantity $V = V(p^*, q^*)$ is termed the value of the game.

The two-person, zero-sum game can be generalized in many directions: more than two players, nonzero-sum payoff structures, imperfect information about opponents' actions, and so forth. For our purposes here, the most interesting extension is to consider payoff structures that are not zero-sum. This means that $a_{ij} + b_{ij} \neq 0$. Such games are often termed *cooperative* or *mixed-motive games.*

It turns out that there are exactly 12 *symmetric* mixed-motive games, in which the two players can exchange roles without changing the outcome of the game. Of these, it can be shown that eight have equilibrium points corresponding to dominant strategies for both players, i.e., courses of action that remain unchanged regardless of what the other player is doing. Such games are, of course, uninteresting from a strategic point of view. So we now briefly look at each of the four remaining types of mixed-motive games.

Suppose, for simplicity, that each player has two actions at his disposal which, for reasons that will become apparent later, we'll label "C" and "D." The general form of the payoff matrix is then

Player II

		C	D
	C	(R, R)	(S, T)
Player I			
	D	(T, S)	(P, P)

where P, R, S, and T are real numbers.

Each of the four "interesting" mixed-motive games has been studied extensively in the literature and can be represented by a prototypical situation capturing the concepts peculiar to that particular type of game. These qualitatively different games can be classified by the relative magnitudes of the numbers P, R, S, and T:

- *Leader*—$(T > S > R > P)$.
- *The Battle of the Sexes*—$(S > T > R > P)$.
- *Chicken*—$(T > R > S > P)$.
- *The Prisoner's Dilemma*—$(T > R > P > S)$.

Since our concern here will be with the last type of game, the Prisoner's Dilemma, we refer the reader wishing a more thorough discussion of the first three types of mixed-motive games to the volumes [1, 2] cited in the References.

By far the most interesting and well-studied mixed-motive game is the famous situation involving two prisoners who are accused of a crime. Each prisoner has the option of concealing information from the police (C) or disclosing it (D). If they both conceal the information (i.e., they cooperate), they will be acquitted with a payoff of 3 units to each. If one conceals while the other "squeals" to the police, the squealer receives the defector's reward of 5 units, while the payoff to the "martyr" is zero, reflecting his role in the obstruction of justice. Finally, if they both talk, they will each be convicted of a lesser crime, thereby receiving a payoff of only 1 unit apiece. The appropriate payoff matrix for the Prisoner's Dilemma game is then

Prisoner II

		C	D
		C	D
	C	(3, 3)	(0, 5)
Prisoner I			
	D	(5, 0)	(1, 1)

(Note: Use of the symbols C and D to represent the possible actions by the players in these games is motivated by the usual interpretation of the actions in the Prisoner's Dilemma game. Here C represents "cooperating" with your pal and not confessing, whereas D signifies "defecting" and giving information to the police.)

The Prisoner's Dilemma is a real paradox. The minimax strategies intersect in the choice of mutual defection, which is also the only equilibrium point in the game. So neither prisoner has any reason to regret a minimax choice if the other also plays minimax. The minimax options are also dominant for both prisoners, since each receives a larger payoff by defecting than by cooperating when playing against either of the other player's choices. Thus, it appears to be in the best interests of each prisoner to defect—*regardless* of what the other player decides to do. But if both prisoners choose this individually rational action, the 2 units they each receive are less than the 3 units they could have obtained if they had chosen to remain silent.

The essence of the paradox in the Prisoner's Dilemma lies in the conflict between individual and collective rationality. According to individual rationality, it's better for a prisoner to defect and give information to the police. But, paradoxically, if both attempt to be "martyrs" and remain silent, they each wind up being better off. What's needed to ensure this better outcome for both players is some kind of selection principle based on their collective interests. Perhaps the oldest and best known principle of this sort is the Golden Rule of Confucius: "Do unto others as you would have them do unto you."

Arguably the most important development in the theory of games since von Neumann's proof of the Minimax Theorem was the notion of an *evolutionary stable strategy (ESS)*, introduced by Maynard Smith and Price in 1973. The basic idea behind an ESS comes from animal ecology.

Suppose we have a population of animals that interact with each other in a sequence of contests or "games." The contests involve two animals fighting for a scarce resource, say food. Each animal has a particular strategy that it always employs, and the result of a particular contest is determined by the strategies used by the two contestants. Following Maynard Smith and Price [3], we suppose there are only two pure strategies in the population, an aggressive strategy called *Hawk (H)*, and a nonaggressive strategy termed *Dove (D)*.

Payoffs in the Hawk-Dove game are measured in units of Darwinian fitness, i.e., an increase or decrease in the expected number of offspring. Suppose the victor of a contest receives V units of fitness, while an injury to either party reduces fitness by W units. Finally, there is the time and energy wasted in a "war of attrition," which costs each participant T units.

Each contest is one of three possible types: Hawk-Hawk, Hawk-Dove, or Dove-Dove. The outcomes are determined by the following rules:

• *Hawk-Hawk:* In this case we assume that the two participants each have an equal chance of winning the contest or getting injured. Since such a contest will be short, there is a negligible amount of time and energy wasted. Thus, an individual's expected payoff is $\frac{1}{2}(V - W)$. Generally, we assume that the cost of injury W is greater than the gain from victory V, so that there is some nontrivial danger associated with escalation.

• *Hawk-Dove:* The Dove will flee immediately in this type of encounter, thereby receiving a payoff of zero. The Hawk, on the other hand, then wins the contest without injury, receiving thereby a gain of V units of fitness.

• *Dove-Dove:* Again, in this case, the two participants each have equal chances of winning. But the contest is likely to be a long one, with a lot of time and energy expended in posturing and maneuvering, before the issue is settled. Thus, the expected payoff to each party is $\frac{1}{2}V - T$.

The following payoff matrix summarizes the overall Hawk-Dove situation:

Animal II

		H	D
	H	$\frac{1}{2}(V-W)$	V
Animal I			
	D	0	$\frac{1}{2}V-T$

Note that here the matrix entries express the payoffs to Animal I, omitting by convention an explicit listing of the payoffs to Animal II— even though this is not a zero-sum game. But since the game is the same for both players, it's easy enough to fill-in the payoffs to Animal II if need be.

We have already assumed that any given *individual* is either a Hawk or a Dove. So the outcome of every encounter is rigidly fixed by the above rules. With this payoff structure, the problem is to determine the optimal distribution of strategies within the overall population, since it's clear that from an evolutionary point of view either a population of all Hawks or all Doves can easily be invaded by a mutant playing the other strategy. This is because too much fitness is lost due to injuries if there are too many Hawk-Hawk encounters, while too much time and energy are expended in protracted contests if the dominant encounters are between Doves. The evolutionary stable strategy (ESS) addresses the question of how the population should distribute itself in order to achieve an optimal balance between the two pure strategies.

Underlying the ESS is the following simple question: Is there a strategy such that if most members of the population adopt it, no mutant strategy can invade the population by natural selection? Any such strategy is what we call an ESS, since no mutant strategy confers greater Darwinian fitness on the individuals adopting it; consequently, the ESS cannot be invaded by any competing strategy. For this reason, an ESS is the strategy we would expect to see in Nature.

In the Hawk-Dove game, neither of the pure strategies is an ESS. To see this explicitly, a population of Hawks can be invaded by Doves because the expected payoff to the winner, $\frac{1}{2}(V-W)$, is less than zero, the payoff to a Dove battling a Hawk. Similarly, a population of Doves can be invaded by Hawks, since the payoff V to a Hawk fighting a Dove is greater than $\frac{1}{2}(V-T)$, the payoff to the winner of a Dove-Dove battle. So an ESS for this game must be a mixture of Hawk and Dove.

To formalize the ESS concept, assume that there are n pure strategies available and that p_i represents the probability that the animal plays pure strategy i, where, of course, $\sum_{i=1}^{n} p_i = 1$. Further, let

$A \in R^{n \times n}$ be the payoff matrix for the game. Under these circumstances, the probability vector $p = (p_1, p_2, \ldots, p_n)$ is an ESS strategy if it satisfies the following equilibrium and stability conditions:

1. p is as good a reply against itself as any other strategy x,

and

2. if x is a best reply against p, then p is a better reply against x than x itself.

Mathematically, we can state these two conditions succinctly as

I. $(p, Ap) \geq (x, Ap)$ for all $x \in S^n$ (equilibrium),

II. If $(p, Ap) = (x, Ap)$ for $x \neq p$, then $(p, Ax) > (x, Ax)$ (stability).

Using these definitions, I'll leave it to the reader to verify that the mixed strategy

$$p_H = 1 + \frac{V - W}{2T + W},$$
$$p_D = \frac{W - V}{2T + W} \ (= 1 - p_H),$$

constitutes an ESS for the Hawk-Dove game. Inspection of this strategy offers mathematical support for the fairly obvious fact that as the reward for winning a contest outpaces the loss in fitness from injuries, it becomes increasingly more advantageous to be a Hawk.

With these game-theoretic ideas at our disposal, let's turn now to a consideration of how they have been employed to study how cooperative behavior can emerge in an evolutionary context.

3. TIT FOR TAT and the Evolution of Cooperation

The cornerstone of sociobiological reasoning is the claim that human behavior patterns, including what look on the surface like selfless acts of altruism, emerge out of genetically selfish actions. The relevance of the Prisoner's Dilemma game for sociobiology is evident. In the context of the Prisoner's Dilemma, we can translate the sociobiological thesis into the statement that the individually rational act of defection will always be preferred to the collectively rational choice of cooperation. The question then becomes: Can that situation ever lead to a population of cooperators? If there is no way for cooperative acts to emerge naturally out of self-interest, it's going to be very difficult for the sociobiologists to support their case.

Put in game-theoretic terms, the strategy of defecting on every play (ALL D) is an evolutionary stable strategy, since players who deviate from this strategy can never make inroads against a population of defectors. Or can they? Are there situations in which a less cutthroat course of action can ultimately get a foothold in a population of defectors? This was the question that political scientist Robert Axelrod set out to answer in one of the most intriguing psychological experiments carried out in recent years. The separate issues that Axelrod wanted to address were: (1) How can cooperation get started at all in a world of egoists? (2) Can individuals employing cooperative strategies survive better than their uncooperative rivals? (3) Which cooperative strategies will do best, and how will they come to dominate?

Axelrod's key observation was to note that while ALL D, the strategy of always defecting, is uninvadable for a sequence of Prisoner's Dilemma interactions that is of known, fixed, and finite duration, there may be alternative ESS strategies if the number of interactions is not known by both parties in advance. So after having played a round of the Prisoner's Dilemma, if there is a nonzero chance that the game might continue for another round, then perhaps there is a nice strategy that is also ESS. Here by "nice" we mean a strategy that would not be the first to defect.

To put this speculation to the test, Axelrod invited a number of psychologists, mathematicians, political scientists, and computer experts to participate in a contest pitting different strategies against one another in a computer tournament [4]. The idea was for each participant to supply what he or she considered to be the best strategy for playing a sequence of Prisoner's Dilemma interactions, with the different strategies then competing against each other in a round-robin tournament. Fourteen competitors sent in strategies, which were in the form of computer programs. The ground rules allowed the programs to make use of any information whatsoever about past plays of the game. Furthermore, the programs didn't have to be deterministic, but were allowed to arrive at their choice of what to do by some kind of randomizing device if the player so desired. The only condition imposed was that the program ultimately come to a definite decision for each round of play: C or D. In addition to the strategies submitted, Axelrod included the strategy RANDOM, which decided whether to cooperate or defect by, in effect, flipping a coin. In the tournament itself, every program was made to engage every other (including a clone of itself)

200 times, the entire experiment being carried out five times in order to smooth out statistical fluctuations in the random-number generator used for the nondeterministic strategies.

Using the payoff values $R = 3, S = 0, T = 5$, and $P = 1$, the strategy that won the tournament turned out to be the simplest. This was the three-line program describing the strategy TIT FOR TAT. It was offered by game theorist Anatol Rapoport, and consisted of the two rules: (1) cooperate on the first play; (2) thereafter, do whatever your opponent did on the preceding round. That such a simple, straightforward strategy could prevail against so many seemingly far more complex and sophisticated rules for action seems nothing short of miraculous. The central lesson of this tournament was that in order for a strategy to succeed, it should be both nice and forgiving, i.e., it should be willing both to initiate and to reciprocate cooperation. Following a detailed analysis of the tournament, Axelrod decided to hold a second tournament to see if the lessons learned the first time around could be put into practice to develop even more effective cooperative strategies than TIT FOR TAT.

As prelude to the second tournament, Axelrod packaged up all the information and results from the first tournament and sent it to the various participants, asking them to submit revised strategies. He also opened up the tournament to outsiders by taking out ads in computer magazines, hoping to attract some programming fanatics who might take the time to devise truly ingenious strategies. Altogether Axelrod received 62 entries from around the world, including one from John Maynard Smith, mentioned earlier as the developer of the ideas of the evolutionary game and the ESS. The winner? Again it was Rapoport with TIT FOR TAT!

Even against this supposedly much stronger field, Rapoport's game-theoretic version of the Golden Rule was the hands-down winner. The general lesson that emerged from the second tournament was that not only is it important to be nice and forgiving, but it's also important to be both provocable and recognizable, i.e., you should get mad at defectors and retaliate quickly but without being vindictive, and you should be straightforward, avoiding the impression of being too complex. After extensive study of the results, Axelrod summarized the success of TIT FOR TAT in the following way:

> TIT FOR TAT won the tournaments not by beating the other player but by eliciting behavior from the other player that allowed both to do well. ... so in a non-zero sum world, you do not have to do better than the other player to do well for yourself. This is especially true when you

are interacting with many different players. . . . The other's success is virtually a pre-requisite for doing well yourself.

So what are the implications of these results for sociobiology and evolution of cooperative behavior?

If we think of the total points amassed by a strategy during the course of the tournament as its "fitness," interpret "fitness" to mean "the number of progeny in the next generation," and let "next generation" mean "next tournament," then what happens is that each tournament's results determine the environment for the next tournament. The fittest strategies then become more heavily represented in the population of strategies fighting it out in the next tournament. This interpretation leads to a kind of ecological adaptation without evolution, since no *new* species come into existence. Sociobiologists can take heart in this sort of interpretation of Axelrod's experiments because it shows that it's possible for phenotypically altruistic (cooperative) behavior to emerge out of individually selfish motives. It's important to emphasize here, though, that these results say nothing about the actual causal factors at work generating the individual motives. They **could** be genetic, as hard-core sociobiologists would love to argue, but there is nothing in Axelrod's work to say that they are. Nevertheless, the experiments do offer some support to the sociobiological explanation of cooperative behavior by means of reciprocal altruism.

Following his work on the evolution of cooperation, Axelrod carried out another set of experiments that also give succor to the sociobiologist's claim for an evolutionary development of standards of behavior, i.e., cultural norms. The basic idea is to use a souped-up version of the Prisoner's Dilemma in which the players had the choice not only of cooperation or defection, but also of punishing a defection or letting it pass. Players in the Norms Game are characterized by two qualities: **Boldness (B)**, which measures the risk they are willing to run in defecting; and **Vengefulness (V)**, a measure of their inclination to punish defection. Strategies were assigned randomly to 20 players, with the first round of play lasting until each player had four opportunities to defect. At the end of the first generation, a strategy was given one offspring if its score was near average, two offspring if its score was at least one standard deviation above the mean, and no offspring if its score was more than one standard deviation below the mean. Furthermore, Axelrod allowed for the emergence of new strategies through a process of mutation in such a way that about one new strategy emerged in each generation.

The results of the simulation showed that with enough time, all populations eventually converge to the collapse of the norm, i.e., V approaching zero. The reason appears to be that the players lack sufficient incentive to punish the defectors, i.e., nobody wants to play sheriff. As one way of enforcing the norm, Axelrod suggests a *metanorm:* direct vengeance not only against those who defect, but also against those who refuse to punish them. This is the kind of procedure we see in some totalitarian countries, where when a citizen is accused by the authorities of some real or imagined ideological transgression, others are called upon to pile their own denunciations onto the back of the hapless offender.

While these results are still in the preliminary stage, the Evolution of Cooperation Game and the Norms Game both provide some theoretical evidence in support of the idea that cooperative social behavior can emerge as the result of evolutionary processes involving individually selfish agents.

4. Beyond TIT FOR TAT

The overwhelming success of TIT FOR TAT in Axelrod's experiments raised a number of intriguing questions surrounding the stability of this particular strategy. For the most part, these questions can be divided into two qualitatively different types:

A. Asking about whether there are strategies for playing the game *under the original rules* that are even better than TIT FOR TAT.

B. Asking about the degree to which TIT FOR TAT retains its preeminent position if we change the ground rules of the game by, for example, admitting uncertainty into the identification of what the other player has actually done, changing the actual numbers used in the payoffs, or allowing the players to have more than two courses of action.

Here we examine studies that have looked at both of these questions.

In an attempt to discover strategies that perform even better than TIT FOR TAT under the original Axelrod rules of the game, Martin Nowak and Karl Sigmund reported an improved strategy for playing the game in [5]. This was the strategy they termed PAVLOV, which can be described very simply as "win-stay, lose-shift." What this means is that a PAVLOV player cooperates if and only if both players chose the same action on the preceding round; otherwise, the player defects. The terminology "Pavlov" comes from the fact that this strategy embodies an almost conditioned reflex-like response to the payoff: it repeats its previous move if it was rewarded with the R or T payoff, while it changes its action if it was punished by receiving only the S or P payoff.

The reasoning leading up to the discovery of PAVLOV's superiority to TIT FOR TAT starts by observing that TIT FOR TAT can suffer from random perturbations in two ways. First, a TIT FOR TAT population can be degraded by the random appearance of unconditional cooperators (ALL C), whose appearance then allows the growth of exploiters. This problem arises because TIT FOR TAT is **not** an evolutionary stable strategy. (Note: It can be shown, in fact, that *no* pure strategy can be ESS in the iterated Prisoner's Dilemma game.) And second, an occasional mistake between two TIT FOR TAT players can result in a prolonged run of mutual backbiting, as the players get "out of synch" with each other and begin engaging in a sequence of mutual defections until another mistake occurs that neutralizes the previous misunderstanding.

PAVLOV has two major advantages over TIT FOR TAT: (1) a misunderstanding between two PAVLOV players causes only a single round of mutual defection followed by a return to joint cooperation; (2) PAVLOV cannot be invaded by unconditional cooperators (ALL C), thereby allowing exploiters to gain a foothold in the population. To the contrary, PAVLOV has no qualms about fleecing a sucker, once it has discovered (after a misunderstanding) that it need not fear retaliation. So in this sense, cooperation based on PAVLOV is a safer bet than cooperation arising from TIT FOR TAT, since such cooperation is less prone to exploitation by the intermediary of an ALL C invasion.

In an extensive series of computer simulations, the work of Nowak and Sigmund uncovered the fact that over relatively long time horizons (number of generations in the population), there is a pronounced tendency for populations to move toward generally cooperative behavior. Moreover, as the time horizon gets longer, the cooperative behavior is composed exclusively of two strategies: PAVLOV and "Generous TIT FOR TAT," a strategy that cooperates after the opponent cooperates, but also cooperates with a certain probability after a defection, as well. As an example, after 1 million rounds, approximately 60 percent of the population consisted of cooperative strategies, with PAVLOV accounting for about two-thirds of the cooperators and Generous TIT FOR TAT another 5 percent or so. And when the number of generations reaches 10 million, nearly 90 percent of the population is cooperators, with 80 percent of the population playing PAVLOV and 10 percent

Generous TIT FOR TAT. Readers wishing more details of these fascinating experiments should consult the original article [5].

We have just noted the problems that can arise in "no frills" TIT FOR TAT when there is a nonzero chance that the players will misunderstand the action taken by their opponent, thereby opening up the possibility for a long sequence of mutual defections. If a misunderstanding of this type occurs, the two players will alternate between three modes of behavior: (a) First, they will play the ordinary TIT FOR TAT actions (C, C), but when a misunderstanding takes place, (b) they will shift to alternating (C, D) and (D, C). Finally, the third possibility is (c) sequences of mutual defections (D, D). The average likelihoods of the three modes (a), (b), and (c) are 1/4, 1/2, and 1/4, respectively, leading to an average payoff of 9/4.

In an attempt to develop strategies that are more resistant to misunderstanding, Kristian Lindgren employed genetic algorithms in [6] as a way of "evolving" strategies for playing the iterated Prisoner's Dilemma game. What he discovered is that there exist deterministic (i.e., pure) strategies that are robust against noise (misunderstandings), unexploitable (in the sense that they cannot be invaded), and reach an average score of nearly 3—substantially better than ordinary TIT FOR TAT.

In Lindgren's experiments, he coded different finite-memory strategies using the following binary scheme. An m-length history is given by the vector

$$h_m = (a_{m-1}, \ldots, a_1, a_0),$$

where each a_i is either 0 or 1. Thus, since a deterministic strategy of memory m associates an action C or D to each history h_m, we can specify the strategy by a binary sequence

$$S = (A_0, A_1, \ldots, A_{n-1}).$$

This strategy is like the "genetic code" for the strategy that chooses action A_k when history h_k turns up. So the length of such a "genome" is $n = 2^m$.

As an example of this scheme for strategies of memory length $m = 1$, the histories are labeled 0 or 1, corresponding to the opponent's having defected or cooperated, respectively. Thus, the four memory-1 strategies are: $S_1 = (00), S_2 = (01), S_3 = (10)$, and $S_4 = (11)$. The strategy S_1 always defects (ALL D), S_2 cooperates only if history 1 turns up (TIT FOR TAT), S_3 does the opposite of S_2 (ANTI-TIT FOR TAT), and S_4 always cooperates (ALL C).

The graph in Figure 1 shows a simulation of 30,000 generations, starting with a population equally distributed among the four memory-1 strategies. Here we see that the system reaches periods of stasis, where there is coexistence between TIT FOR TAT and ANTI-TIT FOR TAT. This stasis is then punctuated by a number of memory-2 strategies, and after a period of unstable behavior, the system slowly stabilizes when the strategy $S = (1001)$ increases in the population. (Note: The perceptive reader will recognize this strategy as PAVLOV: It cooperates only if both players took the same action in the previous round.) Another period of stasis then ensues, before being broken up by memory-3 and finally, memory-4 strategies.

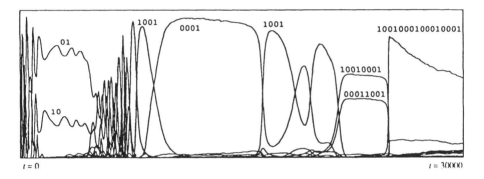

Figure 1. Evolution of population strategies.

Figure 2 shows the average score s and number of genotypes n for the population generated in the simulation shown in Figure 1. Here we see that when the exploiting memory-2 strategy dominates, the average score drops down close to 1. But the final stasis populated by the ESS memory-4 strategy reaches a score of 2.91, close to the maximum possible score of 3 that's attained by the best strategies in a noise-free environment.

There are many conclusions that we can draw from Lindgren's experiments, one of the most important being that when there is a nonzero chance for a misunderstanding, there exists an unexploitable strategy that is cooperative. But the simulations suggest that the minimal memory needed for such a strategy is 4, i.e., the player should take account of both parties actions for the previous two rounds. Basically, defecting twice in response to a single defection is a strategy that cannot be exploited by intruders. Lindgren offers many other insights into the structure of "good" strategies for the Prisoner's Dilemma that

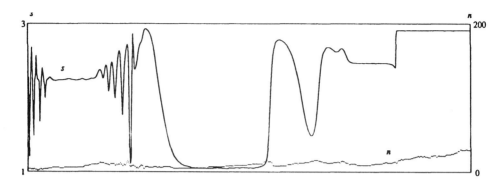

Figure 2. Average scores and number of genotypes in the population.

we have no space to recount here. So again, the reader is referred to the source [6] for these details.

For the most part, the foregoing results have with minor exceptions adhered pretty much to Axelrod's original ground rules for playing the iterated Prisoner's Dilemma. Now we consider briefly a cluster of experiments involving major changes in the rules of the game, starting with work in which the players retain a memory of which players they have faced in the past and can choose whether or not to play the game again with particular players.

In [7], Stanley, Ashlock, and Tesfatsion set up an iterated Prisoner's Dilemma ecology in which the agents are represented by finite automata. These agents interact with each other in the usual Prisoner's Dilemma way, with the notable exception that players choose and refuse potential game partners on the basis of continually updated expected payoffs. This choice mechanism gives players a way to increase their chance of encountering other cooperative players, as well as offering protection against defectors without having to become defectors themselves.

Under the choice/refusal option, ostracism of defectors occurs endogenously as an increasing number of players refuse their game offers. However, choice and refusal also permits clever "ripoff" players to home-in quickly on exploitable players and to establish parasitic partnerships. Simulation studies indicate that the emergence of cooperation tends to be accelerated by the choice/refusal option, but that the rate of emergence is a highly nonlinear function of the parameters characterizing the choice/refusal mechanism. Unfortunately, to chronicle the details of this mechanism and the resulting simulation studies would involve a bit more space than this chapter can accommodate, so we again refer the reader to the paper [7] for the full story.

In the survey paper [8], Axelrod and Dion have summarized a number of studies in which the original iterated Prisoner's Dilemma game has been modified in one direction or another. Here is a brief account of their findings:

- *Interactions*—The original game specifies interactions between pairs of players. In some situations, though, it makes more sense to assume that interactions are among more than two players. In these games, players make a choice that they then play with all other players simultaneously. This leads to what's termed the *n-person iterated Prisoner's Dilemma.*

It has been shown that increasing the number of players makes cooperation more difficult. In both evolutionary and nonevolutionary settings, cooperation can be part of an equilibrium of the game—but only if the "shadow of the future" is long and/or the number of cooperators is large.

- *Choices*—If the players are allowed more than the two choices of cooperation or defection, the outcome of the game can also be altered significantly. For example, we have already seen the effect of allowing players the option of playing or not playing a particular partner. In a related study, it was shown that the stable equilibria of a nonsimultaneous move game are either ALL D or a form of conditional cooperation.

- *Payoffs*—Experiments with changing the payoff matrix also lead to interesting departures from the standard Axelrod results. One such result involved having the players play both the iterated Prisoner's Dilemma and Chicken. It turned out that under these conditions, TIT FOR TAT does well in this situation—but not as well in Chicken as the strategy PERMANENT RETALIATION, under which the player cooperates until the other player defects and then defects for the remainder of the game.

- *Noise*—In the work of Nowark and Sigmund, as well as in the studies of Lindgren, we have already seen the rather dramatic effect of adding faulty transmission of strategy choices (i.e., noise) into the picture. In one study it was shown that in the presence of any amount of noise, two TIT FOR TAT players will in the long run achieve the same average payoffs as two interacting RANDOM players. Nevertheless, if the amount of noise is small, TIT FOR TAT can still perform well. For instance, when the noise level is only 1 percent, TIT FOR TAT still turned out to be the best strategy in one tournament. But when the noise level went up to 10 percent, TIT FOR TAT dropped down to sixth place among 21 strategies.

But there are many ways that noise can be introduced into the Prisoner's Dilemma setup, and its difficult to make any general statements about when TIT FOR TAT can be "souped up" to still win out in these noisy environments. About the best that can be said is that if the amount of noise is small, unilateral generosity seems to be the best response. But for larger amounts of noise, there is a trade-off between the unnecessary conflict that can be avoided by being generous and the exploitation that generosity invites.

• *Shadow of the Future*—The expected length of the game plays a crucial role in strategic thinking in the Prisoner's Dilemma situation. So when the probability of the game continuing for one more round w is modified, so are the winning strategies in the computer tournaments we've been talking about here. The work by Stanley, Ashlock, and Tesfatsion mentioned earlier involving the changing likelihood of encountering a given player again is an example of this sort of behavior-dependent play.

Other experiments have focused on changing the shadow of the future by using a stochastic payoff matrix. For instance, in some moves of this kind of game, the players face a normal Prisoner's Dilemma, while in other moves cooperation is the dominant choice. If the probability of surviving the current move is a function of the total payoff received to that time, an egoist may cooperate on the Prisoner's Dilemma moves of the game to ensure the presence of a "congenial" partner in other stages of play.

In general, it seems to be the case that extending the shadow of the future by whatever means tends to enhance the likelihood of cooperative behavior substantially. So as a guiding principle, if you want to promote cooperation, change the rules of the game so as to increase the probability that two players will meet again in future rounds of play.

• *Population Dynamics*—It turns out that if a population can be invaded by multiple mutants (i.e., those playing strategies different from the population as a whole), then there is no single strategy the population can adopt that is ESS. In other words, if there is some chance for future interaction, then for each strategy X, there is always a set of strategies $Z(X)$ such that X is not an ESS against Z. Moreover, no pure strategy whose behavior is determined solely by the history of the game is an ESS if the future is important enough. More specifically, if the shadow of the future $w > \min\{(T-R)/(T-P), (P-S)/(R-S)\}$, then there can be no ESS for the iterated Prisoner's Dilemma.

This kind of result places even greater emphasis on the analysis of sets of strategies, since they show that the evolutionary pathway leading to cooperation can be understood only by looking at the set of possible competitors to a population playing a given strategy. The results considered earlier by Lindgren using genetic algorithms is one approach to studying this kind of issue. But there are others. What one tends to find is that strategies similar to TIT FOR TAT often emerge within a few dozen generations. But not always. Occasionally, strategies like WIN STAY, LOSE CHANGE appear that outperform TIT FOR TAT in a given environment appear as well. In general, about all one can say is that the situation is an unstable one, in the sense that what's best against one population can be worst against another. There's just no universal strategy that can do well under all circumstances.

• *Population Structure*—In his original studies, Axelrod found that cooperation can invade a population of egoists *if* the cooperative strategies invade in clusters. This raises the question of whether cooperation can evolve without such a population structure.

Part of the answer comes from considering an ESS. Since no strategy is ESS if the shadow of the future is important enough, it is possible for cooperation to emerge without clustering if the "right" combination of strategies is present. For example, if we have what's termed "assortative mating and meeting," in which players pick only fellow cooperators and shun defectors, then cooperation can emerge without clustering. In another direction, it has been shown that cooperation can get started without clustering if the payoffs to cooperative acts depend on the frequency with which those acts are performed.

This is just the briefest of reviews of the many ways in which the Prisoner's Dilemma game can be extended. For full details, together with references to the literature on these and many other extensions of the game, we again refer the reader to the survey article [8].

5. Testing the Theory in the Real World

The proof of any pudding is in the eating. And the various theories of cooperation embodied in the iterated Prisoner's Dilemma game is no exception: If we're going to take seriously the idea that cooperation can emerge in a world of egoists, we have to validate the theory with examples of such behavior in the real world of animals and humans.

This section discusses several studies of this sort. So on to the real world of cabbages and kings, or in this case, the world of spiders.

One of the most interesting tests of the ESS in nature was carried out by Susan Riechert, who studied the behavior of the common grass spider *A. aperta* in settling territorial disputes [9]. Riechert studied these spiders in two habitats that differed greatly in the availability of suitable locations for building webs—a desert grassland in New Mexico and a desert riparian area consisting of a woodland bordering a stream in Arizona, a region offering many more favorable locations for webs. While there is no room here to go into the details of how Riechert determined the actions available to the spider and assigned the various payoffs, her final conclusions are worth pondering. She discovered that the contest behavior for web sites in the riparian regions deviated substantially from the ESS predicted by the game-theoretic model. In particular, contrary to theory, a riparian spider does not withdraw from occupied territory when it encounters the owner of the web. Rather, they engage in a dispute that escalates to potentially injurious behavior. On the other hand, the behavior of grassland spiders does follow the ESS as predicted by the theory, with the time and energy they expend in fights varying with their probability of emerging victorious.

So, while the riparian spiders are less aggressive than their desert grassland cousins, just as ESS theory predicts, they are still somewhat more aggressive than they should be. This leads us to ask: Why does the behavior in these territorial disputes differ from the ESS for riparian spiders and not for their grassland cousins? Riechert gives an answer that will gladden the heart of any sociobiologist. She states:

> If one assumes that the model is correct—that it has taken into account all the important parameters and includes all possible set of strategies— then there must be some biological explanation for the observed deviation. . . . One possibility is that the release from strong competition is a recent event and that there just has not been sufficient time for natural selection to operate on the behavioral traits to complete the expected change. . . . Finally, a major change in the wiring of *A. aperta*'s nervous system might be required to achieve the new ESS, and such a mutant may simply not have arisen yet.

Another example of the ESS in nature was reported by Maynard Smith in [10]. This involves six stickleback fish who were fed at both ends of a tank, the rate of feeding being twice as great at one end of the tank as at the other. The ESS is when no fish can gain by moving from one end of the tank to the other. This occurs when there are four fish at the end with the higher food supply and two fish at the other end.

In a set of experiments in which the fish are all dropped into the tank at random locations, Milinsky found that the fish distribute themselves statistically in accordance with the ESS. Individual fish, on the other hand, continue to move from one end of the tank to the other. This is to be expected, since in nature the relative attractiveness of patches of food will not remain constant over time. And when the experimenters switched the rate of supply between the two ends during the course of a given experiment, the fish again distributed themselves statistically as predicted by the ESS.

The spawning of sea bass provides a final example in nature of how cooperative behavior can emerge in a world of egoists. These fish have both male and female sexual organs, form pairs, and can be said to take turns at being the high investment (laying eggs) and the low investment (providing sperm to fertilize the eggs) partner. Up to ten spawnings occur in a single day, with only a few eggs provided each time. Pairs of fish tend to break up if sex roles are not divided evenly. From studies in the wild, it appears that this type of cooperative behavior got started at a time when the sea bass was scarce as a species, leading to a high level of inbreeding. Such inbreeding implies a relatedness in the pairs, which would have initially promoted cooperation without the need for further relatedness. A more complete account of this phenomenon is found in the book [4]. Now let's turn our attention from general cooperation and the ESS to the specific case of real-world laboratory experiments involving the iterated Prisoner's Dilemma.

By now, there have been well over a thousand experiments based on the Prisoner's Dilemma and related games. So it would be futile to try to discuss even a small fraction of them here. We refer the reader to an excellent review article on the topic [11], contenting ourselves with only a brief summary of a few of the more interesting cases.

Without a doubt, the most striking general finding of the Prisoner's Dilemma experiments is the *DD lock-in effect*. When the game is repeated many times, there is a pronounced tendency for long runs of D choices by both players to take place. Of course, even though the ALL D strategy is dominant—and even minimax—for this game, both players are better off if they both chose C instead of D. Strangely, though, in their influential book *Games and Decisions* (Wiley, New York, 1957), Luce and Raiffa stated that "We feel that in most cases an unarticulated collusion between the players will develop." That this prediction has *not* been borne out in hundreds of experiments is a fact of which many game theorists still remain largely ignorant.

We can get some insight into the DD lock-in effect from the work of Rapoport and Chammah [12]. They found that three phases typically occur in a long series of plays of the Prisoner's Dilemma game: Initially, the proportion of C choices is typically slightly higher than 1/2. But this is followed by a rapid decline in cooperation, the so-called "sobering period." This period usually lasts for about 30 rounds, after which cooperative choices begin to increase slowly in frequency (the "recovery period"). The frequency of C choices generally then reaches about 60 percent by round 300.

The moderately high fraction of initial C choices is often attributed to an initial reservoir of goodwill or simply to a naïve lack of comprehension on the part of the subjects of the strategic structure of the game. The sobering period then reflects a decline in trust and trustworthiness, an increase in competitiveness, or perhaps just a dawning understanding of the structure of the payoff matrix. The recovery period can be thought of then as reflective of the slow and imperfect growth of the "unarticulated collusion" mentioned by Luce and Raiffa.

Another interesting outcome of experiments with human subjects in Prisoner's Dilemma situations involves programmed strategies. Experiments in this area pit subjects against another "player," who in reality is just a confederate of the experimenter programmed in advance to make predetermined sequences of choices.

Certainly the simplest of these kinds of experiments are those in which the confederate chooses either C or D on every trial. In the Prisoner's Dilemma situations, an ALL C strategy elicits much higher frequencies of C choices from subjects than ALL D. Of course, this is in line with expectations, since self-defense considerations force a subject to choose D against an ALL D player in order to avoid the sucker's payoff. What is surprising, however, is the large proportion of subjects who seize the opportunity to exploit an ALL C player by choosing to defect. And, in fact, the tendency for many subjects to exploit "pacifist strategies" has been confirmed in mixed-motive games other than the Prisoner's Dilemma.

One of the most striking and unexpected findings to emerge from early experiments with games like the Prisoner's Dilemma is the apparent tendency for females to exhibit much lower frequencies of C choices than males. In light of traditional sex roles, one might have expected just the opposite. As an illustration of this effect, in games involving two male players, the frequency of C choices was 59 percent. In the same study, it was only 34 percent when the two players were both female.

There are now over a hundred experiments on record showing this pronounced sex difference in mixed-motive games. But the effect is still shrouded in mystery. It has been claimed that the sex difference is an artifact, and there is some evidence that the effect disappears when the experimenter is a female. At present, no one seems to have been able to come up with a convincing explanation of the effect, but it's not unreasonable to assume that in Western industrial cultures, males are brought up to behave more boldly than females and with a greater willingness to take risks. This would then account for their greater likelihood of choosing the non-minimax C choice in the Prisoner's Dilemma in spite of the danger of receiving the sucker's payoff.

While there are many other effects of the above sort scattered throughout the experimental gaming literature, space constraints demand closing our discussion here. Again, the interested reader is urged to consult the literature on these matters cited in the chapter references.

References

[1] Colman, A. *Game Theory and Experimental Games.* Oxford: Pergamon Press, 1982.

[2] Casti, J. *Reality Rules–II: Picturing the World in Mathematics.* New York: Wiley, 1992.

[3] Maynard Smith, J. and G. Price. "The Logic of Animal Conflicts." *Nature,* 246 (1973), 15–18.

[4] Axelrod, R. *The Evolution of Cooperation.* New York: Basic Books, 1984.

[5] Nowak, M. and K. Sigmund. "A Strategy of Win-Stay, Lose-Shift that Outperforms Tit-for-Tat in the Prisoner's Dilemma Game." *Nature,* 364 (1993), 56–58.

[6] Lindgren, K. "Evolutionary Phenomena in Simple Dynamics," in *Artificial Life–II,* C. Langton et al., eds. Redwood City, CA: Addison-Wesley, 1992, pp. 295–312.

[7] Stanley, A., D. Ashlock, and L. Tesfatsion. "Iterated Prisoner's Dilemma with Choice and Refusal," preprint, June 1992.

[8] Axelrod, R. and D. Dion. "The Further Evolution of Cooperation." *Science,* 242 (1988), 1385–1390.

[9] Riechert, S. "Spider Fights as a Test of Evolutionary Game Theory." *American Scientist,* 74 (1986), 604–610.

[10] Maynard Smith, J. *Evolution and the Theory of Games.* Cambridge: Cambridge University Press, 1982.

[11] Schlenker, B. and T. Bonoma. "Fun and Games: The Validity of Games for the Study of Conflict." *Journal of Conflict Resolution,* 22 (1978), 7–38.

[12] Rapoport, A. and A. Chammah. *Prisoner's Dilemma: A Study in Conflict and Cooperation.* Ann Arbor, MI: University of Michigan Press, 1965.

CHAPTER 4

Randomness in Arithmetic and the Decline and Fall of Reductionism in Pure Mathematics*

GREGORY J. CHAITIN

Computer Science Department
IBM Research Division
Box 704
Yorktown Heights, NY 10598, USA

1. Hilbert on the Axiomatic Method

Last month I was a speaker at a symposium on reductionism at Cambridge University where Turing did his work. I'd like to repeat the talk I gave there and explain how my work continues and extends Turing's. Two previous speakers had said bad things about David Hilbert. So I started by saying that in spite of what you might have heard in some of the previous lectures, Hilbert was not a twit!

Hilbert's idea is the culmination of two thousand years of mathematical tradition going back to Euclid's axiomatic treatment of geometry, going back to Leibniz's dream of a symbolic logic and Russell and Whitehead's monumental *Principia Mathematica.* Hilbert's dream was to clarify once and for all the methods of mathematical reasoning. Hilbert wanted to formulate a formal axiomatic system which would encompass all of mathematics.

Hilbert emphasized a number of key properties that such a formal axiomatic system should have. It's like a computer programming

---

*This is the transcript of a lecture given at the University of New Mexico in October 1992. It follows closely the presentation made at the Abisko meeting.

Cooperation and Conflict in General Evolutionary Processes, edited by John L. Casti and Anders Karlqvist.

ISBN 0–471–59487–3 ©1994 John Wiley & Sons, Inc.

language. It's a precise statement about the methods of reasoning, the postulates, and the methods of inference that we accept as mathematicians. Furthermore, Hilbert stipulated that the formal axiomatic system encompassing all of mathematics that he wanted to construct should be "consistent" and it should be "complete."

Formal Axiomatic System

$\longrightarrow$ consistent

$\longrightarrow$ complete

$\longrightarrow$

Consistent means that you shouldn't be able to prove an assertion and the contrary of the assertion.

Formal Axiomatic System

$\longrightarrow$ consistent $A \neg A$

$\longrightarrow$ complete

$\longrightarrow$

You shouldn't be able to prove A and not A. That would be very embarrassing.

Complete means that if you make a meaningful assertion you should be able to settle it one way or the other. It means that either A or not A should be a theorem, should be provable from the axioms using the rules of inference in the formal axiomatic system.

Formal Axiomatic System

$\longrightarrow$ consistent $A \neg A$

$\longrightarrow$ complete $A \neg A$

$\longrightarrow$

Consider a meaningful assertion A and its contrary not A. Exactly one of the two should be provable if the formal axiomatic system is consistent and complete.

A formal axiomatic system is like a programming language. There is an alphabet and rules of grammar, in other words, a formal syntax. It's a kind of thing that we are familiar with now. Look back at Russell and Whitehead's three enormous volumes full of symbols and you'll feel you're looking at a large computer program in some incomprehensible programming language.

Now there's a very surprising fact. Consistent and complete means only truth and all the truth. They seem like reasonable requirements. There's a funny consequence, though, having to do with something called the decision problem. In German it's the *Entscheidungsproblem.*

Formal Axiomatic System

$\longrightarrow$ consistent $A \neg A$

$\longrightarrow$ complete $A \neg A$

$\longrightarrow$ decision problem

Hilbert ascribed a great deal of importance to the *Entscheidungsproblem,* or what in English is called the decision problem.

HILBERT

Formal Axiomatic System

$\longrightarrow$ consistent $A \neg A$

$\longrightarrow$ complete $A \neg A$

$\longrightarrow$ decision problem

Solving the decision problem for a formal axiomatic system is giving an algorithm that enables you to decide whether or not any given meaningful assertion is a theorem. A solution of the decision problem is called a decision procedure.

HILBERT

Formal Axiomatic System

$\longrightarrow$ consistent $A \neg A$

$\longrightarrow$ complete $A \neg A$

$\longrightarrow$ decision procedure

This sounds weird. The formal axiomatic system that Hilbert wanted to construct would have included all of mathematics: elementary arithmetic, calculus, algebra, everything. If there's a decision procedure, then mathematicians are out of work. This algorithm, this mechanical procedure, can check whether something is a theorem or not, can check whether it's true or not. So to require that there be a decision procedure for this formal axiomatic system sounds like you're asking for a lot.

However, it's very easy to see that if it's consistent and it's complete, that implies that there must be a decision procedure. Here's how you do it. You have a formal language with a finite alphabet and a grammar. And Hilbert emphasized that the whole point of a formal axiomatic system is that there must be a mechanical procedure for checking whether or not a purported proof is correct, whether it obeys the rules. That's the notion that mathematical truth should be objective so that everyone can agree whether or not a proof follows the rules.

So if that's the case you run through all possible proofs in size order, and look at all sequences of symbols from the alphabet one character long, two, three, four, a thousand, a thousand and one . . . a hundred thousand characters long. You apply the mechanical procedure which is the essence of the formal axiomatic system, to check whether each proof is valid. Most of the time, of course, it'll be nonsense, it'll be ungrammatical. But you'll eventually find every possible proof. It's like a million monkeys typing away. You'll find every possible proof, though only in principle of course. The number grows exponentially and this is something that you couldn't do in practice. You'd never get to proofs that are one page long.

But in principle you could run through all possible proofs, check which ones are valid, see what they prove, and that way you can systematically find all theorems. In other words, there is an algorithm, a mechanical procedure, for generating one by one every theorem that can be demonstrated in a formal axiomatic system. So if for every meaningful assertion within the system, either the assertion is a theorem or its contrary is a theorem, only one of them, then you get a decision procedure. To see whether or not an assertion is a theorem you just run through all possible proofs until you find the assertion coming out as a theorem or you prove the contrary assertion.

So it seems that Hilbert actually believed that he was going to solve once and for all, all mathematical problems. It sounds amazing, but apparently he did. He believed that he would be able to set down a consistent and complete formal axiomatic system for all of mathematics and from it obtain a decision procedure for all of mathematics. This is just following the formal, axiomatic tradition in mathematics.

But I'm sure he didn't think that it would be a practical decision procedure. The one I've outlined would only work in principle. It's exponentially slow, it's terribly slow! Totally impractical. But the idea was that if all mathematicians could agree whether a proof is correct and be consistent and complete, in principle that would give a decision procedure for automatically solving any mathematical problem. This was Hilbert's magnificent dream, and it was to be the culmination of Euclid and Leibniz, and Boole and Peano, and Russell and Whitehead.

Of course the only problem with this inspiring project is that it turned out to be impossible!

2. Gödel, Turing, and Cantor's Diagonal Argument

Hilbert is indeed inspiring. His famous lecture in the year 1900 is a call to arms to mathematicians to solve a list of twenty-three difficult problems. As a young kid becoming a mathematician you read that list of twenty-three problems and Hilbert is saying that there is no limit to what mathematicians can do. We can solve a problem if we are clever enough and work at it long enough. He didn't believe that in principle there was any limit to what mathematics could achieve.

I think this is very inspiring. So did John von Neumann. When he was a young man he tried to carry through Hilbert's ambitious program. Because Hilbert couldn't quite get it all to work, in fact he started off just with elementary number theory, 1, 2, 3, 4, 5, ... , not even with real numbers at first.

And then in 1931 to everyone's great surprise (including von Neumann's), Gödel showed that it was impossible, that it couldn't be done.

– Gödel 1931 –

This was the opposite of what everyone had expected. Von Neumann said it never occurred to him that Hilbert's program couldn't be carried out. Von Neumann admired Gödel enormously, and helped him to get a permanent position at the Institute for Advanced Study.

What Gödel showed was the following. Suppose that you have a formal axiomatic system dealing with elementary number theory, with 1, 2, 3, 4, 5 and addition and multiplication. And we'll assume that it's consistent, which is a minimum requirement—if you can prove false results it's really pretty bad. What Gödel showed was that if you assume that it's consistent, then you can show that it's incomplete. That was Gödel's result, and the proof is very clever and involves self-reference. Gödel was able to construct an assertion about the whole numbers that says of itself that it's unprovable. This was a tremendous shock. Gödel has to be admired for his intellectual imagination; everyone else thought that Hilbert was right. However I think that Turing's 1936 approach is better.

– Turing 1936 –

Gödel's 1931 proof is very ingenious, it's a real tour de force. I have to confess that when I was a kid trying to understand it, I could read it and follow it step by step but somehow I couldn't ever really feel that I was grasping it. Now, Turing had a completely different approach.

Turing's approach I think it's fair to say is in some ways more fundamental. In fact, Turing did more than Gödel. Turing not only got as a corollary Gödel's result, he showed that there could be no decision procedure.

You see, if you assume that you have a formal axiomatic system for arithmetic and it's consistent, from Gödel you know that it can't be complete, but there still might be a decision procedure. There still might be a mechanical procedure which would enable you to decide if a given assertion is provable or not within the system. That was left open by Gödel, but Turing settled it. The fact that there cannot be a decision procedure is more fundamental and you get incompleteness as a corollary.

How did Turing do it? I want to tell you how he did it because that's the springboard for my own work. The way he did it, and I'm sure all of you have heard about it, has to do with something called the halting problem. In fact, if you go back to Turing's 1936 paper you will not find the words "halting problem." But the idea is certainly there.

People also forget that Turing was talking about "computable numbers." The title of his paper is "On computable numbers, with an application to the Entscheidungsproblem." Everyone remembers that the halting problem is unsolvable and that comes from that paper, but not as many people remember that Turing was talking about computable real numbers. My work deals with computable and dramatically uncomputable real numbers. So I'd like to refresh your memory as to how Turing's argument goes.

Turing's argument is really what destroys Hilbert's dream, and it's a simple argument. It's just Cantor's diagonal procedure (for those of you who know what that is) applied to the computable real numbers. That's it, that's the whole idea in a nutshell, and it's enough to show that Hilbert's dream, the culmination of two thousand years of what mathematicians thought mathematics was about, is wrong. So Turing's work is tremendously deep.

What is Turing's argument? A real number, you know something like $3.1415926\ldots$, is a length measured with arbitrary precision, with an infinite number of digits. And a computable real number, said Turing, is one for which there is a computer program or algorithm for calculating the digits one by one. For example, there are programs for π, and there are algorithms for solutions of algebraic equations with integer coefficients. In fact, most of the numbers that you actually find in analysis are computable. However, they're the exception, if you know set theory, because the computable reals are denumerable and the reals are nondenumerable (you don't have to know what that means). That's the essence of Turing's idea.

The idea is this. You list all possible computer programs. At that time there were no computer programs, and Turing had to invent the Turing machine, which was a tremendous step forward. But now you just say, imagine writing a list with every possible computer program.

p_1 **Gödel 1931**

p_2 **Turing 1936**

p_3

p_4

p_5

p_6

$\vdots$

If you consider computer programs to be in binary, then it's natural to think of a computer program as a natural number. And next to each computer program, the first one, the second one, the third one, write out the real number that it computes if it computes a real (it may not). But if it prints out an infinite number of digits, write them out. So maybe it's 3.1415926 and here you have another and another and another:

p_1 $3.1415926\ldots$ **Gödel 1931**

p_2 $\cdots$ **Turing 1936**

p_3 $\cdots$

p_4 $\cdots$

p_5 $\cdots$

p_6 $\cdots$

$\vdots$

So you make this list. Maybe some of these programs don't print out an infinite number of digits, because they're programs that halt or that have an error in them and explode. But then there'll just be a blank line in the list.

p_1 3.1415926...   **Gödel 1931**

p_2 ...   **Turing 1936**

p_3 ...

p_4 ...

p_5

p_6 ...

$\vdots$

It's not really important—let's forget about this possibility.

Following Cantor, Turing says go down the diagonal and look at the first digit of the first number, the second digit of the second, the third ... .

p_1 $-.\underline{d_{11}}d_{12}d_{13}d_{14}d_{15}d_{16}\ldots$ **Gödel 1931**

p_2 $-.d_{21}\underline{d_{22}}d_{23}d_{24}d_{25}d_{26}\ldots$ **Turing 1936**

p_3 $-.d_{31}d_{32}\underline{d_{33}}d_{34}d_{35}d_{36}\ldots$

p_4 $-.d_{41}d_{42}d_{43}\underline{d_{44}}d_{45}d_{46}\ldots$

p_5

p_6 $-.d_{61}d_{62}d_{63}d_{64}d_{65}\underline{d_{66}}\ldots$

$\vdots$

Well, actually, it's the digits after the decimal point. So it's the first digit after the decimal point of the first number, the second digit after the decimal point of the second, the third digit of the third number, the fourth digit of the fourth, the fifth digit of the fifth. And it doesn't matter if the fifth program doesn't put out a fifth digit, it really doesn't matter.

What you do is you change these digits. Make them different. Change every digit on the diagonal. Put these changed digits together into a new number with a decimal point in front, a new real number. That's Cantor's diagonal procedure. So you have a digit which you choose to be different from the first digit of the first number, the second

digit of the second, the third of the third, and you put these together
into one number.

$$p_1 \quad -.\underline{d_{11}}d_{12}d_{13}d_{14}d_{15}d_{16}\ldots \qquad \textbf{Gödel 1931}$$

$$p_2 \quad -.d_{21}\underline{d_{22}}d_{23}d_{24}d_{25}d_{26}\ldots \qquad \textbf{Turing 1936}$$

$$p_3 \quad -.d_{31}d_{32}\underline{d_{33}}d_{34}d_{35}d_{36}\ldots$$

$$p_4 \quad -.d_{41}d_{42}d_{43}\underline{d_{44}}d_{45}d_{46}\ldots$$

$$p_5$$

$$p_6 \quad -.d_{61}d_{62}d_{63}d_{64}d_{65}\underline{d_{66}}\ldots$$

$$\vdots$$

$$.\neq d_{11} \neq d_{22} \neq d_{33} \neq d_{44} \neq d_{55} \neq d_{66}\ldots$$

This new number cannot be in the list because of the way it was
constructed. Therefore, it's an uncomputable real number. How does
Turing go on from here to the halting problem? Well, just ask yourself
why can't you compute it? I've explained how to get this number
and it looks like you could almost do it. To compute the Nth digit of
this number, you get the Nth computer program (you can certainly do
that) and then you start it running until it puts out an Nth digit, and
at that point you change it. Well what's the problem? That sounds
easy.

The problem is, what happens if the Nth computer program never
puts out an Nth digit, and you sit there waiting? And that's the halting
problem—you cannot decide whether the Nth computer program will
ever put out an Nth digit! This is how Turing got the unsolvability of
the halting problem. Because if you could solve the halting problem,
then you could decide if the Nth computer program ever puts out an
Nth digit. And if you could do that then you could actually carry out
Cantor's diagonal procedure and compute a real number which has to
differ from any computable real. That's Turing's original argument.

Why does this explode Hilbert's dream? What has Turing proved?
That there is no algorithm, no mechanical procedure, which will decide
if the Nth computer program ever outputs an Nth digit. Thus there
can be no algorithm which will decide if a computer program ever halts
(finding the Nth digit put out by the Nth program is a special case).
Well, what Hilbert wanted was a formal axiomatic system from which
all mathematical truth should follow, only mathematical truth, and
all mathematical truth. If Hilbert could do that, it would give us a
mechanical procedure to decide if a computer program will ever halt.
Why?

You just run through all possible proofs until you either find a proof that the program halts or you find a proof that it never halts. So if Hilbert's dream of a finite set of axioms from which all of mathematical truth should follow were possible, then by running through all possible proofs checking which ones are correct you would be able to decide if any computer program halts. In principle you could. But you **can't** by Turing's very simple argument, which is just Cantor's diagonal argument applied to the computable reals. That's how simple it is!

Gödel's proof is ingenious and difficult. Turing's argument is so fundamental, so deep, that everything seems natural and inevitable. But of course he's building on Gödel's work.

3. The Halting Probability and Algorithmic Randomness

The reason I talked to you about Turing and computable reals is that I'm going to use a different procedure to construct an uncomputable real, a much more uncomputable real than Turing's.

$$
\begin{array}{lll}
p_1 & -.\underline{d_{11}}d_{12}d_{13}d_{14}d_{15}d_{16}\ldots & \textbf{Gödel 1931} \\
p_2 & -.d_{21}\underline{d_{22}}d_{23}d_{24}d_{25}d_{26}\ldots & \textbf{Turing 1936} \\
p_3 & -.d_{31}d_{32}\underline{d_{33}}d_{34}d_{35}d_{36}\ldots & \text{uncomputable reals} \\
p_4 & -.d_{41}d_{42}d_{43}\underline{d_{44}}d_{45}d_{46}\ldots & \\
p_5 & & \\
p_6 & -.d_{61}d_{62}d_{63}d_{64}d_{65}\underline{d_{66}}\ldots & \\
\vdots & & \\
\end{array}
$$

$$. \neq d_{11} \neq d_{22} \neq d_{33} \neq d_{44} \neq d_{55} \neq d_{66}\ldots$$

And that's how we're going to get into much worse trouble.

How do I get a much more uncomputable real? (And I'll have to tell you how uncomputable it is.) Well, not with Cantor's diagonal argument. I get this number, which I like to call Ω, like this:

$$\Omega = \sum_{p \text{ halts}} 2^{-|p|}.$$

This is just the halting probability. It's sort of a mathematical pun. Turing's fundamental result is that the halting problem is unsolvable— there is no algorithm that'll settle the halting problem. My fundamental result is that the halting probability is algorithmically irreducible or algorithmically random.

What exactly is the halting probability? I've written down an expression for it:

$$\Omega = \sum_{p \text{ halts}} 2^{-|p|}.$$

Instead of looking at individual programs and asking whether they halt, you put all computer programs together in a bag. If you generate a computer program at random by tossing a coin for each bit of the program, what is the chance that the program will halt? You're thinking of programs as bit strings, and you generate each bit by an independent toss of a fair coin, so if a program is N bits long, then the probability that you get that particular program is 2^{-N}. Any program p that halts contributes $2^{-|p|}$, two to the minus its size in bits, the number of bits in it, to this halting probability.

By the way, there's a technical detail which is very important and didn't work in the early version of algorithmic information theory. You couldn't write this:

$$\Omega = \sum_{p \text{ halts}} 2^{-|p|}.$$

It would give infinity. The technical detail is that no extension of a valid program is a valid program. Then this sum,

$$\sum_{p \text{ halts}} 2^{-|p|},$$

turns out to be between zero and one. Otherwise, it turns out to be infinity. It took only ten years until I got it right. The original 1960s version of algorithmic information theory is wrong. One of the reasons it's wrong is that you can't even define this number:

$$\Omega = \sum_{p \text{ halts}} 2^{-|p|}.$$

In 1974 I redid algorithmic information theory with "self-delimiting" programs and then I discovered the halting probability Ω.

Okay, so this is a probability between zero and one,

$$0 < \Omega = \sum_{p \text{ halts}} 2^{-|p|} < 1,$$

like all probabilities. The idea is you generate each bit of a program by tossing a coin and ask what is the probability that it halts. This number Ω, this halting probability, is not only an uncomputable real—Turing already knew how to do that. It is uncomputable in the worst possible way. Let me give you some clues as to how uncomputable it is.

Well, one thing is that it's algorithmically incompressible. If you want to get the first N bits of Ω out of a computer program, if you want a computer program that will print out the first N bits of Ω and then halt, that computer program has to be N bits long. Essentially you're only printing out constants that are in the program. You cannot squeeze the first N bits of Ω. This

$$0 < \Omega = \sum_{p \text{ halts}} 2^{-|p|} < 1$$

is a real number; you could write it in binary. And if you want to get out the first N bits from a computer program, essentially you just have to put them in. The program has to be N bits long. That's irreducible algorithmic information. There is no concise description.

Now that's an abstract way of saying things. Let me give you a more concrete example of how random Ω is. Émile Borel at the turn of this century was one of the founders of probability theory.

$$0 < \Omega = \sum_{p \text{ halts}} 2^{-|p|} < 1$$

(Questions from the audience)

Q: Can I ask a very simple question before you get ahead?

A: Sure.

Q: I can't see why Ω should be a probability. What if the two one-bit programs both halt? I mean, what if the two one-bit programs both halt and then some other program halts. Then Ω is greater than one and not a probability.

A: I told you that no extension of a valid program is a valid program.

Q: Oh right, no other programs can halt.

A: The two one-bit programs would be all the programs there are. That's the reason this number,

$$0 < \Omega = \sum_{p \text{ halts}} 2^{-|p|} < 1,$$

can't be defined if you think of programs in the normal way.

(End of questions)

So here we have Émile Borel, and he talked about something he called a normal number.

Émile Borel—Normal Reals

What is a normal real number? People have calculated π out to a billion digits, maybe two billion. One of the reasons for doing this, besides that it's like climbing a mountain and having the world's record, is the question of whether each digit occurs the same number of times. It looks like the digits 0 through 9 each occur 10% of the time in the decimal expansion of π. It looks that way, but nobody can prove it. I think the same is true for $\sqrt{2}$, although that's not as popular a number to ask this about.

Let me describe some work Borel did around the turn of the century when he was pioneering modern probability theory. Pick a real number in the unit interval, a real number with a decimal point in front, with no integer part. If you pick a real number in the unit interval, Borel showed that with probability one it's going to be "normal." Normal means that when you write it in decimal each digit will occur in the limit exactly 10% of the time, and this will also happen in any other base. For example, in binary, 0 and 1 will each occur in the limit exactly 50% of the time. Similarly with blocks of digits. This was called an absolutely normal real number by Borel, and he showed that with probability one if you pick a real number at random between zero and one, it's going to have this property. There's only one problem. He didn't know whether π is normal, he didn't know whether $\sqrt{2}$ is normal. In fact, he couldn't exhibit a single individual example of a normal real number.

The first example of a normal real number was discovered by a friend of Alan Turing's at Cambridge called David Champernowne, who is still alive and who's a well-known economist. Turing was impressed with him—I think he called him "Champ"—because Champ had published this in a paper as an undergraduate. This number is known as

Champernowne's number. Let me show you Champernowne's number.

Émile Borel—Normal Reals
Champernowne
0.01234567891011121314 ... 99100101 ...

It goes like this. You write down a decimal point, then you write 0, 1, 2, 3, 4, 5, 6, 7, 8, 9, then 10, 11, 12, 13, 14 until 99, then 100, 101. And you keep going in this funny way. This is called Champernowne's number and Champernowne showed that it's normal in base ten, only in base ten. Nobody knows if it's normal in other bases, I think it's still open. In base ten, though, not only will the digits 0 through 9 occur exactly 10% of the time in the limit, but each possible block of two digits will occur exactly 1% of the time in the limit, each block of three digits will occur exactly 0.1% of the time in the limit, etc. That's called being normal in base ten. But nobody knows what happens in other bases.

The reason I'm saying all this is because it follows from the fact that the halting probability Ω is algorithmically irreducible information that this

$$0 < \Omega = \sum_{p \text{ halts}} 2^{-|p|} < 1$$

is normal in any base. That's easy to prove using ideas about coding and compressing information that go back to Shannon. So here we finally have an example of an absolutely normal number. I don't know how natural you think it is, but it is a specific real number that comes up and is normal in the most demanding sense that Borel could think of. Champernowne's number couldn't quite do that.

This number Ω is in fact random in many more senses. I would say it this way. It cannot be distinguished from the result of independent tosses of a fair coin. In fact, this number,

$$0 < \Omega = \sum_{p \text{ halts}} 2^{-|p|} < 1,$$

shows that you have total randomness and chaos and unpredictability and lack of structure in pure mathematics! This is in the same way that all it took for Turing to destroy Hilbert's dream was the diagonal argument. You just write down this expression

$$0 < \Omega = \sum_{p \text{ halts}} 2^{-|p|} < 1,$$

and this shows that there are regions of pure mathematics where reasoning is totally useless, where you're up against an impenetrable wall. This is all it takes. It's just this halting probability.

Why do I say this? Well, let's say you want to use axioms to prove what the bits of this number Ω are. I've already told you that it's uncomputable—right?—like the number that Turing constructs using Cantor's diagonal argument. So we know there is no algorithm that will compute digit by digit or bit by bit this number Ω. But let's try to prove what individual bits are using a formal axiomatic system. What happens?

The situation is very, very bad. It's like this. Suppose you have a formal axiomatic system which is N bits of formal axiomatic system (I'll explain what this means more precisely later). It turns out that with a formal axiomatic system of complexity N, that is, N bits in size, you can prove what the positions and values are of at most $N + c$ bits of Ω.

Now what do I mean by formal axiomatic system N bits in size? Well, remember that the essence of a formal axiomatic system is a mechanical procedure for checking whether a formal proof follows the rules or not. It's a computer program. Of course in Hilbert's days there were no computer programs, but after Turing invented Turing machines you could finally specify the notion of computer program exactly, and of course now we're very familiar with it.

So the proof checking algorithm which is the essence of any formal axiomatic system in Hilbert's sense is a computer program, and just see how many bits long this computer program is. That's essentially how many bits it takes to specify the rules of the game, the axioms and postulates and the rules of inference. If that's N bits, then you may be able to prove, say, that the first bit of Ω in binary is 0, that the second bit is 1, that the third bit is 0, and then there might be a gap, and you might be able to prove that the thousandth bit is 1. But you're only going to be able to settle N cases if your formal axiomatic system is an N-bit formal axiomatic system.

Let me try to explain better what this means. It means that you can only get out as much as you put in. If you want to prove whether an individual bit in a specific place in the binary expansion of the real number Ω is a 0 or a 1, essentially the only way to prove that is to take it as a hypothesis, as an axiom, as a postulate. It's irreducible mathematical information. That's the key phrase that really gives the whole idea.

– Irreducible Mathematical Information –

$$0 < \Omega = \sum_{p \text{ halts}} 2^{-|p|} < 1$$

Émile Borel—Normal Reals

Champernowne

0.01234567891011121314 ... 99100101 ...

Okay, so what have we got? We have a rather simple mathematical object that completely escapes us. Ω's bits have no structure. There is no pattern, there is no structure that we as mathematicians can comprehend. If you're interested in proving what individual bits of this number at specific places are, whether they're 0 or 1, reasoning is completely useless. Here mathematical reasoning is irrelevant and can get nowhere. As I said before, the only way a formal axiomatic system can get out these results is essentially just to put them in as assumptions, which means you're not using reasoning. After all, anything can be demonstrated by taking it as a postulate that you add to your set of axioms. So this is a worst possible case—this is irreducible mathematical information. Here is a case where there is no structure, there are no correlations, there is no pattern that we can perceive.

4. Randomness in Arithmetic

Okay, what does this have to do with randomness in arithmetic? Now we're going back to Gödel—I skipped over him rather quickly, and now let's go back.

Turing says that you cannot use proofs to decide whether a program will halt. You can't always prove that a program will or will not halt. That's how he destroys Hilbert's dream of a universal mathematics. I get us into more trouble by looking at a different kind of question, namely, can you prove that the fifth bit of this particular real number,

$$0 < \Omega = \sum_{p \text{ halts}} 2^{-|p|} < 1,$$

is a 0 or a 1, or that the eighth bit is a 0 or a 1? But these are strange-looking questions. Who had ever heard of the halting problem in 1936? These are not the kind of things that mathematicians normally worry about. We're getting into trouble, but with questions rather far removed from normal mathematics.

Even though you can't have a formal axiomatic system which can always prove whether or not a program halts, it might be good for everything else and then you could have an *amended* version of Hilbert's dream. And the same with the halting probability Ω. If the halting problem looks a little bizarre, and it certainly did in 1936, well, Ω is brand new and certainly looks bizarre. Who ever heard of a halting probability? It's not the kind of thing that mathematicians normally do. So what do I care about all these incompleteness results?

Well, Gödel had already faced this problem with his assertion which is true but unprovable. It's an assertion which says of itself that it's unprovable. That kind of thing also never comes up in real mathematics. One of the key elements in Gödel's proof is that he managed to construct an *arithmetical* assertion which says of itself that it's unprovable. It was getting this self-referential assertion into a statement in elementary number theory that took so much cleverness.

There's been a lot of work building on Gödel's work, showing that problems involving computations are equivalent to arithmetical problems involving whole numbers. A number of names come to mind. Julia Robinson, Hilary Putnam, and Martin Davis did some of the important work, and then a key result was found in 1970 by Yuri Matijasevič. He constructed a diophantine equation, which is an algebraic equation involving only whole numbers, with a lot of variables. One of the variables, K, is distinguished as a parameter. It's a polynomial equation with integer coefficients and all of the unknowns have to be whole numbers—that's a diophantine equation. As I said, one of the unknowns is a parameter. Matijasevič's equation has a solution for a particular value of the parameter K if and only if the Kth computer program halts.

In the year 1900 Hilbert had asked for an algorithm which will decide whether a diophantine equation, an algebraic equation involving only whole numbers, has a solution. This was Hilbert's tenth problem. It was tenth is his famous list of twenty-three problems. What Matijasevič showed in 1970 was that this is equivalent to deciding whether an arbitrary computer program halts. So Turing's halting problem is exactly as hard as Hilbert's tenth problem. It's exactly as hard to decide whether an arbitrary program will halt as to decide whether an arbitrary algebraic equation in whole numbers has a solution. Therefore, there is no algorithm for doing that and Hilbert's tenth problem cannot be solved—that was Matijasevič's 1970 result.

Matijasevič has gone on working in this area. In particular there is a piece of work he did in collaboration with James Jones in 1984. I can use it to follow in Gödel's footsteps, to follow Gödel's example. You see, I've shown that there's complete randomness, no pattern, lack of structure, and that reasoning is completely useless, if you're interested in the individual bits of this number:

$$0 < \Omega = \sum_{p \text{ halts}} 2^{-|p|} < 1.$$

Following Gödel, let's convert this into something in elementary number theory. Because if you can get into all this trouble in elementary number theory, that's the bedrock. Elementary number theory, 1, 2, 3, 4, 5, addition and multiplication, that goes back to the ancient Greeks and it's the most solid part of all of mathematics. In set theory you're dealing with strange objects like large cardinals, but here you're not even dealing with derivatives or integrals or measure, only with whole numbers. And using the 1984 results of Jones and Matijasevič I can indeed dress up Ω arithmetically and get randomness in elementary number theory.

What I get is an exponential diophantine equation with a parameter. "Exponential diophantine equation" just means that you allow variables in the exponents. In contrast, what Matijasevič used to show that Hilbert's tenth problem is unsolvable is just a polynomial diophantine equation, which means that the exponents are always natural number constants. I have to allow X^Y. It's not known yet whether I actually need to do this. It might be the case that I can manage with a polynomial diophantine equation. It's an open question, I believe that it's not settled yet. But for now, what I have is an exponential diophantine equation with 17,000 variables. This equation is 200 pages long and again one variable is the parameter.

This is an equation where every constant is a whole number, a natural number, and all the variables are also natural numbers, that is, positive integers (actually, *nonnegative* integers). One of the variables is a parameter, and you change the value of this parameter—take it to be 1, 2, 3, 4, 5. Then you ask, does the equation have a finite or an infinite number of solutions? My equation is constructed so that it has a finite number of solutions if a particular individual bit of Ω is a 0, and it has an infinite number of solutions if that bit is a 1. So deciding whether my exponential diophantine equation in each individual case

has a finite or an infinite number of solutions is exactly the same as
determining what an individual bit of this

$$0 < \Omega = \sum_{p \text{ halts}} 2^{-|p|} < 1$$

halting probability is. And this is completely intractable because Ω is
irreducible mathematical information.

Let me emphasize the difference between this and Matijasevič's
work on Hilbert's tenth problem. Matijasevič showed that there is a
polynomial diophantine equation with a parameter with the following
property: You vary the parameter and ask, does the equation have a
solution? That turns out to be equivalent to Turing's halting problem,
and therefore escapes the power of mathematical reasoning, of formal
axiomatic reasoning.

How does this differ from what I do? I use an exponential diophantine equation, which means that I allow variables in the exponent.
Matijasevič allows only constant exponents. The big difference is that
Hilbert asked for an algorithm to decide if a diophantine equation has
a solution. The question I have to ask to get randomness in elementary
number theory, in the arithmetic of the natural numbers, is slightly
more sophisticated. Instead of asking whether there is a solution, I ask
whether there are a finite or an infinite number of solutions—a more
abstract question. This difference is necessary.

My 200 page equation is constructed so that it has a finite or infinite number of solutions, depending on whether a particular bit of
the halting probability is a 0 or a 1. As you vary the parameter, you
get each individual bit of Ω. Matijasevič's equation is constructed so
that it has a solution if and only if a particular program ever halts.
As you vary the parameter, you get each individual computer program.

Thus even in arithmetic you can find Ω's absolute lack of structure,
Ω's randomness and irreducible mathematical information. Reasoning
is completely powerless in those areas of arithmetic. My equation shows
that this is so. As I said before, to get this equation I use ideas that
start in Gödel's original 1931 paper. But it was Jones and Matijasevič's
1984 paper that finally gave me the tool that I needed.

So that's why I say that there is randomness in elementary number
theory, in the arithmetic of the natural numbers. This is an impenetrable stone wall, it's a worst case. From Gödel we knew that we couldn't
get a formal axiomatic system to be complete. We knew we were in

trouble, and Turing showed us how basic it was, but Ω is an extreme case where reasoning fails completely.

I won't go into the details, but let me talk in vague information-theoretic terms. Matijasevič's equation gives you N arithmetical questions with yes/no answers which turn out to be only $\log N$ bits of algorithmic information. My equation gives you N arithmetical questions with yes/no answers which are irreducible, incompressible mathematical information.

5. Experimental Mathematics

Okay, let me say a little bit in the minutes I have left about what this all means.

First of all, the connection with physics. There was a big controversy when quantum mechanics was developed, because quantum theory is nondeterministic. Einstein didn't like that. He said, "God doesn't play dice!" But as I'm sure you all know, with chaos and non-linear dynamics we've now realized that even in classical physics we get randomness and unpredictability. My work is in the same spirit. It shows that pure mathematics, in fact even elementary number theory, the arithmetic of the natural numbers, 1, 2, 3, 4, 5, is in the same boat. We get randomness there too. So, as a newspaper headline would put it, God not only plays dice in quantum mechanics and in classical physics, but even in pure mathematics, even in elementary number theory. So if a new paradigm is emerging, randomness is at the heart of it. By the way, randomness is also at the heart of quantum field theory, as virtual particles and Feynman path integrals (sums over all histories) show very clearly. So my work fits in with a lot of work in physics, which is why I often get invited to talk at physics meetings.

However, the really important question isn't physics, it's mathematics. I've heard that Gödel wrote a letter to his mother who stayed in Europe. You know, Gödel and Einstein were friends at the Institute for Advanced Study. You'd see them walking down the street together. Apparently Gödel wrote a letter to his mother saying that even though Einstein's work on physics had really had a tremendous impact on how people did physics, he was disappointed that his work had not had the same effect on mathematicians. It hadn't made a difference in how mathematicians actually carried on their everyday work. So I think that's the key question: How should you really do mathematics?

I'm claiming that I have a much stronger incompleteness result. If so, maybe it'll be clearer whether mathematics should be done the ordinary way. What is the ordinary way of doing mathematics? In spite of the fact that everyone knows that any finite set of axioms is incomplete, how do mathematicians actually work? Well suppose you have a conjecture that you've been thinking about for a few weeks, and you believe it because you've tested a large number of cases on a computer. Maybe it's a conjecture about the primes and for two weeks you've tried to prove it. At the end of two weeks you don't say, well obviously the reason I haven't been able to show this is because of Gödel's incompleteness theorem! Let us therefore add it as a new axiom! But if you took Gödel's incompleteness theorem very seriously, this might in fact be the way to proceed. Mathematicians will laugh, but physicists actually behave this way.

Look at the history of physics. You start with Newtonian physics. You cannot get Maxwell's equations from Newtonian physics. It's a new domain of experience—you need new postulates to deal with it. As for special relativity, well, special relativity is almost in Maxwell's equations. But Schrödinger's equation does not come from Newtonian physics and Maxwell's equations. It's a new domain of experience and again you need new axioms. So physicists are used to the idea that when you start experimenting at a smaller scale, or with new phenomena, you may need new principles to understand and explain what's going on.

Now, in spite of incompleteness mathematicians don't behave at all like physicists do. At a subconscious level they still assume that the small number of principles, postulates, and methods of inference that they learned early as mathematics students are enough. In their hearts they believe that if you can't prove a result it's your own fault. That's probably a good attitude to take rather than to blame someone else, but let's look at a question like the Riemann Hypothesis. A physicist would say that there is ample experimental evidence for the Riemann Hypothesis and would go ahead and take it as a working assumption.

What is the Riemann Hypothesis? There are many unsolved questions involving the distribution of the prime numbers that can be settled if you assume the Riemann hypothesis. Using computers people check these conjectures and they work beautifully. They're neat formulas but nobody can prove them. A lot of them follow from the Riemann hypothesis. To a physicist this would be enough: It's useful, it explains a lot of data. Of course, a physicist then has to be prepared to say

"Oh oh, I goofed!" because an experiment can subsequently contradict a theory. This happens very often.

In particle physics you throw up theories all the time and most of them quickly die. But mathematicians don't like to have to backpedal. But if you play it safe, the problem is that you may be losing out, and I believe you are.

I think it should be obvious where I'm leading. I believe that elementary number theory and the rest of mathematics should be pursued more in the spirit of experimental science, and that you should be willing to adopt new principles. I believe that Euclid's statement that an axiom is a self-evident truth is a big mistake. The Schrödinger equation certainly isn't a self-evident truth! And the Riemann hypothesis isn't self-evident either, but it's very useful.

So I believe that we mathematicians shouldn't ignore incompleteness. It's a safe thing to do but we're losing out on results that we could get. It would be as if physicists said, okay no Schrödinger equation, no Maxwell's equations, we stick with Newton, everything must be deduced from Newton's laws. (Maxwell even tried it. He had a mechanical model of an electromagnetic field. Fortunately they don't teach that in college!)

I proposed all this twenty years ago when I started getting these information-theoretic incompleteness results. But independently a new school on the philosophy of mathematics is emerging called the "quasi-empirical" school of thought regarding the foundations of mathematics. There's a book of Tymoczko's called *New Directions in the Philosophy of Mathematics* (Birkhäuser, Boston, 1986). It's a good collection of articles. Another place to look is *Searching for Certainty* by John Casti (Morrow, New York, 1991), which has a good chapter on mathematics. The last half of the chapter talks about this quasi-empirical view. By the way, Lakatos, who was one of the people involved in this new movement, happened to be at Cambridge at that time. He'd left Hungary.

The main schools of mathematical philosophy at the beginning of this century were Russell and Whitehead's view that logic was the basis for everything, the formalist school of Hilbert, and an "intuitionist" constructivist school of Brouwer. Some people think that Hilbert believed that mathematics is a meaningless game played with marks of ink on paper. Not so! He just said that to be absolutely clear and precise as to what mathematics is all about, we have to specify the rules determining whether a proof is correct so precisely that they become mechanical. Nobody who thought that mathematics is meaningless would have been so energetic and done such important work and been such an inspiring leader.

Originally, most mathematicians backed Hilbert. Even after Gödel and, even more emphatically, Turing showed that Hilbert's dream did not work, in practice mathematicians carried on as before, in Hilbert's spirit. Brouwer's constructivist attitude was mostly considered a nuisance. As for Russell and Whitehead, they had a lot of problems getting all of mathematics from logic. If you get all of mathematics from set theory you discover that it's nice to define the whole numbers in terms of sets (von Neumann worked on this). But then it turns out that there's all kinds of problems with sets. You're not making the natural numbers more solid by basing them on something which is more problematical.

Now everything has gone topsy-turvy. It's gone topsy-turvy, not because of any philosophical argument, not because of Gödel's results or Turing's results or my own incompleteness results. It's gone topsy-turvy for a very simple reason—the computer!

The computer as you all know has changed the way we do everything. The computer has enormously and vastly increased mathematical experience. It's so easy to do calculations, to test many cases, to run experiments on the computer. The computer has so vastly increased mathematical experience, that in order to cope, people are forced to proceed in a more pragmatic fashion. Mathematicians are proceeding more pragmatically, more like experimental scientists do. This new tendency is often called "experimental mathematics." This phrase comes up a lot in the field of chaos, fractals, and nonlinear dynamics.

It's often the case that when doing experiments on the computer, numerical experiments with equations, you see that something happens, and you conjecture a result. Of course, it's nice if you can prove it. Especially if the proof is short. I'm not sure that a 1,000-page proof helps too much. But if it's a short proof, it's certainly better than not having a proof. And if you have several proofs from different viewpoints, that's very good.

But sometimes you can't find a proof and you can't wait for someone else to find a proof, and you've got to carry on as best you can. So now mathematicians sometimes go ahead with working hypotheses on the basis of the results of computer experiments. Of course, if it's physicists doing these computer experiments, then it's certainly okay; they've always relied heavily on experiments. But now even mathematicians sometimes operate in this manner. I believe that there's a new journal called the *Journal of Experimental Mathematics.* They should've put me on their editorial board, because I've been proposing this for twenty years based on my information-theoretic ideas.

So in the end it wasn't Gödel, it wasn't Turing, and it wasn't my results that are making mathematics go in an experimental mathematics direction, in a quasi-empirical direction. The reason that mathematicians are changing their working habits is the computer. I think it's an excellent joke! (It's also funny that of the three old schools of mathematical philosophy, logicist, formalist, and intuitionist, the most neglected was Brouwer, who had a constructivist attitude years before the computer gave a tremendous impulse to constructivism.)

Of course, the mere fact that everybody's doing something doesn't mean that they ought to be. The change in how people are behaving isn't because of Gödel's theorem or Turing's theorems or my theorems, it's because of the computer. But I think that the sequence of work that I've outlined does provide some theoretical justification for what everybody's doing anyway without worrying about the theoretical justification. And I think that the question of how we should actually do mathematics requires **at least** another generation of work. That's basically what I wanted to say—thank you very much!

Acknowledgment

Some of this material appeared under a different title in the *International Journal of Bifurcation and Chaos,* volume 4, pp. 3–15, 1994.

References

[1] G. J. Chaitin, *Information-Theoretic Incompleteness,* World Scientific, 1992.

[2] G. J. Chaitin, *Information, Randomness & Incompleteness,* second edition, World Scientific, 1990.

[3] G. J. Chaitin, *Algorithmic Information Theory,* revised third printing, Cambridge University Press, 1990.

CHAPTER 5

Narratives of Evolution and the Evolution of Narratives

N. KATHERINE HAYLES

Department of English
University of California
Los Angeles, CA 90024, USA

Evolution is a science; it is also a narrative, or more properly a series of competing and cooperating narratives that emerge out of particular cultural moments. Whether they occur in literature or in science, narratives do not exist in isolation but rather take their shape and significance from larger discursive patterns that I will call grand narratives. These grand narratives are the stories a culture tells to make sense of the world and itself. Like all narratives, grand narratives have beginnings, middles, and ends. The patterns that fulfill these requirements, however, vary from culture to culture. A beginning and an end in a Navajo myth is not the same as a beginning and an end in the Judaeo-Christian tradition. Grand narratives, then, are culturally and historically specific, varying not only in their thematics but also in their underlying dynamics and shapes.[1] A grand narrative important throughout the nineteenth century in Western culture, including evolutionary biology, was the narrative of progress. Its erosion in the wake of World War I precipitated a major crisis whose aftershocks are still being felt. Perhaps the only Western narrative of comparable importance in the modernist period has been the narrative of individualism, which in the United States and elsewhere came under widespread attack after the Vietnam War and underwent various mutations that theorists are

Cooperation and Conflict in General Evolutionary Processes, edited by John L. Casti and Anders Karlqvist.
ISBN 0–471–59487–3 ©1994 John Wiley & Sons, Inc.

still trying to sort out. As these examples illustrate, grand narratives change over time. They evolve in response to cultural pressures and changing historical conditions. Some become moribund and die out; others adapt to the changing environment and continue to survive in altered form.

Claiming that evolution has been constructed as a narrative implies not only that it is embedded within grand narratives but also that it is a story which itself is evolving. Putting the matter this way makes clear that evolutionary narratives are inevitably caught in a reflexive loop. They are stories of change, and they are stories that themselves are changing, so they are in part self-descriptions as well as descriptions of an exterior world. I will return to the significance of this reflexive dimension of evolutionary narratives. For now, enough has been said to allow me to state my thesis. I will argue that theories of evolution have been constructed as narratives, and that these narratives have distinctive features deriving from their embeddedness in language and culture.[2] This realization should not, I will suggest, be cause for despair. Understanding the dynamics of evolutionary narratives can provide rich and powerful insights into how these models are evolving over time, as well as how the narratives themselves are changing. Particularly, it can explain some of the current shifts in presuppositions as narratives of competition are displaced by narratives of cooperation.

Let me begin my own narrative by asking what it implies to say that something is a narrative. (I will leave aside without comment the prior question of why, in this forum, starting with a definitional question should be recognized as a beginning.) To count as a narrative, a story must explicitly or implicitly constitute relationships, usually causal, between the events it relates. The sequence, "Richard Dawkins wrote *The Selfish Gene*. It was criticized by several people. Dawkins responded to his critics. Then he wrote *The Blind Watchmaker*," is only a minimal narrative, if that, because it creates few or no links between events. The accepted term for such a sequence is chronicle rather than narrative. To make the sequence into a full-fledged narrative, I can recast it as follows: "When Richard Dawkins wrote *The Blind Watchmaker*, many of his basic assumptions underwent important shifts because of the criticism he received on *The Selfish Gene*." The second formal quality of narrative is a distinction between the order of narration and the order of events. Note that in recasting my little narrative, I also changed the order of events. Whereas writing *The Selfish Gene* comes first in the chronicle, as it logically would, it appears in the narrative as a flashback. Following the traditional distinction articulated by Lionel Trilling, I will call the logical or temporal sequence of events the story, and the sequence formed by the order of narration the plot. The third formal quality possessed by all narratives is a narrator. There has to be someone or something telling or writing the story, and this implies a perspective from which the story is told. When plot and story diverge, as they often do, the divergences are important clues to the narrator's positionality.

Using these criteria, how does Darwin's theory of evolution in *Origin of the Species* stack up as a narrative? It constructs causal relationships between events; it orders events logically, creating a story; it is told in a certain order, that is, emplotted within a story; and it is written by a narrator who is culturally and historically positioned. It is significant for the narrative Darwin constructed that he was reading Malthus, and that Malthus in turn was writing a narrative embedded within the larger narratives that were producing and produced by an emerging capitalism. Gillian Beer, in her fine study *Darwin's Plots* (1983), demonstrates that the narrative and rhetorical formulations available to Darwin from his culture affected what he said as well as how he said it, particularly his attempts to get away from teleology. There are two points I want to underscore with this example. The first you have heard before, that the stories of evolutionary biology are narratives in dynamic interplay with cultural grand narratives. The second is that the particular linguistic formulations at work do not merely contain narratives but actively constitute them.

To illustrate these claims, I want to turn now to Richard Dawkins' work, especially *The Selfish Gene* (first ed. 1976, revised ed. 1989) and *The Blind Watchmaker* (1986). Dawkins is an interesting writer for my purposes because he is an unusually skillful rhetorician, keenly aware of the value of a good story. At the same time, he espouses a view of language that falls squarely in the mainstream of practicing scientists—what I will call the giftwrap model of language. This view sees language as a wrapper that one puts around an idea to present it to someone else. I wrap an idea in language, hand it to you, you unwrap it and take out the idea. It does not really matter what the language is—whatever the wrapping, the idea is conveyed intact. Despite the mountains of research, demonstration, and argument within the humanities testifying otherwise, my conversations with a range of scientists from a variety of disciplines suggest that it is no exaggeration to say that this view still represents a consensus among scientists. Whatever differences Dawkins has with other evolutionary biologists in his style of writing and in his ideas, he shares with them the premise that the language he uses is neutral. If it can be demonstrated that this is not the case with his texts, then we will have established a basis for considering how broadly the intertwining of language and idea in narrative extends beyond the kind of popularized writing that Dawkins does to other kinds of scientific discourse.

The Selfish Gene tells the story of how genes manipulate the "lumbering robots" who serve them as transmission vehicles. The temporal vector of the story is consistent with shifting the relevant unit of selection from the individual to the gene. Arguing that genes last for "thousands and millions of years" while individuals last only a few decades (p. 34), the story relates events that stretch through eons. The causal connections tying these events together must therefore be constituted through entities capable of lasting for these periods of time. In this way the narrative form, by presupposing causal connections between events, helps to constitute part of the argument. To emphasize the constructedness of the narrative, I will hereafter call the authorial voice the narrator, reserving Dawkins for the historical author.

The places where the narrator puts forth the giftwrap model of language are not randomly dispersed throughout the text. Rather, they come precisely at the strategic points in the argument where crucial transitions are being made. For example, when the narrator is defining the gene, we read this claim: "[T]here is nothing sacred about definitions. We can define a word how we like for our own purposes, provided we do so clearly and unambiguously" (p. 28). The definition that follows, emphasizing the gene's longevity, basically sets up the entire argument by calling the gene "any portion of chromosomal material that potentially lasts for enough generations to serve as a unit of natural selection" (p. 28). The definition presupposes that natural selection works upon genes rather than upon organisms, which is a central enabling claim for the text's narratives (p. 33). To say that one has the right to whatever definition one chooses, as long as one is clear, implies that claims are separate from definitions; one can be granted a definition and denied a claim. What this formulation sidesteps is the recognition that the definition and claim here are mutually entwined. The narrator acknowledges how important the definition is when he points out, "What I have now done is to *define* the gene in such a way that I cannot really help being right!" (p. 33). Yet at the same time, his giftwrap view of language makes the definition seem as if it were only the opening move in an argument, not the argument itself.

Narratives emerge from this definition once the genes are set into motion as autonomous actors. At the critical juncture where the narrator is switching the effective unit of selection from the individual to the gene, we find another assertion of the language's neutrality. "At times, gene language gets a bit tedious, and for brevity and vividness we shall lapse into metaphor. But we shall always keep a skeptical eye on our metaphors, to make sure they can be translated back into gene language if necessary" (p. 45). Subsequently, the narrator uses language that endows genes with purpose and will, thus enabling them to become selfish. Contrary to the narrator's assertion, the metaphors are not expendable, for they do more than simply express the same idea in another form. By creating the genes as actors, they constitute the subjects who can perform the actions that will comprise the narrative. Without actors, the narrative could not exist.

What kind of narrative is constructed from these actors? Greg Myers in *Writing Biology* (1990), comparing popular science writing to scientific journal articles, notes that when scientific findings are rewritten for a popular audience, the nature of the narrative changes. In a technical context, the story typically is constituted as what he calls a narrative of science. Citations of supporting evidence, rebuttals of opposing views, presentation of experimental methods, and interpretation of data emphasize the mediated, problematic nature of the phenomena under investigation. Implicitly, the actors are the practitioners, even though the use of passive voice and impersonal pronouns partially mask their presence. In popular science writing, by contrast, the mediation tends to be underplayed or to disappear altogether through language that renders natural phenomena as if they were experienced directly. In a technical journal article about unisex lizards, for example, the actors are the biologists who maintain the lizards, observe their behavior, and interpret their observations.[3] Rewritten for a popular magazine like *Scientific American,* the results are cast into a different kind of narrative in which the lizards themselves become the actors. Rewriting the story for a popular audience thus does more than add contextualization and replace technical vocabulary with accessible explanations. It transforms the story's meaning, metamorphizing a narrative of science into a narrative of nature.

Myers draws on this idea in his analysis of Wilson's *Sociobiology* (Harvard University Press, 1975) to explain this text's distinctive features and rhetorical force. Wilson uses natural history anecdotes throughout—"Just-So Stories," one of his critics calls them—but Myers notes that Wilson systematically transforms these anecdotes into behavior through devices, such as classification and quantitative analysis, that strip away anecdotal particularities. What Wilson really wants, Myers writes, is to "make behavioral biology as systematic and quantitative as physics or molecular biology—that is, he wants to remove entirely the narrative elements of particular places, times, and actors" (p. 206). But he also wants to animate his narrative through natural history subjects that can then be inscribed into species-centered narratives; these species-centered narratives are in turn enfolded into the "grand evolutionary narrative that structures the whole book, a Great Chain of Behaving" (pp. 206–7). How to tell a story and still achieve the systematic schema of a quantitative behavioral biology? Myers argues that Wilson achieves this double goal by filling out the "very abstract narrative structures with the actors of natural history" (p. 207). Thus the story can still unfold as a narrative—that is, as actors performing events that are structured meaningfully through time—but the story is told in such a way that the actors perform only on a microscale, while the macroscale is dominated by an actorless schema and quantitative classification.

Dawkins shares with Wilson the problem of how to create a narrative without actors, but he takes a different approach. As even a casual reader must notice, Dawkins works through metaphors and analogies that anthropomorphize the genes, attributing human intentions and desires to them. Throughout, he insists that such language is only a vivid shorthand for more cumbersome technical terms, and that it is always possible to "translate" into more "respectable" language. When he is about to script the individual into the gene's evolutionary stable strategies, a crucial transition in his argument, we read this disclaimer: "Remember that we are picturing the animal as a robot survival machine with a pre-programmed computer controlling the muscles. To write the strategy out as a set of simple instructions in English is just a convenient way for us to think about it" (p. 69). When he wishes to explain altruism, a sticky but essential point for his argument, we read the following: "If we allow ourselves the license of talking about genes as if they had conscious aims, always reassuring ourselves that we could translate our sloppy language back into respectable terms if we wanted to, we can ask the question, what is a single selfish gene trying to do?" (p. 88).

The simultaneous use of a story constituted through the gene's intentionality and disclaimers about the story's language allows Dawkins to write two kinds of narratives at once. *The Selfish Gene* is a narrative of nature because the phenomena themselves—the genes—are the actors. In this narrative the analogies are not just incidental to the plot, for without them the story could not be told. The actions of the genes make sense to the reader, become plausible and understandable, because they are inscripted into a human narrative of competition and selfishness. If, as Dawkins repeatedly insists, selection operates on the organism rather than the group, the source of narrative movement must be located within the organism. Dawkins's contribution, which he himself calls a shift in perspective so dramatic as to amount to a new paradigm, is to identify an internal actor distinct from the individual, who is re-conceptualized as a remote-control mechanism operated by the actor. If this actor did not possess intentionality—if the selfish gene were only a metaphor that could be discarded at will—the motive force driving the narrative would collapse. Without this narrative, there are only shifts in populations that can be statistically measured but not causally explained. The entire argument, then, depends on the narrative of nature in which the selfish gene, far from being a mere rhetorical flourish, is the constitutive actor.

There is also a sense in which *The Selfish Gene* is a narrative of science. Here too the language is not incidental but fundamentally constitutive of the story. Like other narratives of science, *The Selfish Gene* emphasizes mediation. The mediating agents who present nature to the readers are not researchers, however. Rather, they are the analogies that constitute the genes as actors. In a conventional narrative of science, the voices of the researchers are partially suppressed through passive constructions and impersonal pronouns. In Dawkins' text, by contrast, the presence of the analogies is highlighted—but only in order to claim that these actors are merely window-dressing for the "real" narrative of science that lies behind them. Thus the analogies are linguistic actors whose peculiar function it is to deny that they are actors and to point beyond themselves to the "respectable" language that could, if it existed, constitute the researchers as actors in their place. This "real" narrative of science of course does not exist, except insofar as it is called into a kind of virtual existence by the language that claims it is merely a wrapper for what lies within. But the giftwrap does not present the idea. Rather, the idea is present only in and through the giftwrap. By constituting a distinction between wrapper and content, the giftwrap creates a virtual content that stands, the narrator claims, always ready to displace the analogies-as-actors. Thus the story is told through actors who erase their own agency even as they speak.

The dynamic of an actor who denies his agency and points beyond or behind himself to the "real" actor has a familiar ring. Like many rhetorically canny texts, *The Selfish Gene* creates a narrative structure that *performs* as well as relates the story. The story is about displaced agency, about a subjectivity that has the illusion of control while the real locus of control lies with another agent who inhabits the subject and uses him for its own ends. The analogies creating this story also enact it by insisting they are inhabited by a "respectable" discourse that is the real agent telling the story. Dawkins' rhetorical strategy is thus considerably more complex than Wilson's in *Sociobiology*. Rather than confining his actors to a microscale, Dawkins creates a reflexive structure in which the story of mobile, long-lived genes who inhabit lumbering robots is told through analogies presenting themselves as deft, nimble robots inhabited by a long-lived, lumbering language.

How does this reflexively enfolded narrative relate to grand cultural narratives of the period? It is no accident that Dawkins wrestled with creating actors for his narrative who would not possess traditional agency and intention. Darwin faced a similar problem when he struggled with expressing a sense of design that nevertheless did not imply a conscious, intentional designer. By the late twentieth century, the question of intentionality had percolated down from a macroscale global designer to microscale local actors who enact or create designs. The idea of a coherent subjectivity who possesses independent agency, long a pillar of American individualism, had by the 1970s been attacked on a number of fronts, from object relations psychology to the critique of the unified self in critical theory and poststructuralist linguistics. Among the most influential of these attacks was the work of Michel Foucault. From *The Order of Things* (1970) through *The History of Sexuality* (1980), Foucault argued that culture is not created by autonomous thinkers inspired by great thoughts. Rather, he saw subjectivity as an effect produced by particular kinds of discursive systems. In his discursive model, the locus of control is taken away from individuals and displaced into articulated systems of power and knowledge.

Read in this context, Dawkins' theory can be seen as a narrative that partly concedes and partly contests the Foucaultian model of cultural inscription. Like Foucault's archeologies, Dawkins' text posits a shift in the locus of control away from the individual, with a consequent erosion of the autonomy of a human being. Unlike Foucault, however, Dawkins' rhetoric recuperates the basic formation of autonomous individualism by attributing it to the gene. Dawkins' text can thus be seen as an adaptive strategy in a particular intellectual climate. Self-interested individuals may have become moot, but self-interested individualism is alive and kicking, ready to be transmitted into the next generation of culturally situated narratives and narrators.

Although Dawkins does not locate his narrative in this wider cultural context, he implicitly places himself within his narrative in a chapter entitled, appropriately, "The Long Reach of the Gene" (pp. 234–266). When a writer situates himself within his own plot, the move is always potentially reflexive, for his position then becomes ambiguous. On the one hand, he can be seen as outside the story's frame, a free agent bringing the story into existence through the mediated agency of the narrator. On the other hand, he is also positioned inside the story as one of the actors constituted by the drama the narrator tells. In this chapter, Dawkins occupies a reflexive position from which he at once tells the story and is also told by the story. He creates such a position when the narrator speculates that the "long reach of the gene" may extend to units of discourse as well as units of DNA. "The gene . . . happens to be the replicating entity that prevails on our own planet," the narrator writes. "There may be others. If there are, provided certain other conditions are met, they will almost inevitably tend to become the basis for an evolutionary process" (p. 192). These discursive units the narrator calls memes, which can be ideas, slogans, or even musical phrases—any cultural artifact small and persistent enough to replicate itself many times over.

A meme is an idea that can survive and replicate itself; the narrator's argument implies that the concept of the selfish gene is just such a replicating unit. Thus the narrative becomes another kind of "lumbering robot" carrying within itself the unit that will become immortal by manipulating its vehicle of transmission. The narrative validates the selfish gene-meme by presenting arguments that make it plausible, and the gene-meme in turn demonstrates the narrative's validity by instantiating the entity that the narrative claims to exist. Reflexively embedding descriptive and enactive levels into each other, the narrative operates in this chapter as a recursive loop that includes the author within its circumference. The loop can also extend to the reader. Recognizing that the narrative can conceivably inoculate the reader with the meme, I am forced to wonder if it has turned me into a lumbering robot as well, making me replicate it in my own essay (as indeed to some extent I have, at least in mutated form).

This triumph of reflexivity is not achieved without a price. Appropriately, it is the same price that the theory itself demands, namely the sacrifice of autonomous identity as the payment for genetic immortality. The point is beautifully illustrated by an anecdote that Brett Cooke tells.[4] He was at a conference where Dawkins' work was being discussed and made a comment alluding to "Dawkins' idea of a meme." Dawkins, present in the audience, rose to correct him. He explained that he did not wish to have his name used in connection with the meme because the editors of the *Oxford English Dictionary* were considering including the word in their next edition, and their rules precluded coinages too closely associated with a single person. To achieve the kind of immortality that the OED offers, Dawkins was forced to separate his identity from the meme. It could get in only if it became an autonomous entity in its own right, selfishly surviving after its creator's name was deleted.

At this point you may feel that I am beating a dead meme. Perhaps you will object that Dawkins is atypical in his use of flamboyant language, and that in any event the selfish gene is no longer a credible idea within the community of evolutionary biology. Dawkins' rhetorical style may be unusual, but the processes at work within his text are not unusual. D. R. Croaker (1981) has shown that anthropomorphizing, far from being a flourish added to popular texts to make them more interesting (which is Dawkins' claim about his own rhetorical practices), is rather an unavoidable part of ethology that is more or less successfully concealed in scientific publications. It is no mystery why cognitive understanding should be based upon the empathic mechanism of anthropomorphism. We can understand the world as different from ourselves only if we first compare it to ourselves, which necessarily involves an implicit anthropomorphism. In the dance of similarity and difference, similarity comes first and makes difference possible.

Nor is the constitution of argument through language unusual in the discourse of evolutionary biology, as Evelyn Fox Keller has shown in "Language and Ideology in Evolutionary Theory" (1991). Keller notes multiple instances where "competition" is used as a putatively technical term, even when the proposed definitions are strikingly at odds with its colloquial use. Asking what the stakes are in retaining the term, she notes that scarcity has traditionally been assumed to imply competition. Embedded within the coupling of scarcity and competition, she argues, lies a series of nested assumptions that serve to direct thought in a certain direction while tacitly making other possibilities seem counterintuitive. Among these assumptions are the proposition that "resource competition is ... a zero-sum game"; that a "resource can be defined and quantitatively assessed independent of the organism"; that "each organism's utilization of this resource is independent of other organisms"; and that consumption of resources therefore "has a kind of de facto equivalence to murder" (p. 92). Competition is a term that cannot easily be relinquished, her argument implies, because it carries with it a network of assumptions central to evolutionary theory. Dawkins thus shares with the larger community the strategy of using language that evokes these assumptions even while claiming that the specific words do not matter. Keller summarizes the strategic effect: "The colloquial connotations lead plausibly to one set of inferences and close off others, while the technical meaning stands ready to disclaim responsibility if challenged" (p. 92).

Although the stress on competition somewhat diminishes in Dawkins' later work, he does not abandon the strategies that are so evident in *The Selfish Gene*. In *The Blind Watchmaker*, for example, he explains that lions and zebras are enemies because they compete for the same resource, namely the zebra's body. "The role of the zebra in the relationship seems too innocent and wronged to warrant the putative 'enemy.' But individual zebras do everything in their power to resist being eaten by lions, and from the lions' point of view this is making life harder for them" (p. 179). From this it follows that these "enemies" are engaged in an "arms race." The usage serves to naturalize competition and also to inscribe Dawkins' 1986 narrative within the larger cultural narratives of capitalism and communism in which competition was a highly charged and politicized term.

To illustrate how far from inevitable are the assumptions embedded within the formula "scarcity implies competition," Keller points out that there are interactions that generate resources. These are marginalized, however, into the "special case" categories of cooperative, mutualist, and symbiotic interactions. The most important of these mutualist interactions is sexual reproduction, which by definition, Keller observes, cannot be a special case for sexually reproducing species. Fully taking sexual reproduction into account would challenge the widespread assumption that "intrinsic properties of individual units are primary to any description of evolutionary phenomena" (pp. 93–4). The language of competition, Keller concludes, is inextricably bound up with the language of reproductive autonomy. To invoke the one is to reinforce the other.

It is significant, then, that evolutionary narratives concerned with competition tend to ignore the effects of fertility on differential survival. When the focus is shifted to fertility, as in E. Sober and R. Lewontin or F. B. Christiansen, Keller notes that the context for fitness is "determined by the genotype of the mating partner rather than by the complementary allele" (p. 98). Dawkins moves toward this position in *The Blind Watchmaker* in his discussion of female preference for long tails in peahens and widow birds (pp. 200–210), but he continues to inscribe female choice within a continuing narrative of competition. Surely it is no accident that he interjects into his discussion of female choice the alleged human female preference for long penises, a staple of masculine competitive fantasies projected onto women. The example illustrates that the regressive pull of a narrative of competition remains strong in *The Blind Watchmaker*. Nevertheless, some cracks appear on its surface. In a decade in which debates over women's reproductive rights frequently dominated the headlines, female choice was an obvious factor to take into consideration. The fissured, intertextual nature of the chapter suggests that as a narrative begins to shift, it is constituted less as a coherent unity than as an anisotropic field in which the arrows do not all point in the same direction. With characteristic rhetorical mastery, Dawkins hints that his narrative is undergoing just such a transition when he emphasizes the extreme subtlety and dangerous novelty of the ideas he introduces in this chapter.

Yet perhaps he is not so iconoclastic as he presents himself, for as Keller observes, even the shift to the mating pair is insufficient to account for the complexities of the situation. "Mating pairs do not reproduce themselves any more than do individual genotypes," she writes (p. 99). To take full account of sexual reproduction, fitness should be seen not as an "individual property but a composite of the entire interbreeding population, including, but certainly not limited to, genic, genotypic, and mating pair contributions. By undermining the reproductive autonomy of the individual organism, the advent of sex undermines the possibility of locating the causal efficiency of evolutionary change in individual properties. At least part of the 'causal engine' of natural selection must be seen as distributed throughout the entire population of interbreeding organisms" (p. 99). Her argument implies that the autonomy of the individual that was displaced onto the gene in *The Selfish Gene* would have to be re-thought from the ground up if the interdependencies of sexual reproduction were taken fully into account. In an important sense, Dawkins' gene is asexual as well as selfish. I would add to Keller's analysis that it is not just individual words, phrases, and metaphors that function as she describes, but the inter- and intratextual narratives that weave these units together to make persuasive stories. Persuasion does not exist in isolation, any more than do evolutionary fitness and sexual reproduction.

In this evolving narrative of narrative evolution, I want to turn now to more recent work by Dawkins and others that registers a shift in basic premises as well as in intertextual reference. Dawkins' article "The Evolution of Evolvability" appeared in *Artificial Life*, edited by Chris Langton (1989). Although it covers much the same ground as Chapter 3 in The *Blind Watchmaker* explaining how the computer program of the same name works, there are significant differences in tonality and conclusions that I want to address. Dawkins begins his piece by acknowledging a shift in his position. "A title like 'The Evolution of Evolvability' ought to be anathema to a dyed-in-the-wool neo-Darwinian like me," he comments (p. 201). He explains that in constructing the Blind Watchmaker program, he was intuitively led to try recursive local rules that would lead to emergent behavior. "[M]y intuition proved to have been a considerable understatement," he exults. "I was genuinely astonished and delighted at the richness of morphological types that emerged before my eyes as I bred" (p. 208). Wait a minute—"bred"? From the author of *The Selfish Gene*, whose narrator consistently elided the contingencies of sexual reproduction? The verb choice indicates that a quite different pattern of imagery is emerging, along with new kinds of narrative rhythms and embeddings.

The pattern becomes clear as the narrator explains that there are two kinds of changes in genetic programs: those that change phenotypes, and those that can change the programs themselves. Of the latter, the most significant and rare are the "watershed events," so-called because they "open the floodgates to further evolution" (p. 219). The narrator consistently associates watershed events with images of fertility, especially pregnancy. Segmentation, for example, "represented a change in embryology that was pregnant with evolutionary potential" (p. 218). There is in natural selection, the narrator speculates, "a kind of ratchet such that changes in embryology that happen to be relatively fertile, evolutionarily speaking, tend to be still with us" (p. 218). Selection thus works not only for survivability but also for evolvability; "an embryology that is pregnant with evolutionary potential is a good candidate for a higher-level property of just the kind we must have before we allow ourselves to speak of species or higher-level selection" (p. 219).

As if to prove Keller's point that sexual reproduction, if really considered, would force a redefinition of interlinked levels of selection across a full range of groupings, the narrator chooses to cast his concession that higher-level selection might be valid in the gendered language of sexual reproduction. The narrative that emerges from this gendered language envisions a male programmer mating with a female program to create progeny whose biomorphic diversity surpasses the father's imagination. These biomorphs inherit their mother's remarkable fertility, at once instantiating and transmitting a pregnant embryology that promises to bridge the gap between natural and artificial life—a move foretold in the interspecies mating of human and computer program that comprises the central narrative of this essay.

Into what grand cultural narratives is this narrative interpolated? Chris Langton's introductory essay "Artificial Life" (1991) in the same volume provides clues. The principal assumption of the artificial life program, Langton writes, "is that the 'logical form' of an organism can be separated from its material base of construction, and that 'aliveness' will be found to be a property of the former, not the latter" (p. 11). As Stefan Helmreich (1992) has pointed out, throughout the discourse of artificial life are gendered dichotomies that associate logical form with a male seed and reproductive capacity with a female ground. In describing the work of L. S. Penrose, for example, Langton uses "seed" to describe a pair of hooked blocks that, when dropped into a box of unhooked blocks, begin a process of self-organization. The metaphors of a male seed and female ground have a long history in Western culture, stretching back at least to Aristotle and informing biological models of reproduction well into the seventeenth century. It was not until the late eighteenth century that the female contribution to sexual reproduction was recognized, and even then the belief persisted that the male sperm carried the human form within itself as a homunculus, with the egg merely providing a fertile ground for the sperm's activity. Long after these ideas ceased to be accepted within biology, the gendered patterns of language that assigned active form to the male and passive receptivity to the female persisted, as Ruth Hubbard's study of contemporary biology textbooks has shown. Having given the metaphors renewed vivacity by transforming them from *in vivo* to *in silico,* the aim of the artificial life program is to transport them back from the computer into the biological realm. "Since we know that it is possible to abstract the logical form of a machine from its physical hardware," Langton writes, "it is natural to ask whether it is possible to abstract the logical form of an organism from its biochemical wetware" (p. 21).

Even as they take their imagery and resonance from the grand cultural narratives of patriarchy, however, the A-Life programs are pushing them in a new direction. For the female ground of the computer program is not merely passive. On the contrary, it is capable of producing forms so unpredictable and surprising that they astonish their male creator. The key to this fertility is the application of a rule to the seed structure, which results in embedded self-reproducing loops that function like a computer program. "This 'program' within the loop computer is also applied recursively to growing structure," Langton explains. "Thus, this system really involves a double level of recursively applied rules ... This system makes use of the signal propagation capacity to embed a structure that itself computes the resulting structure, rather than the 'physics' being directly responsible for developing the final structure from a passive seed" (p. 30).

The tangled hierarchy of A-Life that embeds one level in another recalls the reflexive structure of *The Selfish Gene,* although the concerns that animated *The Selfish Gene* stand in significant contrast to A-Life. Whereas the focus of Dawkins' argument was a debate about which level of the biological unit is most important for natural selection, in artificial life these questions are rendered moot, for the local and global levels mutually and cooperatively determine each other, producing through their feedback mechanisms emergent behavior that cannot be predicted analytically. "Thus, local behavior supports global dynamics," Langton summarizes, "which shapes local context, which affects local behavior, which supports global dynamics, and so forth" (p. 31). Similarly, anxiety about where the locus of control resides is also diffused. Whereas in *The Selfish Gene* considerable rhetorical energy was expended to suggest that control resided in the gene rather than the individual, that is to say at the local rather than the global level, in artificial life control is distributed between levels because of the recursive loops that connect them.

The differences between the narratives of the selfish gene and the breeding A-Life program can now be brought into clearer focus. As we have seen, the reflexive looping of *The Selfish Gene* performed as well as constituted a narrative of displaced agency underwritten by two imperatives: preserving the capacity for autonomous action at some level, and insisting that this level was not coextensive with the traditional Western subject. The looping had the effect of reconstituting agency so that a subject previously considered autonomous (a human being, the text's language, the author, the reader) was encircled within another agent's intentionality (the selfish gene, "respectable" discourse, the meme). Reflexivity thus carried a double valence of anxiety and reassurance—even though one's own agency was coopted, agency itself was preserved. The key to the narrative was autonomy. To qualify as a "real" actor in the drama, an agent had to have the ability to preserve its own identity and defend itself against encroaching foreign elements. The winners were those actors who could use reflexivity to penetrate another's boundaries while keeping their own intact. In the A-Life narrative, by contrast, reflexivity breaks out of this scenario of anxiety and displaced control to celebrate the fertility of a reproductive partner whose exuberant productions are seen as the co-creation of male programmer and female computer. Autonomy gives way to interdependence, preservation of the individual agent to cooperative procreation, anxiety about keeping boundaries intact to a feminine recursivity whose enfolded structure, rather than closing in on itself, opens outward toward complexity and prolific progeny.

Someone once said that the deep questions of an era are never really answered. Rather, they are rendered irrelevant by new developments, fading into the wings as other kinds of questions are energized and take center stage. One indication that the nature of the questions has shifted is the appearance in the discourse of artificial life of metaphors of mating and sexual reproduction. As long as the important questions were concerned with individual autonomy, sexual reproduction tended to be elided because it posed a potential challenge to the model's presuppositions. In artificial life, by contrast, the important questions focus on the relation of artificial to natural life. It is therefore not surprising in this context to encounter metaphors of mating between human and computer, for they reinforce rather than challenge the model.

What does it mean that this is a mating between two profoundly different species, one encoded in protein, the other in silicon? The A-Life narratives are part of the grand narrative of cybernetics, in which human beings face intelligent machines that are constructed as their evolutionary partners or, more threateningly, their evolutionary successors. Narratives of competition and cooperation assume new significance in this context, for the stakes are not just survival of individual autonomy but survival of the species. The bifurcation between competition and cooperation becomes more ambiguous, for competition between species may require cooperation among species members (humans against computers), just as cooperation between species may result in competition among species members (cyborgs against unaltered humans). Interspecies mating of the kind Dawkins imagines points toward still another scenario, in which cooperation between species facilitates a move toward redefining gender roles in evolutionary narratives so that females become coequal creators with males. The narratives of A-Life are thus subtly transforming the possibilities of the grand narrative that lends them plausibility, even as the grand narrative shapes and constrains the stories they can tell.

What should our response be to these complex interactions between language and narratives, scientific theories and cultural grand narratives? One possibility is to see natural language as a treacherous, ambiguous swamp from which we should strive to extricate ourselves as much as possible, retreating to the solid high ground of mathematical symbols. Although it is beyond the scope of this essay to address this question, there are good reasons to think that mathematical models do not in fact leave language behind. Brian Rotman (1987, 1988, 1993) has argued that the semiotic structures of natural languages are re-encoded in mathematical discourse. In addition to this deep encoding, natural language plays a constitutive role in the presuppositions that constrain and direct how models will evolve. Moreover, after models are formulated, language inevitably enters at moments of interpretation, informing the meaning and significance attributed to them (see Hayles, 1990, 1992). If language cannot be avoided, then ignoring it simply blinds us to the influences that it exerts upon thought. Given that we cannot move beyond language, a better strategy is to move through it, for example by using narrative to understand narrative, which is what I have tried to do in this essay. To say there are connections between ideas and language, theories and narrative does not invalidate ideas and theories. It does, however, suggest that ideas and theories do not exist in a Platonic purity above and beyond culture. They are always already positioned, relevant to, and emerging from a variety of interacting discourses.

There is a deeply ingrained belief in our culture that objectivity requires distance between subject and object; we assume that we know the world because we are separate from it. It is at least as true, however, that we know the world because we are connected to it. If objectivity requires subjects who are capable of a "view from nowhere," as Thomas Nagel (1986) put it, then none of us can be objective, for we are always situated within particular times, places, and cultures that both enable and limit what we can imagine, think, and know. To acknowledge and explore the conditions that produce knowledge need not destroy the possibilities for objectivity. Rather, as many feminist theorists and social constructivists have argued, including Donna Haraway (1988), Sandra Harding (1986, 1993), Bruno Latour (1987), and Greg Myers (1990), taking positionality into account produces a *stronger* version of objectivity than is possible if we suppose that knowledge is constructed by autonomous subjects who exist outside language and culture. To say that evolutionary theory is informed by narrative structures and positioned discourses, among other factors, is not to deny that it can produce reliable knowledge. Although models can never be true in an absolute sense, they can be consistent with our experiences of reality in specified contexts. Understanding how knowledges are positioned makes them more rather than less reliable, for then we are better able to see the limits beyond which they may cease to be useful.

In this context, Dawkins' work can be seen as marking an important transition between an era in which reproductive autonomy, individual agency, and competition were mutually entwined and reinforcing in evolutionary narratives, to a period in which cooperation, sexual reproduction, and symbiotic alliances between species are foregrounded. Given where advanced capitalism and information technologies are at the moment, the knowledges produced and implied by the latter scenario are likely to be at least as useful as that articulated by *The Selfish Gene.*

Notes

[1] For an analysis of the grand narratives important in postmodernism, see Jean-Francois Lyotard (1984).

[2] For a structuralist approach to evolution as narrative, see Misia Landau (1991). In her analysis, Landau relies on techniques introduced by Vladimir Propp and other structuralists to show that underlying mythic elements inform the theories of T. H. Huxley, Ernst Haeckel, Charles Darwin, Arthur Keith, and Grafton Elliot Smith. My approach differs substantially from Landau's in that it is cultural and contextual rather than structuralist and mythic.

[3] These articles are discussed in Myers, pp. 141–192.

[4] The anecdote is included in Brett Cooke, "The Biopoetics of Immmortality: A Darwinist Perspective on Science Fiction," presented at the Eaton Science Fiction Conference, May 1992.

References

Beer, Gillian. 1983. *Darwin's Plots: Evolutionary Narratives in Darwin, George Eliot, and Nineteenth-Century Fiction.* London: Routledge and Kegan Paul.

Croaker, D. R. 1981. "Anthropomorphism: Bad Practice, Honest Prejudice?" *New Scientist* (16 July): 159–62.

Dawkins, Richard. 1986. *The Blind Watchmaker.* New York: W. W. Norton and Company.

Dawkins, Richard. 1989. "The Evolution of Evolvability," pp. 201–220 in *Artificial Life,* Vol. VI, Santa Fe Institute Studies in the Sciences of Complexity. Redwood City, CA: Addison-Wesley.

Dawkins, Richard. 1989. *The Selfish Gene.* Oxford: Oxford University Press. First ed. 1976.

Foucault, Michel. 1970. *The Order of Things: An Archaeology of the Human Sciences.* New York: Pantheon.

Foucault, Michel. 1980. *A History of Sexuality. Vol. 1, An Introduction.* Trans. Robert Hurley. New York: Random House.

Haraway, Donna. 1988. "Situated Knowledges: The Science Question in Feminism as a Site of Discourse on the Privilege of Partial Perspective," *Feminist Studies,* 14: 575–99.

Harding, Sandra. 1986. *The Science Question in Feminism.* Ithaca, NY: Cornell University Press.

Harding, Sandra. 1993. "Standpoint Epistemology Revisited: What Is 'Strong Objectivity'?" in *Feminist Epistemologies,* Linda Alcoff and Elizabeth Potter, eds. New York: Routledge.

Hayles, N. Katherine. 1990. *Chaos Bound: Orderly Disorder in Contemporary Literature and Science.* Ithaca, NY: Cornell University Press.

Hayles, N. Katherine. 1992. "Gender Encoding in Fluid Mechanics: Masculine Channels and Feminine Flows," *Differences,* 4 (Summer): 16–44.

Helmreich, Stefan. 1992. "The Historical and Epistemological Ground of von Neumann's Theory of Self-Reproducing Automata and Theory of Games," unpublished manuscript.

Keller, Evelyn Fox. 1991. "Language and Ideology in Evolutionary Theory: Reading Cultural Norms into Natural Law," in *The Boundaries of Humanity: Humans, Animals, Machines,* James J. Sheehan and Morton Sosna, eds. Los Angeles: University of California Press.

Landau, Misia. 1991. *Narratives of Human Evolution.* New Haven, CT: Yale University Press.

Langton, Christopher G. 1989. "Artificial Life," pp. 1–48 in *Artificial Life,* Vol. VI, Santa Fe Institute Studies in the Sciences of Complexity. Redwood City, CA: Addison-Wesley.

Latour, Bruno. 1987. *Science in Action: How to Follow Scientists and Engineers Through Society.* Milton Keynes, UK: Open University Press.

Lyotard, Jean-Francois. 1984. *The Postmodern Condition: A Report on Knowledge.* Trans. Geoff Bennington and Brian Massumi. Minneapolis, MN: University of Minnesota Press.

Myers, Greg. 1990. *Writing Biology: Texts in the Social Construction of Scientific Knowledge.* Madison, WI: University of Wisconsin Press.

Nagel, Thomas. 1986. *The View from Nowhere.* New York: Oxford University Press.

Rotman, Brian. 1987. *Signifying Nothing: The Semiotics of Zero.* New York: St. Martin's Press.

Rotman, Brian. 1988. "Towards a Semiotics of Mathematics," *Semiotica,* 72: 1–35.

Rotman, Brian. 1993. *Ad Infinitum ... The Ghost in Turing's Machine: Taking God Out of Mathematics and Putting the Body Back In: An Essay in Corporeal Semiotics.* Palo Alto, CA: Stanford University Press.

CHAPTER 6

Biologically Bound Behavior, Free Will, and the Evolution of Humans

PHILIP LIEBERMAN

Department of Cognitive and Linguistic Sciences
Brown University
Providence, RI 02912, USA

1. Introduction

Whether free will exists has long been debated. But it is certainly the case that many individuals believe that they can exercise free will and in some measure control their destiny. Moreover, the concept of free will is central to many of the world's religious and philosophic systems. However, the focus of this paper is not whether free will exists, but the biological and evolutionary basis for this belief. There must be a basis for the belief that free will exists, unless we postulate that it derives from divine intervention, which is outside the domain of evolutionary biology. The proposition that I would like to advance is that we believe that free will exists because as human beings we can exercise voluntary control over the actions and decisions that we make during the ordinary course of life.

In other words, humans have sufficient voluntary control over most of their actions and decisions to foster the belief that free will exists. In particular, we can execute voluntary and arbitrary maneuvers that reflect cognitive activity—for example, dance, speech, and song—using the muscles of our bodies, feet, hands, and most especially the organs of speech. It is this ability, the human capacity to exercise voluntary

Cooperation and Conflict in General Evolutionary Processes, edited by John L. Casti and Anders Karlqvist.
ISBN 0-471-59487-3 ©1994 John Wiley & Sons, Inc.

control over motoric acts that express our cognitive capacity, that is the basis for a belief in free will. Indeed, I will argue that the neurophysiological bases of those aspects of human cognitive ability that allow us to develop and express the concept of free will, derive from the neurological mechanisms involved in the voluntary motoric activity of human language—speech and signing. Recent data and theories[1] point to a link between the evolution of fine manual control, speech production, and some aspects of human linguistic and cognitive ability (Lieberman, 1984, 1985, 1991; Kimura, 1979; MacNeilage et al., 1987).

– Preadaptation –

The key to the theory that I will develop is Darwin's (1859, p. 190) observation that an "organ might be modified for some other and quite distinct purpose." This observation, which has been codified as the mechanism of "preadaptation" (Mayr, 1982), is a central element of evolutionary theory. Darwin understood that the gradual, small changes effected by natural selection could not account for many of the abrupt transitions evident in nature. He also implicitly noted the distinction that must be drawn between small changes in the physical structure of an organ and the behavioral consequences of these physical changes.

Darwin illustrated predaptation in his discussion of the evolution of the lungs of terrestrial animals. Gradual improvements in the gills and other mechanisms that fish use to extract air from water could only result in fish better adapted for life in water. They could not account for the evolution of terrestrial, air-breathing animals. The solution hinged on the distinction between structure and behavior. Many species of fish have swim bladders, internal bladders that allow them to adjust their specific gravity to that of the water at a particular depth. The fish normally pumps air extracted from the water by its gills into the elastic swim bladders, expanding them to adjust its specific gravity to that of the water at some particular depth. Swim bladders were the "predaptive," fortuitous starting point for a respiratory system that enabled air to oxygenate the bloodstream. A gap in the palate of the lungfish which was sealed under water by a primitive larynx (Negus, 1949) allowed air to enter the swim bladder from the fish's mouth when it was stranded on dry land. Reverse flow supplied the bloodstream with oxygen. A series of small changes effected by natural selection acted on the lungfish swim bladder, refining the system to ultimately yield the mammalian lung. As Darwin (1859, p. 190) put it, an "organ originally constructed for one purpose, namely flotation, may be converted into one for a wholly different purpose, namely respiration."

– Neural Mechanisms –

The thesis of this brief essay is that the preadaptive basis of free will was the evolution of the neural mechanisms that allow humans to organize and effect novel patterns of motoric activity. Humans, more than any other animals, are not constrained to stereotyped motor-response patterns. I shall discuss neurophysiologic and comparative data that are consistent with the hypothesis that neural mechanisms that were adapted initially to allow hominids to transcend the limits imposed by stereotyped motor-response patterns, also underlie the "abstract" thought processes that are exemplified in human cognitive and linguistic ability. The particular neurophysiologic thesis that I will propose derives from recent data and theories that link the activity of neocortical prefrontal cortex and subcortical pathways linking prefrontal cortex to other parts of the brain to precise manual and speech motor control, and what Kurt Goldstein (1948) presciently termed the "abstract capacity." The morphological distinctions between human beings and other species resides in the quantitative expansion of prefrontal cortex and subcortical pathways, and perhaps specialized prefrontal cortical mechanisms—the point of this paper being a neurophysiologic theory that can be tested.

2. On the Expression of Free Will

Though some other species may display free will to a limited degree, in its developed human form it is probably the distinguishing characteristic of human beings. Free will is defined, in a broad sense, as behavior that is not a stereotyped response to a specific situation, not a "bound behavior" that may be genetically programmed. Many aspects of human and animal behavior are similar in that they involve a sequence of actions and cognitive decisions that constitute a frozen form that is automatically executed in response to a particular situation or stimulus. Human beings are able to free the particular atomic units that constitute the inventory of automatized bound behaviors; we can combine these atomic elements to form novel responses and novel thoughts.

Although we usually think of major life-defining decisions in reference to free will, the basic concept applies to other less august, indeed mundane, levels of behavior. If one reflects on the probable evolutionary processes that would have yielded the ability of human free will, it would have involved aspects of behavior closer to the daily lives of our hominid ancestors. Therefore, the evolutionary roots of free will are more likely to rest in the ability to modify patterns of behavior relative to daily life than in abstract moral behavior. Free will is, in fact, evident in humans in motoric acts that are closely linked to cognition such as speech and dance:

1. Speech—where the articulatory gestures that specify particular sounds can be combined in novel sequences to form words that are not simply stereotyped response patterns triggered by our emotional state, or particular events or objects.

2. Dance—where the motor activity of the legs, arms, etc., the entire body, are combined in novel patterns that transcend the needs of the Darwinian "struggle for existence."

Free will also takes expression in virtually every aspect of human behavior. In our relations with others—conflict and cooperation—as well as shaping our own lives. Free will can act to enhance or to oppose innately determined "bioprograms" postulated by socio-biological theory (Wilson, 1975). In reduced degree some of these socio-behavioral manifestations of free will can be observed in the activities of non-human primates such as chimpanzees (De Waal, 1982). Chimpanzees also exhibit in their manual dexterity and tool-making and tool-using activities a reduced form of the motor manifestations of free will that may derive from the first stages of the neurological adaptations that ultimately are the basis of human free will.

3. Comparative Evidence Concerning Free Will

Inferences concerning cognitive behavior may be derived from the archaeological and fossil record, but direct access to the past is impossible, so we must rely on comparative studies that study the behavior and morphology of humans and other species. Many other living species show signs of intelligence and, in reduced form, attain some of the attributes that in a developed form differentiate human culture from the behavior of other species.

Chimpanzees, who are genetically closest to humans, for example, use and make tools (e.g., Goodall, 1986). Other species also make tools, but comparison with chimpanzees is most likely to yield insights concerning hominid evolution, since we presumably share more common, "primitive" features with chimpanzees than with, say, sea lions (who also use tools). The tools that chimpanzees fabricate are simple compared to the complicated devices that people made 2,000 years ago, but they nonetheless appear to have been invented by chimpanzees without human tutelage. Goodall (1986), for example, notes a number of tools—termite sticks, leaf sponges, stone missiles, branch clubs, and so forth. Different chimpanzee tool traditions appear to exist; Boesch and Boesch (1981, 1991) document chimpanzees using stone and wood hammers and stone anvils to crack nuts in the Ivory Coast, an activity absent in Goodall's chimpanzees in Tanzania.

When humans intervened and demonstrated stone-working techniques in a laboratory setting, the bonobo *(Pan paniscus)* Kanzi at the Yerkes Field Station was able to imitate work and produce stone tools that resemble some of the implements often associated with early hominids (Toth, personal communication). For that matter, chimpanzees in their native habitat have been observed teaching juveniles how to make and use tools (Boesch and Boesch, 1991). Chimpanzees in their native state also carry out some of the less laudatory activities that we associate with humans, organized warfare and cannibalism (Goodall, 1986). And when raised under conditions that approximate those of human infants in a normal linguistic and social environment, they can attain some aspects of human language (Gardner and Gardner, 1984; Gardner et al., 1989; Rumbaugh and Savage-Rumbaugh, 1992; Savage-Rumbaugh et al., 1985, 1986).

It is in the cognitive domain of language that enhanced free will most obviously differentiates human beings from chimpanzees and other living species. The linguistic abilities of chimpanzees are very impressive when one reflects on the fact that they are *not* human beings. Human children commonly acquire language when they are exposed to a normal linguistic environment. Hearing-impaired children obviously must be exposed to languages that do not depend on sounds, but virtually any child exposed to language used in a normal communicative fashion will acquire his or her first language without overt instruction. There are some limits, however. Severely retarded people do not acquire a language; contrary to the assertions of linguistic theory they are unable to speak and have limited comprehension (Wills, 1973). Children also must be exposed to languages used in a productive manner. Dutch children exposed to German television programs, for example, do not acquire German (Bowerman, 1987).

A certain amount of tutoring and shaping is involved in the acquisition of language by children, though the adult tutors may not be fully aware of their role as teachers. Fernald (1982), for example, shows that adults typically use a different register of voice when they talk to children. Tomasello and Farrar (1986) note that the process of "joint attention" wherein an adult draws the child's attention to some object or activity while discussing it, plays an important part in language acquisition. Certain aspects of human language may be innately specified, i.e., genetically transmitted. The early expression of the intonation contours that human languages use to segment the flow of speech into sentence-like segments appears to be innate (Lieberman, 1967).

Other aspects of speech production and speech perception also appear to involve innate neural mechanisms which are of particular interest because they reflect some of the motoric manifestations of free will that can be traced in the fossil record of hominid evolution—namely factors relating to the process by which human speech differs from the communications of non-human primates.

4. Human Speech

Until the 1960s it was not realized that speech is a crucial component of human linguistic ability. Linguists thought that any set of arbitrary sounds would suffice to transmit words. The research of Liberman and his colleagues (Liberman et al., 1967) demonstrated that the sounds of speech had a special status. Speech allows us to transmit phonetic "segments" (which are approximated by the letters of the alphabet) at an extremely rapid rate, up to 25 segments per second. This leads to a seeming mystery. Miller (1956) showed that human beings cannot identify nonspeech sounds at rates that exceed 7 to 9 items per second. How then can human beings possibly understand speech which is typically transmitted at a rate of about 15 to 25 sounds per second? The answer turns out to be a set of brain mechanisms that allow us to "decode" speech signals in a very special way, and specialized anatomy and brain mechanisms that allow us to make these speech sounds. The high transmission rate of speech is essential to human linguistic ability, as it allows complex thoughts to be transmitted within the constraints of short-term memory. Although sign language can also achieve a high transmission rate, the signer's hands cannot be used for other tasks. Nor can viewers see the signer's hands except under restricted conditions. Vocal language represents the continuation of the evolutionary trend towards freeing the hands for carrying and tool use that started with upright bipedal hominid locomotion. The contribution to biological fitness is obvious since language pervades all aspects of human culture. Human speech also has some lesser selective advantages; the sounds that are specific to human speech, i.e., those that only human beings can produce, are less susceptible to perceptual confusion than the sounds that living nonhuman primates can make and those that certain extinct hominids could have made.

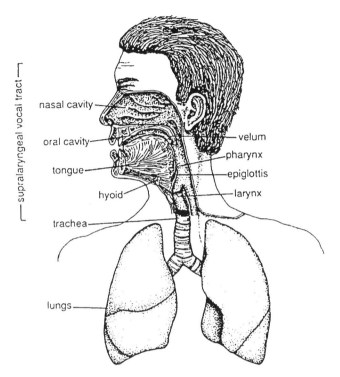

Figure 1. The human supralaryngeal vocal tract. The tongue, which forms the lower margin of the oral cavity and the anterior margin of the pharynx, allows humans to produce the abrupt changes in supralaryngeal vocal tract shape that are necessary to produce sounds like the vowels of the words *tea* and *too*. The velum also can seal off the nasal cavity from the rest of the supralaryngeal airway, yielding nonnasal speech. These speech production qualities facilitate the process of speech perception. After Lieberman (1991).

– The Physiology of Speech –

One of the key biological mechanisms that is necessary for human speech is the supralaryngeal vocal tract (SVT) sketched in Figure 1. The human SVT consists of most of the airway above the larynx. Extending from the lungs upwards is the trachea, capped by the larynx. Above the larynx the airway connects with the esophagus, which connects the stomach to the mouth. This region consists of the pharynx, where the food and air tubes come together. The pharynx branches into two passages—the oral and nasal cavities. As far as the specific properties of human speech are concerned, the important parts of the supralaryngeal vocal tract are the pharynx and oral and nasal cavities. The velum, the soft flexible part of the palate that can close off the nose to the mouth, the tongue, lips, and larynx (which can move upwards or downwards) work together to change the shape of the supralaryngeal vocal tract. A series of changes have occurred in the last 150,000 years or so, which we will briefly discuss below, adapted the SVT for speech production at the expense of vegetative functions such as breathing and preventing choking to death on food (Lieberman, 1968, 1975, 1984, 1989, 1991; Lieberman and Crelin, 1971; Lieberman et al., 1972, 1992b; Laitman et al. 1978, 1979; George, 1978; Grosmangin, 1979).

– Innate Neural Perceptual Mechanisms "Matched" to the SVT –

The process by which human speech attains its rapid data transmission rate is quite complex and has never been successfully modeled on any computer system. It involves at least three innately determined stages of processing that appear to be "matched" to the human SVT:

a. recovering formant frequency patterns,

b. normalizing them,

c. and decoding the "encoded" formant patterns to recover phonetic sequences.

These processes are matched to the physiological constraints of the human SVT; they inherently involve "knowledge" of these constraints and the voluntary motor commands that humans alone can effect—the motor commands that may be the basis of the evolution of the concept of human free will.

– Recovering the Formant Frequency Patterns of Speech –

Human speech is generated by exciting the airway above the larynx, the supralaryngeal vocal tract (SVT), with a source or sources of acoustic energy (Fant, 1960). In the case of vowels or other voiced sounds the larynx produces a series of quasi-periodic puffs of air—the *glottal source*. The *fundamental frequency of phonation* (F0) is the rate at which these puffs of air occur; the average pitch of a person's voice is the perceived consequence of the average F0. The intonation of speech is determined by the temporal pattern of F0 as well as the amplitude of the glottal source. Many animals make use of differing patterns of

F0 to convey information, e.g., frogs (Capranica, 1965) and monkeys (Peterson et al., 1978).

However, the *formant frequencies* of speech that specify many of the phonetic distinctions between different "segmental" vowels and consonants are determined by the supralaryngeal vocal tract, the SVT, which acts as an acoustic *filter*. When the SVT is positioned between a source and the listener's ear, it acts in a manner similar to the tube of a woodwind instrument. The transmission of acoustic energy is impeded, i.e., *attenuated*, as a function of frequency. At certain frequencies minimum attenuation occurs. These frequencies are termed *formant frequencies*. Research since the time of Hellwag (1781) shows that formant frequency patterns are one of the primary determinants of phonetic quality. The effect is no different from that of an optical filter. The dyes in the glass of a section of a stained glass window impede, i.e., *attenuate* the light energy impinging on the window at certain frequencies. At other frequencies light energy passes through that section of the stained glass window with minimum attenuation—the color that we perceive depends on the frequencies of the light passing through.

The recovery of the formant frequencies by human listeners appears to involve "knowledge" of the particular filtering characteristics of the human SVT. The uppermost graph of Figure 2 shows the spectrum of the phonation source that would result if F0 = 500 Hz. The middle graph shows the filter function of the vowel [i] (the vowel of the word *tea*) for a speaker who has a 17-cm-long SVT. The local maxima in the filter function—i.e., the formant frequencies, F1, F2, and F3, are 0.3, 2.1, and 3.1 kHz, respectively. Note that these formant frequencies are not harmonically related. The formant frequencies and fundamental frequency of phonation and its harmonics are independent. Note that for this example, no harmonics coincide with any of the formant frequencies. The spectrum of the resulting speech signal is sketched in the bottom graph. Note that the formant frequencies (marked by circled X's) are not directly manifested in the acoustic signal. This is the usual case; recovering formant frequencies from the acoustic signal is one of the many problems encountered in the recognition of speech by machine. Formant frequencies clearly cannot be recovered by selecting peaks in the short-term acoustic spectrum. Indeed, their recovery is not impeded when the overall spectrum is dramatically modified, e.g., when the "tone controls" of an audio system are manipulated. The process by which humans internally compute these formant frequency patterns appears to involve a process of "analysis by synthesis," in which the listener compares the received acoustic signal with knowl-

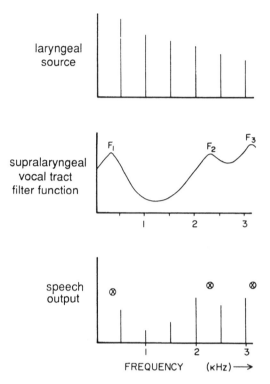

Figure 2. Graphs illustrating the problems inherent in "recovering" formant frequencies from the acoustic signal. The horizontal axis of each graph shows frequency in kHz.

Top—The acoustic energy that would be present in the laryngeal source for a person phonating at 500 Hz (0.5 kHz). The vertical lines show that energy is present only at discrete frequencies.

Middle—The vertical axis shows the relative amount of acoustic energy that would pass through the SVT producing the vowel [i] (the vowel of **bee**). The peaks are the formant frequencies Fl, F2, F3 that specify the phonetic quality of the vowel.

Bottom—The result of the SVT filtering the laryngeal output of the top graph. Note that no energy is present at the X's that mark the formant frequencies. However, human listeners recover the formant frequencies by means of a perceptual process that has implicit "knowledge" of the constraints of the SVT. After Lieberman (1991).

edge of the effects of the inventory formant frequency patterns that a human SVT could generate (Bell et al., 1961).

Human listeners must perform a second stage of speech processing that also appears to be innately specified—SVT "normalization." We must estimate the probable length of a speaker's supralaryngeal airway in order to assign a particular formant frequency pattern to a particular speech sound. The length of the human SVT differs greatly; those of young children are half the length of those of adults. Different length SVTs will have different formant frequencies; a short vocal tract will produce speech sounds that have higher formant frequencies than those of a long vocal tract. The difference is like that between a piccolo and a bassoon, which produce musical notes that have higher or lower "pitches." If we interpreted the formant sounds of speech like the notes produced by woodwind instruments, then the speech sounds of people who had different vocal tract lengths would not have the same phonetic value. For example, the word *bit* spoken by a large adult male speaker can have the same formant frequency pattern as the word *bet* produced by a smaller male. Yet we "hear" the large person's *bit* as *bit* rather than *bet.*

This process appears to be in place by the age of three months in infants (Lieberman, 1984). Infants at this age start to imitate the speech sounds produced by their mothers and take account of differences in SVT length (Lieberman, 1984, pp. 219–222). They will in certain situations imitate the sustained vowel sounds that their mothers direct towards them. The infant's SVT being half the length of his mother's, the infant is physically incapable of producing the absolute formant frequencies of her vowels. The infant instead produces formant frequency patterns that are proportionate to the ratio between the length of his mother's SVT divided by the length of his SVT. The infant's formant frequency "normalization" is similar to that used by adults when they identify the vowels of speakers who have differing SVTs (Nearey, 1978). Three-month-old infants are able to perform a computation that is beyond the power of most of the speech-recognizing computer systems currently available. They are able to infer the probable length of a speaker's SVT from the acoustic signal and compensate for the effects of overall length on the formant frequency patterns that convey vowel distinctions.

– Formant Frequency Encoding and
Voluntary Speech Gestures –

The articulatory maneuvers that underlie the production of human speech are among the most complex that human beings attain. Until age ten years, normal children cannot attain adult levels of precision for basic maneuvers like the lip positions that are necessary to produce different isolated vowels (Watkins and Fromm, 1984). Moreover, in normal running speech speakers must plan ahead, altering their production of individual sounds according to the phonetic context before and after each particular sound.

For example, the vowel [u] which occurs in the word *two* is produced with "rounded," i.e., protruded and pursed, lips. In contrast, the vowel [i], which occurs in the word *tea,* is produced with your lips retracted. When adult speakers of English produce the word *two* they round their lips 100 msec *before* they start to produce the vowel; they don't wait for the vowel to start before they begin to round their lips. When they produce *tea* they don't round their lips at all. The speakers anticipate the occurrence of the vowel. The speakers plan ahead, rounding their lips for the vowel [u] that they know will come along in the stream of speech. Adult speakers of Swedish, in contrast, coarticulate the [u] between 500 and 100 msec ahead of its occurrence (Lubker and Gay, 1982). The difference clearly is not the result of some genetic difference between speakers of English and Swedish. English-speaking three-year-old children, for example, fail to coarticulate their [u]'s at all (Sereno et al., 1987; Sereno and Lieberman, 1987). Acquiring an English versus a Swedish accent involves learning these different automatized motor control patterns. The effect is general during the production of speech and is an example of *anticipatory,* i.e., *preplanned* "coarticulation."

Some notion of the complexity of the automatized motor control patterns that underlie human speech may be gained by looking at experimental data. Graco and Abbs (1985) devised an experiment in which a small electrical torque motor could apply a force to a speaker's lower lip that would impede its closing. The speakers were asked to produce a series of syllables that started with the sound [b]. The "stop consonant" [b] is produced by closing one's lips. The experimenters first used the motor to impede lip closing ocassionally 40 msec before the point when they would have normally reached a closed position. The speakers compensated for the perturbation within 40 msec and closed their lips by applying more muscle force to their lower lip. The speakers also helped close their lips with a slight downwards movement of their upper lips. The experimenters found that the speakers needed a minimum time of 40 msec to perform these compensating maneuvers. This interval corresponds to the time that it takes for a sensory signal to travel up to the motor cortex and a compensating muscular control signal to travel back to the lips (Evarts, 1973). The experimenters then shortened the time between the perturbing force and normal lip closure to 20 msec. The speakers then compensated with a large downwards upper lip deflection, prolonging the duration of the lip closure an additional 20 msec so that there would be sufficient time to overcome the perturbing force by means of lower lip action. The speakers used two different automatized motor response patterns that had a common linguistic goal: closing the lips to produce the consonantal stop closure. The speakers must have had a neural representation of a linguistic goal, closing one's lips for a [b] and the physiology of the SVT.

These neural representations of motoric acts structured in terms of the capabilities of the human SVT also account for the perception of phonetic segments. The perception of human speech is "special." We appear to "decode" the encoded speech signal by means of implicit knowledge of the constraints of the SVT and the motor acts that underlie human speech. The "motor theory of speech perception" proposed by Liberman and his colleagues (Liberman et al., 1967) accounts for the process by which we perceive the phonetic segments that have been melded together as we talk.[2]

– The Speech and Language Limitations of Nonhuman Primates –

It is easy to characterize the speech-producing abilities of nonhuman primates—they have none. Although a homologue of Broca's area (the part of the human brain traditionally associated with language—see below) can be identified in monkeys on anatomical grounds, massive lesions in or near this area of the neocortex have no effect whatsoever on their vocalizations. Electrophysiologic and ablation studies show that it is nonfunctional with respect to regulation of their vocalizations. It is the case that neocortical lesions in nonhuman primates have no effect on vocalizations. As MacLean and Newman (1988) note, the effects on vocalizations ascribed to the neocortical lesions noted in some primate studies actually were due to damage to the anterior cingulate gyrus (the "paleomammalian" limbic system, cf. MacLean, 1985). The vocalizations of nonhuman primates are not regulated by neocortical structures. Though parts of their neocortex resemble Broca's area anatomically, it is not involved in regulating their vocalizations. Their vocalizations are instead controlled by the cingulate cortex, the "old" motor cortex that evolved with the earliest mammals, the basal ganglia, and midbrain structures (Sutton and Jurgens, 1988).

Behavioral studies consistently show that non-human primates lack "voluntary" control of their vocal output. For example, acoustic analyses of chimpanzee calls show that they make use of phonetic "features" that play a linguistic role in human speech (Lieberman, 1968,

1975). Chimpanzees produce falling formant frequency transitions by rounding their lips that are similar to those that specify "labial" sounds like [b] and [w] in English. Chimpanzees also have functional larynges and most of their calls are phonated. Therefore, chimpanzees could, in principle, produce the labial stop consonants [b] and [p], which are differentiated by means of the timing between lips and laryngeal output (Lisker and Abramson, 1964). Moreover, chimpanzees have speech-producing anatomy that, in principle, could produce nasalized vowels similar to the human vowels [I], [U], [æ], etc. (Lieberman et al., 1972), and some of their calls involve slightly different vowel-like contrasts (Lieberman, 1968). Many chimpanzees have been raised from infancy in close proximity to humans speaking to them and to other humans and have been actively tutored. However, no chimpanzee has ever produced *voluntary* speech-like vocalizations approximating human words like *bad, pad, bit, pit, bat,* etc.

It is apparent that the acoustic features that constitute chimpanzee calls are *bound* together. Chimpanzees cannot voluntarily dissociate the bonds that link these elements together. Though their stereotyped vocalizations incorporate elements that could approximate many human words, chimpanzees are not able to "free" these elements and produce voluntary speech. The most that nonhuman primates can do is change the rate at which they'll produce some stereotyped utterance. Humans can say whatever they please, whether they're happy, hungry, feeding, frightened, etc. In contrast, as many observers of chimpanzees and other nonhuman primates note, primate calls are part of a stereotyped response pattern that signals a particular affectual or emotional state (Goodall, 1986). Operant conditioning of nonhuman primates that changes their affectual state would also modify their calls. As the primate's affectual state changes, so would its stereotyped calls. The calls would still be involuntary, stereotyped calls that were mediated by the primate's limbic system.

Likewise, the stereotyped vocal responses of infant and adult primates may differ as a result of either maturation or learning, but the end result is a stereotyped call that signals particular threats, rewards, etc. Nothing in the data presented by Cheney and Sayfarth (1990) or other studies of nonhuman primates demonstrates voluntary control of vocalizations. If nonhuman primates had voluntary control of their vocalizations, we would expect to find them imitating human speech when they were raised in human environments. They never produce any speech sounds despite the most intensive training. In contrast, any normal human child raised in any reasonably normal environment acquires speech effortlessly. We can conclude that chimpanzees and other nonhuman primates don't acquire any aspect of speech because they lack the requisite brain mechanisms.

The case is also clear against chimpanzee's having syntactic ability similar to that of normal human beings. Many language-trained chimpanzees clearly have limited lexical abilities and can learn about 200 words using various manual-visual signals. However, they are *not* able to make use of even simple distinctions of word order. They cannot comprehend the semantic distinctions conveyed by word order in sequences like *see me* versus *me see.* Many claims to the contrary have been made; in recent years some investigators have proposed that the linguistic abilities of the bonobo *(Pan paniscus)* surpass those of the "common" chimpanzee *(Pan troglodytes)* with respect to syntax. However, Rumbaugh and Savage-Rumbaugh (1992) in their recent study of the grammatical ability of the bonobo Kanzi conclude, "there is no example in which a difference in [word] order signals a difference in meaning." Chimpanzees do combine words to form new words. For example, they spontaneously form new words by compounding old words forming *waterbird*— duck, from the semantically distinct words *water* and *bird* (Gardner et al., 1989). Chimpanzees, in other words, command some aspects of morphophonemics. However, they cannot handle the syntax of human language with the proficiency of a normal three-year-old human child. Clearly, some qualitative neural distinction sets the syntactic abilities (rather the lack thereof) of chimpanzees apart from those of humans.

5. Some Aspects of the Neurophysiology of Speech Production

The traditional account of the neurophysiological basis of human speech production is largely based on the study of aphasia—damage to the brain that results in deficits in speech production. Paul Broca in 1861 first claimed that damage to part of the neocortex in the dominant hemisphere of the brain can result in aphasia, i.e., a complex of speech and language deficits. The traditional view of Broca's aphasia is that damage localized to this particular area of the brain will result in these deficits, while damage to any other part of the brain won't. This belief is reflected in popularized accounts of how the human brain "works" and in the supposition of many linguists that human beings have a specific, localized "language organ" (Chomsky, 1986). However, that supposition is not true. In the past decade new brain imaging and electrophysiological techniques have revealed that permanent aphasia is the consequence of damage to subcortical "pathways" or "circuits" connecting Broca's area to the parts of the brain that directly control muscles, and to the prefrontal cortex. In fact, subcortical damage that disrupts the connections from Broca's area, *leaving it intact,* can result in aphasia.

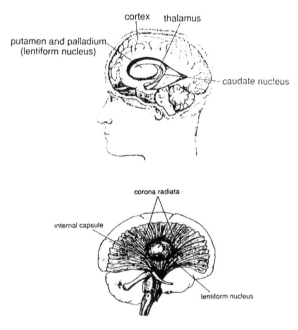

Figure 3. The structures of the basal ganglia. The putamen and palladium are cradled in the descending nerves that converge to form the internal capsule and snake their way down through the basal ganglia. The putamen, palladium, and caudate nucleus connect to various regions of the cortex and to the thalmus through complex pathways that transmit information to and from these structures. After Lieberman (1991).

Damage to the structures sketched in Figure 3, the internal capsule (the bundle of nerve fibers that connect the neocortex to the midbrain), the putamen, and the caudate nucleus (some of the structures of the basal ganglia), can yield impaired speech production and agrammatism similar to that of the classic aphasias as well as other cognitive deficits (Naeser et al., 1982; Alexander et al., 1987). Alexander and his colleagues reviewed 19 cases of aphasia with language impairments that ranged from fairly mild disorders in the patient's ability to recall words, to "global aphasia," in which the patient produced very limited "dysarthric" nonpropositional speech and was not able to comprehend syntax. Dysarthric speech derives from a loss of the coordinated motor control of the respiratory system and larynx—phonation becomes irregular and "breathy." The patients who had the most extreme language and speech deficits also lost control of their right arms and legs. Lesions in the basal ganglia and the nerve connections that run down from the neocortex were noted for the 19 cases. In general, severe language deficits occurred in patients who had suffered the most extensive subcortical brain damage. These patients also suffered paralysis of their dominant right hands.

Studies of the deficits resulting from neurodegenerative diseases like Parkinson's disease (PD) also show the effects of damage to subcortical circuits on cognition and language. These diseases cause major damage to the basal ganglia, sparing the cortex. The primary deficits of subcortical disease are motoric; tremors, rigidity, and repeated movement patterns occur. However, subcortical diseases can also cause linguistic and cognitive deficits. In extreme form the deficits associated with these subcortical diseases constitute a dementia (Albert et al., 1974; Cummings and Benson, 1984). Deficits in the comprehension of syntax have been noted in several independent studies (Lieberman et al., 1990, 1992b; Grossman et al., 1991, 1992; Natsopolous, in press). A pattern of speech production, syntax, and cognitive deficits similar in nature to those typical of aphasia can occur in even moderately impaired PD (Morris et al., 1988; Lieberman, 1992b; Lange et al., in press).

Moreover, studies that make use of CT and PET scans of victims of aphasia show clearly that damage to pathways to prefrontal cortex is implicated in these deficits. Metter et al. (1989), for example, using CT scans found that Broca's patients had subcortical damage to the internal capsule and parts of the basal ganglia. PET scans showed that these patients had vastly reduced metabolic activity in the left prefrontal cortex and Broca's region. Previous studies (Metter et al., 1987) showed a strong correlation between prefrontal and Broca's region metabolic activity and "functional motor loss of the arms and legs, as well as spontaneous speech and writing [and] in normal subjects, a strong correlation between prefrontal cortex and decision making" (Metter et al., 1989, p. 31). They conclude that the behavioral deficits of Broca's aphasia— general "difficulty in motor sequencing and executing motor speech tasks" and "the presence of language comprehension abnormalities"— derive from damage to circuits to the prefrontal cortex.

PET scan studies of subcortical disease reach similar conclusions

(Metter et al., 1984); progressive supranuclear palsy (PSP) patients have less metabolic activity in the prefrontal cortex than that of normal controls. The destruction by PSP of the circuits through the basal ganglia that stimulate the prefrontal cortex is responsible for the reduced activity and concomitant cognitive deficits (D'Antonia et al., 1985). Recent behavioral data show that reduced basal ganglia pathway activity to prefrontal cortex in even moderately impaired PD patients produces cognitive deficits similar in kind to those typical of frontal lesions (Morris et al., 1988: Sahakian et al., 1988; Lieberman, 1992b; Lange et al., in press).

6. What Might Distinguish the Human Brain from Other Brains?

Tracer studies that reveal the detailed wiring diagram of the brain can not be performed on human subjects. However, they have been used to explore the basal ganglia pathways to and from prefrontal cortex in nonhuman primates and other animals (Parent, 1986). Basal ganglia pathways in mammals carry signals between various parts of the cortex. The basal ganglia have become proportionately larger as the cortex has expanded in the course of mammalian evolution; as cortex expands so do basal ganglia. The prefrontal cortex also becomes proportionately larger as the phylogenetic scale is ascended (Brodmann, 1909). The prefrontal cortex is about 200 percent larger in humans than in chimpanzees (Deacon, 1988). It enters into all manner of controlled, planned motoric functions as well as into abstract thought (Stuss and Benson, 1986). The basal ganglia have also become enlarged, elaborated, and differentiated during the course of mammalian evolution. The basal ganglia of rodents, for example, lack the independent caudate nucleus and putamen found in primates (Parent, 1986). There is no reason to suppose that the basal ganglia of humans are structurally identical to those of apes given the fact that the human basal ganglia differ demonstrably from those of apes in size. It is reasonable to propose that the voluntary speech behavior of humans may rest on the enlargement and elaboration of the prefrontal cortex and the basal ganglia circuits connecting it to other parts of the brain.

– Tracing the Evolution of the Human Brain –

The data of recent studies of aphasia and neurodegenerative diseases suggests that the neural circuits for speech and syntax appear to involve independent, though proximate, subcortical pathways. The architecture of these pathways, in all likelihood, reflects the evolutionary history of the human brain (Lieberman, 1991). Neural mechanisms that were initially adapted for speech motor control furnished the starting point for natural selection that ultimately yielded the neural structures and circuits that underlie human syntactic ability. Therefore, it is no accident that chimpanzees lack both voluntary speech and syntax, though they demonstrate some lexical ability. Though it appears to be impossible to make any direct inferences concerning the prefrontal cortex and basal ganglia circuits of fossil hominids by examining their skulls, the strong linkage between the brain mechanisms involved in the production and perception of human speech and the constraints of the supralaryngeal vocal tract (SVT) may furnish some insights.

Therefore, we may be able to trace the evolution of speech in the fossil record by reconstructing the supralaryngeal airways of fossil hominids using the methods of comparative anatomy. The adult human tongue, mouth, and pharynx, which constitute the supralaryngeal airway (the supralaryngeal vocal tract—SVT), are radically different from those of any other animal, including newborn humans (Negus, 1949; Lieberman and Crelin, 1971; Lieberman et al., 1972). Figure 4 shows a chimpanzee's supralaryngeal tract. The larynx in nonhuman species is positioned high, close to the entrance to the nasal cavity. The nonhuman larynx is normally raised while breathing, locking it into the nasal airway to form a sealed pathway from nose to lungs. This sealed breathing path allows animals to breath and drink simultaneously. In contrast, in normal adult-like humans the larynx is positioned low in the neck and food has to be pushed past the opening of the larynx, with an ever-present risk of blocking the airway to the lungs.

Charles Darwin (1859, p. 191) first noticed the peculiar morphology of the human SVT and wondered why it had assumed this odd morphology. The answer is that the human SVT allows us to produce sounds that facilitate the process of formant frequency encoding that yields the high transmission rate of human speech. The human SVT first allows us to produce sounds that are *not* nasalized. We can seal our nose off from the rest of the SVT while we vocalize. Nasalized sounds, which occur when the nose is coupled to the rest of the SVT, reduce speech intelligibility because they obscure the formant frequency patterns that convey the phonetic contrasts of human speech. The human SVT, moreover, allows us to produce sounds like the vowels [u] and [i] (the vowels of *shoe* and *see*) which facilitate the perceptual process of vocal tract normalization that is an essential aspect of the "encoding-decoding" process that yields the high data transmission rate of human speech.

As noted earlier, the formant frequency encoding-decoding process

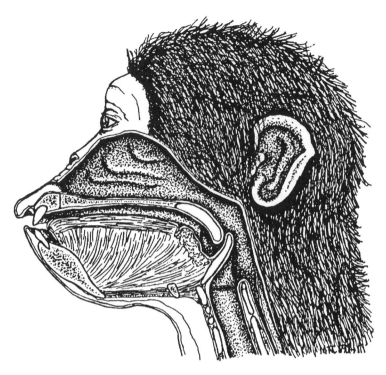

Figure 4. A typical nonhuman supralaryngeal airway, a chimpanzee. The tongue is positioned entirely within the oral cavity; the larynx is positioned high, close to the opening to the nose. The epiglottis and velum overlap to form a watertight seal when the larynx is raised. This allows the chimpanzee to swallow and breath simultaneously. After Lieberman (1991).

exemplified in human speech allows us to circumvent the temporal constraints of the human auditory system; the speech transmission rate can reach 20 phonetic units per second, in contrast to a nonspeech rate of 7 to 9. But the enhanced speech-producing qualities of the human supralaryngeal airway or "tract" (the SVT) would be worse than useless in the absence of the species-specific neural mechanisms that allow us to schedule and produce the necessary articulatory maneuvers. Given the liabilities of the human SVT with respect to breathing and swallowing, as well as the increased risk of infection due to impacted teeth (the human mandible and maxilla have become shorter compared to *Erectus* hominids without a matching reduction in tooth size), the only selective advantage of the human SVT is for speech. Therefore, a human SVT in a fossil hominid indicates the presence of the neural mechanisms that make voluntary speech possible (Lieberman, 1989,

1991). Since these neural mechanisms are also involved in comprehending and producing sentences with moderately complex syntax and in abstract reasoning, we can make some further inferences.

– The "Eve" Hypothesis and Neanderthals –

The different morphology of the human and nonhuman SVT is reflected in the skeletal features of the bottom of the skull and the lower jaw, which makes it possible to reconstruct the SVTs of fossil hominids. The SVT reconstructions that I have discussed elsewhere (Lieberman, 1975, 1984, 1991) are consistent with the "Eve" hypothesis. The Eve hypothesis claims that modern humans originated in Africa in the last 150,000 years or so, and displaced archaic hominids like the European classic Neanderthals (Stringer and Andrews, 1988). The earliest fossils (Qafzeh and Skhul V) that appear to have modern human SVT lived about 100,000 years ago in what is presently Israel. Some of the skeletal features that indicate that they had modern SVTs also appear in earlier African fossils such as those at Broken Hill, about 200,000 years in age. Whether these distant ancestors expressed the concept of free will is something that we shall never know. However, in all likelihood they had a capacity for voluntary speech motor activity similar to ours. Moreover, the archaeological record shows that Qafzeh and Skhul V had advanced cognitive and linguistic ability; they were buried with ritual grave goods, which, in itself, indicates complex linguistic and cognitive ability (Lieberman, 1991). In contrast, classic European Neanderthals lacked a human SVT and hence were not as specialized for vocal communications as our ancestors. No known Neanderthal burial includes grave goods; the items often identified as grave goods in Neanderthal burials could just as likely have been debris (Dibble, 1989).

– Neanderthal Speech –

Neanderthal speech remains a controversial subject. It forms one part of the current debate on the origin of modern human beings. The unflexed basicranium that marks the SVTs of living primates is a primitive feature of *Australopithecine* and *Erectus* hominids (Lieberman, 1975; Laitman and Heimbuch, 1982), and virtually no anthropologist disputes that they lacked human speech capabilities. It is clear to me that the flexed basicranium that marks the human SVT is one of the derived features that differentiates modern humans anatomically from these extinct hominids. Its presence in fossils such as Jebel Qafzeh and Skhul V is an index that demonstrates that these hominids possessed the neural mechanisms necessary for voluntary speech production—a property that differentiates human beings from all other living primates (Lieberman, 1991). In contrast, classic Neanderthal fossils appear to retain the primitive condition—an unflexed basicranium and an SVT maladapted for speech production. Their speech limitations would have served as a genetic isolating mechanism and may account for their ultimate extinction even if they had been in contact with modern humans over an extended period (Lieberman, 1992a). Given the flexed basicranium of the African Broken Hill hominid, the fossil record pertaining to speech is consistent with the "Eve" hypothesis, which proposes an African origin of modern human beings and a concomitant replacement of classic Neanderthals when modern humans migrated from their African homeland.

Recent claims that Neanderthals had the same speech capacity as modern humans, therefore, are part of a larger debate that seeks to refute the Eve hypothesis. Neanderthal and human remains generally are found in association with the same stone tool kit (Bar Yosef et al., 1992). If Neanderthals had the same linguistic and cognitive capacities as modern humans, there would be little reason to expect their replacement. Indeed, opponents of the Eve hypothesis often claim that Neanderthals are among the ancestors of present-day humans. Thus the claim that Neanderthals had modern SVTs transcends the issue of speech ability. However, recent studies claiming that Neanderthals had human SVTs are based on incorrect anatomical premises.

For example, Arensburg et al. (1990) claim that the SVT of the Neanderthal Kebara fossil had a human SVT because its hyoid bone (part of the supporting structure of the larynx) resembles a modern human hyoid bone. Arensburg and his colleagues base this claim on a false anatomical premise. Their argument hinges on the hyoid bone always occupying the same position in humans relative to the vertebral column and mandible. They take this position because the Lieberman and Crelin (1971) reconstruction of the La Chapelle-aux-Saints Neanderthal SVT was based on the similarities that exist between the Neanderthal and newborn human skulls and certain aspects of their mandibles. These skeletal features, which differ from those of normal adult humans, yield a Neanderthal SVT that is similar to that of a newborn human. The reconstructed Neanderthal tongue is positioned in the oral cavity of the La Chapelle-aux-Saints fossil with the larynx in a high position relative to the vertebral column. However, the Arensburg et al. (1990) argument is falsified by all available data (e.g., Negus, 1949; Bosma, 1975; Lieberman and Crelin, 1971; Lieberman et al., 1972; George, 1978; Senecail, 1979). In reality, the hyoid bone and larynx shift in the course of normal ontogentic development in humans from a high neonatal position and a SVT incapable of normal speech production to a low position in the adult human SVT. The Kebara hyoid could have been positioned anywhere in the neck of the fossil hominid during life (Lieberman et al., 1992b; Lieberman, in press).

Other recent claims for Neanderthal speech ability equivalent to that of modern humans rest on equally shaky anatomical foundations. Houghton (1993), for example, proposes a Neanderthal reconstruction with a tongue that would not allow the creature to swallow or speak. The reconstructed tongue is absurdly small; Houghton failed to use radiographic data that show the actual size and position of the human tongue during speech though such data have been available for more than a half a century (Carmody, 1937). Houghton's reconstructed Neanderthal tongue can neither propel food down the pharynx, nor form the constrictions necessary for producing speech sounds like the vowels [i], [u], [a], stop consonants, or fricatives. Moreover, the Neanderthal tongue proposed by Houghton would yield a larynx positioned in the creature's chest (an impossible position), though neither Houghton nor the editors who reviewed his paper appear to realize this defect. Houghton, moreover, claims that the basicranium of the Neanderthal skull (the area relevant to the SVT) does not differ from modern human skulls, although the comprehensive studies of Howells (1987, 1989) show that the length of the skeletal features that define the mouth are greater than the mean length of the comparable human measurement by more than 5 standard deviations. The publication of Houghton's paper attests to the strong beliefs, transcending rational argument, held by many anthropologists concerning Neanderthals.

In any event, the Neanderthal debate simply addresses the issue of *when* the human SVT evolved. The early *Australopithecine* hominid populations whose skulls and SVTs closely resemble those of living apes (Lieberman, 1975; Laitman and Heimbuch, 1982) must have placed less reliance on vocal communication. They, and later *Erectus* hominids, do not appear to have been as capable of innovation. The "frozen forms" evident over a span of one million years in the almost unchanging *Erectus* stone tool kit attest to a lack of innovation that may reflect a general cognitive deficit. Their neural capacities with regard to voluntary motor activity and voluntary speech production abilities most surely were less developed than ours, and if the thesis that I advance is correct, the issue of free will did not enter into their thoughts.

7. To Conclude

Though scholars are a contentious lot, few would argue that other animals exercise free will. Indeed, free will, or the belief therein, is one of the defining characteristics of humanness; it must have a biological basis in the structure of the human brain. And if we accept the theory of evolution, we must account for the evolution of free will. The evolutionary framework for free will presented in this paper—that free will derives from the neurophysiologic mechanisms that yield voluntary motor control, particularly in the domain of speech, will be debated. However, it appears to be consistent with present knowledge of the neural bases of motor control, language, and thought. Whether it will bear up to further inquiry is the question with which I conclude.

Notes

[1] These include the work of Metter et al. (1984, 1987, 1989); Cummings and Benson (1984); Mesulam (1985); Lieberman (1985, 1989, 1991, 1992b); Parent (1986); Stuss and Benson (1986); Lieberman et al. (1990, 1992b); Morris et al. (1988); Sahakian et al. (1988); Grossman et al. (1991, 1992); Lange et al. (in press), and the studies that will be discussed as we proceed.

[2] Recent modified version of the motor theory no longer claim that humans actually match particular motor sequences to the acoustic signal. The neural devices tuned to perceive speech can implicitly code this information; cf. Lieberman (1984, pp. 179–189).

References

Albert, M. A., R. G. Feldman, and A. L. Willis. 1974. "The 'subcortical dementia' of progressive supranuclear palsy." *Journal of Neurology, Neurosurgery, and Psychiatry,* 37:121–130.

Alexander, M. P., M. A. Naeser, and C. L. Palumbo, 1987. "Correlations of subcortical CT lesion sites and aphasia profiles." *Brain,* 110:961–991.

Arensburg, B., L. A. Schepartz, A. M. Tiller, B. Vandermeersch, H. Duday, and Y. Rak. 1990. "A reappraisal of the anatomical basis for speech in middle palaeolithic hominids." *American Journal of Physical Anthropology,* 83:137–146.

Bar Yosef, O., B. Vandermeersch, B. Arensbyrg, A. Belfer-Cohen, P. Goldberg, H. Laville, L. Meignen, Y. Rak, J. D. Speth, E. Tchernov,

A.-M. Tillier, and S. Weiner. 1992. "The excavations in Kebara Cave, Mt. Carmel." *Current Anthropology*, 33:497–550.

Bell, C. G., H. Fujisaki, J. M. Heinz, K. N. Stevens, and A. S. House. 1961. "Reduction of speech spectra by analysis-by-synthesis techniques." *J. Acoust. Soc. Amer.*, 33:1725–1736.

Boesch, C. and H. Boesch. 1981. "Sex differences in the use of natural hammers by wild chimpanzees: a preliminary report." *Journal of Human Evolution*, 10:585–593.

Boesch, C. and H. Boesch. 1991. "Nutcracking using natural hammers by wild chimpanzees." A film record presented at the Wenner-Gren Conference on Tools and Language, Cascais, Portugal.

Bosma, J. F. 1975. "Anatomic and physiologic development of the speech apparatus," in *Human communication and its disorders*, D. B. Towers, ed., pp. 469–481. New York: Raven.

Bowerman, M. 1987. "What shapes children's grammars?" in *The cross-linguistic study of language acquisition*, D. I. Slobin, ed. Hillsdale, NJ: Lawrence Erlbaum.

Broca, P. 1861. "Remarques sur le siege de la faculté de la parole articulée, suies d'une observation d'aphemie (perte de parole)." *Bulletin de la Société d'Anatomie (Paris)*, 36:330-357.

Brodmann, K. 1909. *Vergleichende histologische Lokalisation der Groshirnrinde in iheren Prinzipien Dargestellt auf Grund des Zellenbaues.* Leipzig: Barth.

Capranica, R. R. 1965. *The evoked vocal response of the bullfrog.* Cambridge, MA: MIT Press.

Carmody, F. 1937. "X-ray studies of speech articulation." *University of California Publications in Modern Philology*, 20:187–237. Berkeley, CA: University of California Press.

Cheney, D. L. and R. M. Sayfarth. 1990. *How monkeys see the world.* Chicago: University of Chicago Press.

Chomsky, N. 1986. *Knowledge of language: Its nature, origin and use.* New York: Praeger.

Cummings, J. L. and D. F. Benson. 1984. "Subcortical dementia: Review of an emerging concept." *Archives of Neurology*, 41:874–879.

D'Antonia, R., J. C. Baron, Y. Samson, M. Serdaru, F. Viader, Y. Agid, and J. Cambier. 1985. "Subcortical dementia: Frontal cortex hypometabolism detected by positron tomography in patients with progressive supranuclear palsy." *Brain*, 108:785–799.

Darwin, C. 1859. *On the origin of species.* Facsimile ed. 1964 Cambridge, MA: Harvard University Press.

Deacon, T. W. 1988. "Human brain evolution–II. Embryology and brain allometry," in *Intelligence and evolutionary biology,* H. J. Jerison and I. Jerison, eds., NATO ASI Series, pp. 383–416. Berlin: Springer.

De Waal, F. 1982. *Chimpanzee politics.* London: Collins.

Dibble, H. 1989. "The implications of stone tool types for the presence of language during the lower and middle Palaeolithic," in *The human revolution: Behavioural and biological perspectives in the origins of modern humans,* P. Mellars and C. B. Stringer, eds., pp. 415–432. Edinburgh: Edinburgh University Press.

Evarts, E. V. 1973. "Motor cortex reflexes associated with learned movement." *Science,* 179:501–503

Fant, G. 1960. *Acoustic theory of speech production.* The Hague: Mouton.

Fernald, A. 1982. "Acoustic determinants of infant preference for 'motherese.'" Ph.D. dissertation, University of Oregon.

Fodor, J. 1983. *Modularity of mind.* Cambridge, MA: MIT Press.

Gardner, R. A. and B. T. Gardner. 1984. "A vocabulary test for chimpanzees *(Pan troglodytes)*." *Journal of Comparative Psychology,* 4:381–404.

Gardner, R. A., B. T. Gardner, and T. E. Van Cantfort. 1989. *Teaching sign language to chimpanzees.* Albany, NY: State University of New York Press.

George, S. L. 1978. "A longitudinal and cross-sectional analysis of the growth of the postnatal cranial base angle." *American Journal of Physical Anthropology,* 49:171–178.

Goldstein, K. 1948. *Language and language disturbances.* New York: Grune and Stratton.

Goodall, J. 1986. *The chimpanzees of Gombe: Patterns of behavior.* Cambridge, MA: Harvard University Press.

Graco, V. and J. Abbs. 1985. "Dynamic control of the perioral system during speech: kinematic analyses of autogenic and nonautogenic sensorimotor processes." *Journal of Neurophysiology,* 54:418–432.

Grosmangin, C. 1979. "Base du crâne et pharynx dans leur rapports avec l'appareil de langage articulé." *Memoires du Laboratoire d'Anatomie de la Faculte de Medicine de Paris,* no. 40–1979.

Grossman, M., S. Carvell, S. Gollomp, M. B. Stern, G. Vernon and H. I. Hurtig. 1991. "Sentence comprehension and praxis deficits in Parkinson's disease." *Neurology,* 41:160–1628.

Grossman, M., S. Carvell, M. B. Stern, S. Gollomp, and H. I. Hurtig. 1992. "Sentence comprehension in Parkinson's Disease: The role of attention and memory." *Brain and Language,* 42:347–384.

Hellwag, C. 1781. *De Formatione Loquelae.* Dissertation, Tubingen, Germany.

Houghton, P. 1993. "Neanderthal supralaryngeal vocal tract." *American Journal of Physical Anthropology,* 90:139–146.

Howells, W. W. 1976. "Neanderthal man: facts and figures," in *Proceedings of the Ninth International Congress of Anthropological and Ethnological Sciences,* Chicago, 1973. The Hague: Mouton.

Howells, W. W. 1989. "Skull shapes and the map; craniometric analyses in the dispersion of modern homo." *Papers of the Peabody Museum of Archaeology and Ethnology.* Cambridge, MA: Harvard University, Volume 79.

Illes, J., E. J. Metter, W. R. Hanson, and S. Iritani. 1988. "Language production in Parkinson's disease: Acoustic and linguistic considerations." *Brain and Language,* 33:146–160.

Kimura, D. 1979. "Neuromotor mechanisms in the evolution of human communication," in *Neurobiology of Social Communication in Primates,* H. D. Steklis and M. J Raleigh, eds. New York: Academic Press.

Laitman, J. T. and R. C. Heimbuch. 1982. "The basicranium of Plio-Pleistocene hominids as an indicator of their upper respiratory systems." *American Journal of Physical Anthropology,* 59:323–344.

Laitman, J. T., R. C. Heimbuch, and E. S. Crelin. 1978. "Developmental changes in a basicranial line and its relationship to the upper respiratory system in living primates." *American Journal of Anatomy,* 152, 467–482.

Laitman, J. T., R. C. Heimbuch, and E. S. Crelin. 1979. "The basicranium of fossil hominids as an indicator of their upper respiratory systems." *American Journal of Physical Anthropology,* 51:15–34.

Lange, K. W., T. W. Robbins, C. D. Marsden, J. M. A. M. Owen, and G. M. Paul. "L-Dopa withdrawal in Parkinson's disease selectively impairs cognitive performance in tests sensitive to frontal lobe dysfunction." *Psychopharmacology,* in press.

Liberman, A. M., F. S. Cooper, D. P. Shankweiler, and M. Studdert-Kennedy. 1967. "Perception of the speech code." *Psychological Review,* 74:431–461.

Lieberman, P. 1967. *Intonation, perception and language.* Cambridge MA: MIT Press.

Lieberman, P. 1968. "Primate vocalizations and human linguistic ability." *Journal of the Acoustical Society of America,* 44:1157–1164.

Lieberman, P. 1975. *On the origins of language: An introduction to the evolution of speech.* New York: Macmillan.

Lieberman, P. 1984. *The biology and evolution of language.* Cambridge, MA: Harvard University Press.

Lieberman, P. 1985. "On the evolution of human syntactic ability: Its pre-adaptive bases—motor control and speech." *Journal of Human Evolution,* 14:657–668.

Lieberman, P. 1989. "The origins of some aspects of human language and cognition," in *The human revolution: Behavioural and biological perspectives in the origins of modern humans,* P. Mellars and C. B. Stringer, eds., pp. 391-414. Edinburgh: Edinburgh University Press.

Lieberman, P. 1991. *Uniquely human: The evolution of speech, thought, and selfless behavior.* Cambridge, MA: Harvard University Press.

Lieberman, P. 1992a. "On Neanderthal speech and Neanderthal extinction." *Current Anthropology,* 33:409–410.

Lieberman, P. 1992b. "Could an autonomous syntax module have evolved?" *Brain and Language,* 43: 768–774.

Lieberman, P. in press. "The Kebara KMH-2 Hyoid and Neanderthal speech." *Current Anthropology.*

Lieberman, P. and E. S. Crelin. 1971. "On the speech of Neanderthal man." *Linguistic Inquiry,* 2:203–222.

Lieberman, P., E. S. Crelin, and D. H. Klatt. 1972. "Phonetic ability and related anatomy of the newborn, adult human, Neanderthal man, and the chimpanzee." *American Anthropologist,* 74:287–307.

Lieberman, P., J. Friedman, and L. S. Feldman. 1990. "Syntactic deficits in Parkinson's disease." *Journal of Nervous and Mental Disease,* 178: 360–365.

Lieberman, P., E. T. Kako, J. Friedman, G. Tajchman, L. S. Feldman, and E. B. Jiminez. 1992a. "Speech production, syntax comprehension, and cognitive deficits in Parkinson's disease." *Brain and Language,* 43:169–189.

Lieberman, P., J. T. Laitman, J. S. Reidenberg, and P. Gannon. 1992b. "The anatomy, physiology, acoustics and perception of speech: Essential elements in analysis of the evolution of human speech." *Journal of Human Evolution,* 23:447–467.

Lisker, L. and A. S. Abramson. 1964. "A cross language study of voicing in initial stops: Acoustical measurements." *Word,* 20:384–442.

Lubker, J. and T. Gay. 1982. "Anticipatory labial coarticulation: Experimental, biological, and linguistic variables." *Journal of the Acoustical Society of America,* 71:437–438.

MacLean, P. D. 1985. "Evolutionary psychiatry and the triune brain." *Psychological Medicine,* 15:219–221.

MacLean, P. D. and J. D. Newman. 1988. "Role of midline frontolimbic cortex in the production of the isolation call of squirrel monkeys." *Brain Research,* 450:111–123.

MacNeilage, P. F., M. G. Studdert-Kennedy, and B. Lindblom. 1987. "Primate handedness reconsidered." *Behavioral and Brain Sciences,* 10:247–303.

Mayr, E. 1982. *The growth of biological thought.* Cambridge, MA: Harvard University Press.

Mesulam, M. M. 1985. "Patterns in behavioral neuroanatomy: Association areas, the limbic system and hemispheric specialization," in *Principles of behavioral neurology,* M. M. Mesulam, ed., pp. 1–70. Philadelphia: F. A. Davis.

Metter, E. J., W. H. Riege, W. R. Hanson, M. E. Phelps and D. E. Kuhl. 1984. "Local cerebral metabolic rates of glucose in movement and language disorders from positron tomography." *American Journal of Physiology,* 246:R897–R900.

Metter, E. J., D. Kempler, C. A. Jackson, W. R. Hanson, W. H. Reige, L. M. Camras, J. C. Mazziotta, and M. E. Phelps. 1987. "Cerebular glucose metabolism in chronic aphasia." *Neurology,* 37:1599–1606.

Metter, E. J., D. Kempler, C. Jackson, W. R. Hanson, J. C. Mazziotta, and M. E. Phelps. 1989. "Cerebral glucose metabolism in Wernicke's, Broca's, and conduction aphasia." *Archives of Neurology,* 46:27–34.

Miller, G. A. 1956. "The magical number seven, plus or minus two: Some limits on our capacity for processing information." *Psychological Review,* 63:81–97

Morris, R. G., J. J. Downes, B. J. Sahakian, J. L. Evenden, A. Heald, and T. W. Robbins. 1988. "Planning and spatial working memory in Parkinson's disease." *Journal of Neurology, Neurosurgery, and Psychiatry,* 51:757–766.

Naeser, M. A., M. P. Alexander, N. Helms-Estabrooks, H. L. Levine, S. A. Laughlin, and N. Geschwind. 1982. "Aphasia with predomininantly subcortical lesion sites; description of three capsular/putaminal aphasia syndromes." *Archives of Neurology,* 39:2–14.

Natsopoulos, D., G. Grouios, S. Bostantzopoulou, G. Mentenopoulos, Z. Katsarou, and J. Logothetis. In press. "Algorithmic and heuristic strategies in comprehension of complement clauses by patients with Parkinson's Disease." *Neuropsychologia.*

Nearey, T. 1978. "Phonetic features for vowels." Bloomington, IN: Indiana University Linguistics Club.

Negus, V. E. 1949. *The comparative anatomy and physiology of the larynx.* New York: Hafner.

Parent, A. 1986. *Comparative neurobiology of the basal ganglia.* New York: John Wiley.

Peterson, M. R., M. D. Deecher, S. R. Zolith, D. B. Moody, and W. C. Stebbens. 1978. "Species-specific perceptual processing of vocal sounds by monkeys." *Science,* 202:324–326.

Rumbaugh, D. M. and E. S. Savage-Rumbaugh. 1992. "Biobehavioral roots of language: Words, apes and a child." Conference paper presented at University of Bielefeld, Germany.

Sahakian, B. J., R. G. Morris, J. L. Evenden, A. Heald, R. Levy, M. Philpot, and T. W. Robbins. 1988. "A comparative study of visuospatial memory and learning in Alzheimer-type dementia and Parkinson's disease." *Brain,* 111:695–718.

Savage-Rumbaugh, S., D. Rumbaugh, and K. McDonald. 1985. "Language learning in two species of apes." *Neuroscience and Biobehavioral Reviews,* 9:653–665.

Savage-Rumbaugh, S., K. McDonald, R. A. Sevcik, W. D. Hopkins, and E. Rubert. 1986. "Spontaneous symbol acquisition and communicative use by pygmy chimpanzees *(Pan paniscus)*." *Journal of Experimental General Psychology,* 115:211–235.

Senecail, B. 1979. "L'Os hyoide; introduction anatomique a l'étude de certains mécanismes de la phonation." *Mémoires du Laboratoire d'Anatomie de la Faculté de Médecine de Paris,* no. 36–1979.

Sereno, J. and P. Lieberman. 1987. "Developmental aspects of lingual coarticulation." *Journal of Phonetics,* 15:247–257.

Sereno, J., S. R. Baum, G. C. Marean, and P. Lieberman. 1987. "Acoustic analyses and perceptual data on anticipatory labial coarticulation in adults and children." *Journal of the Acoustical Society of America,* 81:512–519.

Stringer, C. B. and P. Andrews. 1988. "Genetic and fossil evidence for the origin of modern humans." *Science,* 239: 1263–1268.

Stuss, D. T. and D. F. Benson. 1986. *The frontal lobes.* New York: Raven.

Sutton, D. and U. Jurgens. 1988. "Neural control of vocalization," in *Comparative primate biology,* Vol. 4, H. D. Steklis and J. Erwin, eds., pp. 625–647. New York: Alan R. Liss.

Tomasello, M. and M. J. Farrar. 1986. "Joint attention and early language." *Child Development,* 57:1454–1463.

Watkins, K. and D. Fromm. 1984. "Labial coordination in children: Preliminary considerations." *Journal of the Acoustical Society of America,* 75:629–632.

Wills, R. 1973. *The institutionalized severely retarded.* Springfield, IL: Charles Thomas.

Wilson, E. O. 1975. *Sociobiology: The new synthesis.* Cambridge, MA: Harvard University Press.

CHAPTER 7

A Hierarchy of Complex Behaviors in Microbiological Systems

ERIK MOSEKILDE, HEIDI STRANDDORF,
JESPER SKOVHUS THOMSEN, AND GEROLD BAIER

Department of Physics, Systems Dynamics Group
Technical University of Denmark
DK–2800 Lyngby, Denmark

1. Introduction

During the last decade or two, it has become clear that nonlinear systems can exhibit modes of behavior qualitatively different from those we're familiar with from linear systems. Through repeated stretching and folding of phase space, extremely complicated and random-looking variations can arise, even when the underlying equations of motions are completely deterministic and relatively simple. Well-known examples are the Lorenz equations [1], which were first derived as a simplified description of an atmospheric low-pressure system, and Duffing's equation, which was applied by Ueda [2] to describe the behavior of an electrical resonance circuit with saturation characteristics for the inductor. Since then hundreds of different mechanical, electrical, and chemical systems have been found that display similar types of behavior, and the term "random transitional phenomena," which was originally introduced by Ueda, has been replaced by "deterministic chaos." Besides the erratic appearance and the fact that it never repeats itself, chaotic behavior is characterized by sensitivity to the initial conditions [3]. This means that nearby trajectories diverge exponentially, on the average.

Cooperation and Conflict in General Evolutionary Processes, edited by John L. Casti and Anders Karlqvist.

ISBN 0–471–59487–3 ©1994 John Wiley & Sons, Inc.

Consequently, given observational error of any sort, it is impossible to make long-term predictions.

In contrast to the concept of homeostasis, which has dominated physiological thinking for quite some time, many biological control systems have been found to show chaotic dynamics. For example, the regulation of kidney pressures and flows exhibits regular self-sustained oscillations for rats with normal blood pressure, while rats with genetically induced hypertension tend to have a chaotic regulation [4]. Many hormonal systems are unstable and operate in a pulsatory or oscillatory mode. This is true, for instance, with the release of insulin [5], growth hormone, and luteinizing hormone, and interaction between two or more such oscillatory systems is likely to produce frequency-locking, chaos, and a variety of other complicated nonlinear dynamic phenomena [6]. The question arises whether the information associated with the temporal variation in hormone concentration has significance for the regulatory function [7]. While this problem remains a matter of speculation, it's evident that the biological effect of certain hormones may be increased if they are administered in a rhythmic fashion. It is also evident that disruption of certain hormonal rhythms can be associated with states of disease and that new types of oscillations may appear in connection with other diseases.

Rhythmic signals are also essential in intercellular communication [8]. Besides neurons and muscle cells, which communicate by trains of electrical pulses, many other cells exhibit pulsatory variations in their membrane potential, often with complicated patterns of fast and slow spikes. Interaction between such signals may again produce chaotic dynamics. Heart cells, for instance, have been found to show mode locking and chaos when stimulated by an external periodic signal [9]. Similarly, mode locking and chaotic firing may influence the information flow between nerve cells [10]. While in these instances the oscillatory and chaotic dynamics is associated with extracellular signals, recent observations [11] indicate that signal transduction within the cell may also be associated with oscillations and waves of intracellular messengers.

Physiological control systems typically involve negative feedback regulation. In principle, this type of control should be stable. Due to the built-in delays, however, the regulation is actually unstable in many cases, producing self-sustained oscillations and other complicated nonlinear phenomena. Since biological control systems have evolved under selective pressure, one can speculate about the possible advantages of such unstable behavior. While this question can seldom be answered in detail, it is clear that in certain cases the same structure can perform different functions during the various phases of a cycle. This may be associated with a need for a biological clock to synchronize different physiological processes. In other cases, efficiency may be improved through shifts between an active and a dormant phase [12]. It is also possible that oscillatory excursions may serve to guard the system against long-term drift or to protect it against the maintenance of unhealthy conditions for longer periods of time.

Thus, in many ways it appears that oscillations and chaos are the natural modes of operation for physiological control systems [13]. Mathematically, one can show that interaction between three or more self-oscillatory systems is likely to produce chaos. But there are many pulsatory hormonal systems, and there are many other self-oscillatory systems in the human body. The chaotic mode has the advantage of a swift response to changes in the external conditions. At the same time, chaotic behavior may be the most effective way for a regulatory system to avoid mode locking with another self-oscillatory system. Among the hormones known to be released in pulses, both luteinizing hormone and growth hormone are released at nearly two-hour intervals. Interaction between these two rhythms will tend to lock them together, so that one pulse of luteinizing hormone is released for each pulse of growth hormone or, more generally, so that p pulses of luteinizing hormone are released for every q pulses of growth hormone, where p and q are integers. This would reduce the ability of each of the hormonal systems to respond to the needs of the organism, and the most direct way to circumvent this problem appears to be for the hormonal systems to operate in a chaotic mode with no well-defined period.

Having established in this way that biological control systems can function quite well in a chaotic mode, let us direct our attention towards a different type of biological system. Originating with the work by Lotka and Volterra in the 1920s, population dynamics has grown into a major field of theoretical biology, with applications to the description of ecological, epidemiological, and microbiological systems. Immunological problems associated with the response of the immune system to foreign invaders such as viruses or bacteria may also be dealt with in this framework. By virtue of the positive feedback associated with reproduction, and because of maturation and other delays, these systems contain all the ingredients for complex dynamical behaviors [14].

A variety of different models have shown how ecological systems can exhibit unstable behavior in the form of growth and collapse dynamics or self-sustained oscillations. Deterministic chaos has also been reported for forced predator-prey models [15,16], as well as for three variable predator-prey-vegetation models. In a recent paper, Anderson and May [17] suggested that the response of the immune system to the simultaneous infection by HIV and by another virus that activates the same T-cells could produce chaotic bursts of free virus with intervals on the order of 20 weeks. The occurrence of such bursts, which might be intimately related to the progress of the disease, could have a significant influence on the infectiousness of the individual in particular types of risk behavior.

There has also been some success in extracting explainable dynamics in terms of single-variable, discrete-time maps from experimental data for various ecological and epidemiological systems, particularly for childhood diseases where the available time series are relatively long and complete [18–22]. These studies seem to show that whereas for chickenpox the basic pattern is an annual cycle, for measles and rubella chaotic fluctuations are superimposed onto the yearly cycle. This is particularly true for large first-world cities, whereas the epidemic for small isolated communities typically has a more stochastic character. The infection may completely disappear for a couple of years, to be reintroduced through the arrival of infected individuals from outside.

In all of these studies the concern has been with systems with few interacting species. In principle, higher-dimensional systems could show even more complicated dynamics. The general conception, however, appears to be that increasing the dimension of an ecological system by adding more and more species will increase its stability [23,24]. With more species around, it becomes possible for a predator to substitute a more abundant prey for a less abundant one, thus introducing a negative feedback control on the prey populations. At the same time, competition for a diverse set of prey species may introduce balancing interactions between predator populations. Most of the reported examples of oscillatory predator-prey dynamics also appear to come from small, isolated ecosystems or from regions with cold climates and relatively few species. However, ecological systems do not reveal their secrets so easily. Because of the complex interactions, the sensitivity to external disturbances, the long time horizons, and the inhomogeneous spatial distributions in such systems, in almost all cases the available data are insufficient to determine whether observed fluctuations in population sizes are random or the result of explainable, though complex, autonomous dynamics. And at present it remains an open question whether multispecies systems in general are more stable than systems with only a few species [25].

Infection of bacterial populations by phages (viruses) plays an important role for many biotechnological processes. The homogeneous, well-controlled bacterial cultures used in modern cheese production, for instance, are often quite sensitive to phage attack, and considerable effort is expended in the search for more resistant cultures [26–28]. On top of this, the economic incentive for extensive experimentation, the relative simplicity of microbiological systems, the fact that they can be set up under well-specified but variable conditions, their small volume, and the relatively small time scales involved make these systems ideal for population dynamics studies of more complex, multispecies cultures.

In order to examine the phenomena that may arise in such studies, we have simulated a variety of different growth, competition, and selection processes that can take place in a microbiological reactor [29,30]. Our model considers a culture containing several variants of the same bacterium, each sensitive to a specific phage. The culture grows in a chemostat with continuous supply of nutrients. Surplus bacteria and viruses are removed through dilution. Depending on the rate of dilution, we show how the model can exhibit a variety of different nonlinear dynamic behaviors, including self-sustained oscillations, quasiperiodic behavior, and deterministic chaos.

But chaotic behavior is only one of a multitude of complex dynamical phenomena that can occur in nonlinear systems. For certain ranges of parameter values there may be more than one stationary solution, and the observed dynamics will depend on the initial conditions. For instance, a cyclic sequence of dominant bacterial variants $1 \to 2 \to 3$ in a three-variant culture may coexist with the sequence $1 \to 3 \to 2$. These two solutions are symmetric, but they rotate in opposite directions in phase space. Similarly, two symmetric quasiperiodic or chaotic attractors may exist for the same parameter values. Changing a parameter may lead the two chaotic attractors to merge, so that the direction of rotation in phase space is no longer maintained. Moreover, the basins of attraction for the various attractors, i.e., the sets of initial conditions which lead to each of the steady states, may have fractal boundaries. In that case, the slightest change in initial conditions can switch the trajectory from one solution to another. Although we have constructed the model discussed here to be as realistic an account of the microbiological system as possible, with two bacterial variants it exhibits all of these types of behavior. With more variants, it becomes even more complex.

Among the quantities used to characterize chaotic behavior are the Lyapunov exponents [31] and the fractal dimension [32,33]. The Lyapunov exponents, of which there are as many as there are state variables in the system, measure the average rate of divergence of nearby trajectories in the principal directions in phase space. A positive value of the largest Lyapunov exponent signals sensitivity to initial conditions and chaos. If two or more Lyapunov exponents are positive, we talk about hyperchaos [34]. By now, this type of behavior has been observed for a handful of systems [35–39].

In order to bridge the gap between deterministic three-variable chaos and white noise with infinite degrees of freedom, Rössler [40] introduced the concept of a chaotic hierarchy. It was shown that with increasing degrees of freedom, the maximum complexity of a dynamical system may continue to increase. Starting from a closed periodic orbit as a stationary solution, repeated stretching and folding in phase space may produce a chaotic attractor with an underlying fractal geometry. However, with more dimensions, stretching and folding can be applied in several independent directions, leading to higher-order chaos. In general, systems with N deterministic variables $(N > 3)$ may produce hyperchaotic behavior with $N - 2$ positive Lyapunov exponents.

It turns out that our model of interacting populations of bacteria and phages follows this scheme precisely, providing what is thought to be the first realistic example of a chaotic hierarchy [41]. Each time a new bacterial variant is added to the system, the most complex dynamical behavior attains an additional positive Lyapunov exponent. Moreover, this result is robust in the sense that it remains true for different formulations of the interspecies interactions as well as for a wide range of parameter values.

We suggest that it is the particular type of coupling in the system that allows us to create more and more complicated dynamics. This coupling is characterized by the competition of the various bacterial populations for the same substrate. For sufficiently low dilution rates, the interaction between each individual bacterial population and its corresponding virus population leads to oscillatory dynamics, as known from other predator-prey models. The competition for resources couples these oscillatory subsystems into an overall system of increasingly complex behavior.

To study phenomena related to incomplete mixing in the chemostat, we have also considered a diffusive coupling between different compartments of the system. As an illustration, we have investigated the types of behavior that can arise when we couple a periodic attractor in one part of the chemostat with a chaotic or hyperchaotic attractor in another part. The diffusive coupling does not, in general, lead to increased complexity.

In a viral attack, the phage adsorbs to the bacterial surface and injects its DNA (or RNA) into the cell. This can lead to a lytic response in which the virus programs the bacterial cell to replicate the phage DNA or to a lysogenic response in which the viral DNA is inserted into the bacterial DNA with the result that the cell becomes partly resistant to new attacks. It is also possible that the attack is abortive, the penetrating phage being destroyed by the bacterial restriction enzymes, or the cell dying without producing new phages. Under stress, a lysogenic bacterium may again release its viral DNA, providing in this way for the possibility of renewed infection.

The presence of viruses exerts an evolutionary pressure on the bacterial cells. Through a lysogenic response the cell may acquire partial resistance. Often this resistance is paid for through a slightly slower rate of replication, and the resistance may therefore be lost again if there are no more viruses. Nonresistant bacterial variants simply outgrow resistant variants. The viruses may also change. For example, it is sometimes possible for a virus to penetrate the restriction system of a resistant cell and hereafter to multiply in a form towards which this type of cell is no longer resistant. In this way, viruses and bacteria coevolve.

In order to study such evolutionary processes, we have adapted the general modeling framework proposed by Bruckner et al. [42]. In this framework, one considers a countable set of fields (populations) each characterized by the properties of its occupying elements (individual cells or virus particles). Within and between these fields a variety of processes can take place, including spontaneous generation, self-reproduction, error production, deaths of elements, and transitions from one field to another. Mutation is one example of a process by which individuals of one population are transformed into individuals of another. Infection, by which an individual is transformed from exposed to infected, is another example.

Since the first occupation of a new field necessarily starts with a few individuals, a discrete stochastic approach is needed to simulate evolutionary dynamics. As previously noted, a similar approach is appropriate for many epidemiological problems in which an epidemic may start with a single individual being infected from external sources. However, the transition probabilities associated with the various processes need not be linear. Processes involving interaction between individuals from two different populations typically contain bilinear terms, and higher-order terms may become significant in the presence of heavily populated fields. On top of the stochastic phenomena, the presence of such terms allows the system to exhibit nonlinear dynamic phenomena, such as multistability, self-sustained oscillations, quasiperiodic behavior, deterministic chaos, and coexisting solutions. This provides a different picture of a system operating between low-dimensional chaos and random noise.

As a final example of complex behavior in microbiological systems, we consider a model of DNA replication control in bacterial cells and of the subsequent cell divisions. The main hypothesis is that a certain protein, which has a negative feedback regulation on its own production, is an essential factor in initiating the replication process. The model is stochastic in the sense that the kinetic association and dissociation processes are assumed to take place in accordance with a Poisson process with mean values that match experimentally determined parameters. An important feature of the model is that it shows correlation between the magnitude of the kinetic rate constants and the size, stability, and dynamics of the cell. The model thus allows us to analyze the distribution of cell volumes at the time of initiation of the replication process for different growth rates, association and dissociation constants, and promoter strengths.

In all of this work, computer simulation plays an essential role. Living systems have a great richness of dynamic phenomena that needs to be understood. Simulation models represent a convenient way of expressing dynamic interactions in a fashion that is both precise and complete. If essential features are left out, the models simply will not reproduce observed behavior. In many cases, computer models are the only means of thoroughly testing a biological hypothesis. And even for the most experimentally oriented work, models are useful to ensure that the conditions under which the experiments are performed are properly specified.

2. Multispecies Model of Bacterium-Phage Interaction

Bacteriophages (or simply phages) are viruses that infect bacterial cells, take over their replicatory machinery, and use it to produce more viruses. A bacteriophage typically consists of an icosahedral head, formed out of protein and containing the viral DNA. The head is attached to a tail, which is formed of a hollow core surrounded by a contractile sheath and ending in one or more flexible fibers. The total length of this structure may be 0.1 to 0.3μm. As the Brownian motion causes a virus to pass a bacterial cell, the fibers sweep across the surface and affix the phage to specific protein receptors inserted into the cell wall [43]. The number of binding sites depends on the growth conditions that the cell has experienced and may typically be on the order of 100 to 1,000 per cell. The binding probability may also depend on the presence of various ions, such as Mg^{2+}.

A detailed mathematical description of the mechanisms involved in the adsorption process has been given by Schwartz [43]. He also provides experimental results for the adsorption of bacteriophage λ to two strains of E. coli K12, denoted Hfr Gb and Hfr H. Although evidence is presented that phage adsorption proceeds in at least two steps, involving first a reversible attachment to the cell surface and then an irreversible binding, the kinetics governing the formation of bacterium-phage complexes can still be considered to be controlled by the random encounters of the two particles at a frequency that is jointly proportional to their concentrations.

The theoretical adsorption constant, which is attained if the motion of the particles is controlled by passive diffusion and if all collisions between bacteria and phages lead to irreversible binding, is given in [44] as

$$\alpha_0 = 4\pi(r_c + r_v)(D_c + D_v), \tag{1}$$

where r_c and r_v are the effective cell and virus radii, and D_c and D_v are the corresponding diffusion constants. An estimate of α_0 from (1) is slightly smaller than the adsorption constant measured in practice for cell populations with high receptor densities [43], illustrating the efficiency of the tail sweeping process in locating the receptors. The time a viral particle remains within a distance of the cell comparable to the length of its tail is typically 5×10^{-3} sec, depending, of course, on the size of the particles.

Following the adsorption process, the sheath contracts, driving the viral DNA into the cell. Exploiting the resources and the replicatory machinery of the bacterium, the foreign DNA may then start to multiply, and after a latent period on the order of 30 minutes the cell lyses (bursts), with the release of a large number of new viruses. As noted, this is known as a lytic response to the viral attack. The bursting size, i.e., the number of new viruses released in average by each bursting cell, may typically be on the order of 100.

Some cells may be resistant to the viral attack, either because they lack receptors for the specific virus, or because they secrete a protective layer that prevents adsorption of the phages. Many bacteria also have a two-component immune system that protects them against foreign DNA molecules [45]. One component, the restriction enzymes, destroys the invading DNA molecules by cutting them at particular sites, unless they have been specifically modified to resemble the DNA molecules of the cell. The other component, the modification enzymes, tags "self" to those DNA molecules accepted by the cell. The phages can adapt to such resistant cells through so-called host-range mutations, which alter their adsorption specificity or their ability to utilize particular functions in the cells. Occasionally, unmodified DNA may also pass unnoticed by the cell's restriction enzymes. As the virus completes its lytic cycle, the produced phage progeny will be modified, and the new viruses will be able to attack cells of the same restriction-modification type. This and other modification processes will be discussed in more detail in Section 5. In the present section we shall assume that cells and viruses remain unchanged.

Let us consider a culture containing n variants of the same bacterium with concentrations B_i, $i = 1, 2, \ldots, n$. Each bacterial variant is assumed to be sensitive to a particular type of phage and almost completely resistant to the phages that attack the other variants. Thus we have n phage populations of concentration P_j. The viruses adsorb both to susceptible and to resistant cells with kinetic rate constants α_{ij}, the first index denoting the cell type and the second the virus. The viruses infect the cells with probabilities ω_{ij}. Here we have assumed $\omega_{ii} = 0.8$ and $\omega_{ij} = 0.005$ for $i \neq j$. Phages also adsorb to infected cells, but in this case a new infection does not occur. Infected bacteria are represented as a three-variable delay chain with populations I_{i1}, I_{i2}, and I_{i3}. On the average, this structure provides a delay from infection to lysing corresponding to the latent period τ_i. However, the structure allows for a slight variation from cell to cell in this delay. Phages that fail to infect a cell remain stuck to its surface until the cell dies or is washed out of the habitat, at which time the adsorbed viruses also disappear.

The culture grows in a well-stirred tank reactor (a chemostat) with a continuous supply of substrate at a concentration σ. Reactor fluid containing unused resources, wastes, bacteria and phages is removed at the same rate, which we shall refer to as the rate of dilution ρ. In the absence of phages, the bacterial populations are assumed to grow in accordance with standard Monod kinetics [46]:

$$\left. \frac{dB}{dt} \right|_{\text{growth}} = \frac{\nu_i S B_i}{\kappa_i + S}, \tag{2}$$

where S is the concentration of nutrients in the chemostat, ν_i the bacterial growth rate with unlimited resources, and κ_i the nutrient concentration at which the growth rate is equal to half its maximum value.

In the balance equation for the concentration of nutrients, we assume that there is a resource consumption of γ_i at each cell division.

Altogether, this leads to the following set of differential equations:

$$\frac{dB_i}{dt} = \frac{\nu_i S B_i}{\kappa_i + S} - B_i \left(\rho + \sum_{j=1}^{3} \alpha_{ij}\omega_{ij}P_j \right),$$ (3)

$$\frac{dI_{i1}}{dt} = B_i \sum_{j=1}^{3} \alpha_{ij}\omega_{ij}P_j - \rho I_{i1} - 3\tau_i/I_{i1},$$ (4)

$$\frac{dI_{i2}}{dt} = \frac{3}{\tau_i}(I_{i1} - I_{i2}) - \rho I_{i2},$$ (5)

$$\frac{dI_{i3}}{dt} = \frac{3}{\tau_i}(I_{i2} - I_{i3}) - \rho I_{i3},$$ (6)

$$\frac{dP_j}{dt} = \phi_j - P_j \left[\sum_{i=1}^{3} \alpha_{ij}B_i + \rho + \sum_{i,k=1}^{3} \alpha_{ij}I_{ik} \right] + 3\beta_j I_{j3}/\tau_j,$$ (7)

$$\frac{dS}{dt} = \rho(\sigma - S) - \sum_{i=1}^{3} \frac{\nu_i \gamma_i S B_i}{\kappa_i + S},$$ (8)

where ϕ_j is a small external supply of phages and β is the burst size. While all other variables in the model are treated as deterministically determined mean values, ϕ_j is a Poisson-distributed stochastic variable with an average value of $\phi = 0.1/\text{ml·min}$ for all phage variants. Clearly, ϕ is only significant as a source of infection.

Levin et al. [47] have previously suggested a very similar model. One difference is that in their model the latent period is represented as a discrete delay, leading to differential-delay equations. With the same latent period, such a representation, besides being less realistic, is also a little more unstable than the present formulation in terms of a finite number of delay variables. Levin et al. also allow for different types of first-order resources, whereas we have only one. Both models are characterized by the fact that they encompass three trophic levels: the primary resources, first-order consumers or prey (the bacteria), and predators (the viruses). In addition, encounters between prey and predators are taken to be random and governed by the product of the two population sizes. Finally, the various prey populations interact only through their competition for primary resources.

To ensure realism and specificity, Levin et al. [47] have determined the model parameters through experimentation with particular bacteria and viruses. In addition, they have compared the predictions of the model with experimental results for the growth of one- and two-population cultures. Except for the above-noted small differences, we have used the same assumptions about the habitat, the use of primary resources, the population growth, and the nature of the bacterium-phage interaction. However, whereas Levin et al. direct most of their attention to establishing the conditions for equilibria, our interest is with the complex dynamical properties of systems with unstable equilibrium points. The actual parameters used in the present study are slightly different, adjusted to conditions of interest in the dairy industry. But this has no significant impact on the results. We have taken the various bacterial and phage variants to have identical properties. Except for the infection probability, which is as specified above, the parameters are:

$$\alpha = 10^{-9} \text{ ml/min}, \quad \kappa = 10\mu g/ml, \quad \nu = 0.024/min,$$
$$\gamma = 0.01ng, \quad \tau = 30 \text{ min}, \quad \beta = 100,$$
$$\sigma = 10\mu g/ml.$$

As previously noted, ρ is the rate at which a suspension of nutrients is supplied to the reactor and is measured in turnovers per minute. In the following simulations, ρ is taken as a bifurcation parameter. The rate of dilution is a major determinant of dissipation in the system. High values of ρ are expected to lead to stable behavior, with small viral populations and equilibration between bacteria and nutrients. For lower values of ρ, we encounter many different forms of complex dynamical phenomena, including self-sustained oscillations, coexisting periodic solutions, quasiperiodic behavior, chaos, and hyperchaos. As ρ is reduced, the system eventually tends to break down, with one or more species dying out.

Figure 1 shows the results of a simulation in which the rate of dilution has been gradually reduced from $\rho_0 = 0.011/min$ to $0.007/min$ following an expression of the form

$$\rho(t) = \rho_0 \exp(-t/t_0), \tag{9}$$

with a time constant $t_0 = 40,000$ min ≈ 1 month. The idea is that by slowly reducing ρ, the bacterial and viral populations are allowed to grow larger and larger, while at the same time the pressure that the phages exert upon the cell populations increases. In models that allow for mutations and other forms of modifications of cells and viruses, a simulation like this can be used to study a particular evolutionary

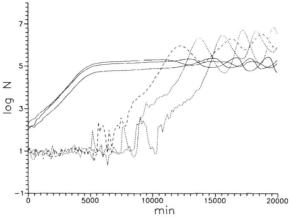

Figure 1. Results of a simulation in which the rate of dilution is gradually reduced from 0.011/min to 0.007/min over a period of 20,000 min. The
vertical axis shows the concentrations of bacteria and viruses on a logarithmic scale. As the viral populations reach macroscopically significant levels,
the system becomes unstable and starts to produce self-sustained oscillations. Solid curves represent the three bacterial variants; dotted curves, the
corresponding viruses.

scenario. In the present case, the main effect of reducing ρ is to decrease
the damping in the system by lengthening the mean residence time in
the chemostat.

Starting with relative high rates of dilution, the model shows a
stable coexistence of the three bacterial variants at levels low enough
for the depletion of primary resources to be nearly negligible. The low
concentrations of bacteria significantly reduce the rate of viral replication, and with the rapid washout of any produced phage the viral
populations are restricted in practice to the level determined by the
random contamination. Without viruses, the three bacterial variants
can be considered identical, and the equilibrium level of the total bacterial population is determined by the equations

$$\frac{dB}{dt} = 0 = \frac{\nu S B}{\kappa + S} - B\rho, \tag{10}$$

$$\frac{dS}{dt} = 0 = \rho(\sigma - S) - \frac{\nu\gamma S B}{\kappa + S}, \tag{11}$$

or

$$B = \frac{1}{\gamma} \left(\sigma - \frac{\kappa \rho}{\nu - \rho} \right), \tag{12}$$

for

$$\rho \leq \frac{\nu \sigma}{\kappa + \sigma}. \tag{13}$$

Here $B = \sum_i B_i$.

The rate of growth of the culture is just sufficient to balance the rate of loss through overflow. If the rate of dilution is reduced a little, the bacterial population increases until a new steady state is established. Of course, there are limits. If the rate of dilution is so high that it requires a growth rate in excess of the rate that the culture can reach, no stable equilibrium exists and cells that may be present in the chemostat will eventually be washed out. Had the bacterial variants been different so that, for instance, one cell type had a lower resource consumption or a higher growth rate than the others, only this variant would survive. This follows from the principle of competitive exclusion [47], according to which the number of prey variants in equilibrium cannot exceed the number of distinct primary resources plus the number of predator variants. The necessary condition for coexistence is that there be different "niches" for the interacting species; i.e., the coexisting populations of predators and preys must exploit their resources and respond to their environment in different ways.

As the rate of dilution is further decreased, the bacterial populations become dense enough and the rate of washout low enough for the viral populations to start growing. Depending on the initial conditions and on the random fluctuations introduced by the infection source, any one of the phage variants may first attain macroscopic significance.

For a monoculture, one typically observes that there is a range of dilution rates in which bacteria and phages can coexist in stable equilibrium. However, as the rate of dilution is reduced beyond a critical value, a Hopf bifurcation takes place and self-sustained oscillations develop. For our three-culture system, such oscillations also arise as soon as the viral populations become large enough to affect the cell populations significantly. Because of the coupling between the bacterial variants via their competition for primary resources, these oscillations may take the form of a cyclic shift in the dominant bacterial variant. If, for instance, bacterial population 1 at a given time happens to be the largest, the corresponding viral population will replicate fastest until it becomes large enough to reduce the bacterial population. With the reduced presence of the first variant, bacteria of type 2 may now grow to become dominant until this population is reduced by its viruses, and so on.

But there is no particular reason why population 2 and not population 3 should follow population 1 as the dominant bacterial variant. And with the symmetry built into the system, it's clear that self-sustained 1–2–3 and 1–3–2 oscillations arise simultaneously in a degenerate Hopf bifurcation [48]. These two periodic attractors are mutually symmetric, but they have opposite directions of rotation in phase space. A very similar situation has been encountered by Sturis and Mosekilde [49] in a model of a simple migratory system, where families from two different ethnic groups can move between three districts of a city. Figure 2 illustrates the two simultaneous periodic solutions that exist for $\rho = 0.0065/\text{min}$. Preliminary results indicate that there is a fractal boundary between the basins of attraction for the two solutions.

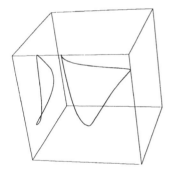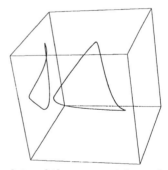

Figure 2. Three-dimensional phase plots of the symmetric periodic solutions which exist for a dilution rate of $\rho = 0.0065/\text{min}$. As axes we have used the three bacterial populations on a logarithmic scale. The existence of two symmetric attractors is characteristic of systems where rotation in one or the other direction is equally possible. An example could be the α-waves of the brain, which can rotate either one way around the head or the other.

As the rate of dilution is further reduced, the viral populations grow at the expense of the bacterial populations. At the same time, the oscillations increase in amplitude and at $\rho \approx 0.0062/\text{min}$ they start to become irregular. Closer examination reveals that we now have two symmetric quasiperiodic attractors. These are shown in Figure 3. Interestingly enough, the simple periodic solutions continue to exist, which implies that the quasiperiodic solutions must have developed

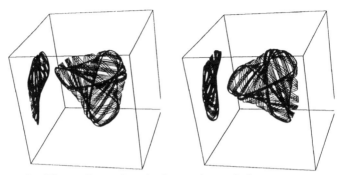

Figure 3. Three-dimensional phase plots of the two symmetric quasi-periodic solutions that exist for $\rho = 0.0055/\text{min}$. Interestingly enough, the periodic solutions shown in Figure 2 still exist as stable attractors.

from another set of presumably unstable solutions. We now have four simultaneous stationary solutions, and the initial conditions determine which of these solutions the system will choose. Reducing ρ further, the oscillations become even more irregular, and for $\rho = 0.0045/\text{min}$ we have two symmetric chaotic solutions. These are shown in Figure 4.

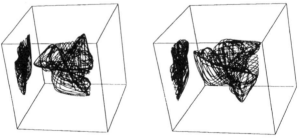

Figure 4. Three-dimensional phase plots of the two symmetric chaotic solutions which exist for $\rho = 0.0045/\text{min}$. The periodic solutions of Figure 2 still exist and can be reached from particular sets of initial conditions. Section 3 describes how the chaotic solutions develop out of the quasiperiodic solutions in Figure 3.

As the rate of dilution is reduced to about $0.0029/\text{min}$, the bacterial populations finally collapse, and hereafter the viruses can no longer replicate. This may be seen in Figure 5, which extends the results of Figure 1 to even lower rates of dilution. Before the collapse occurs, however, the oscillations have become very large and irregular, and the two chaotic attractors of Figure 4 have merged to form a single chaotic attractor that no longer maintains the direction of rotation in phase

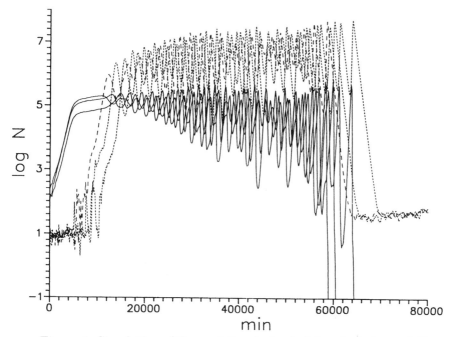

Figure 5. Simulation of the variation in bacterial and viral populations as the rate of dilution is gradually reduced from 0.011/min to 0.0025/min. At the end, the phages kill off the cells, whereafter they can no longer replicate. This may be similar to the breakdown of the ecological system of a lake as the freshwater flow becomes too small.

space. This whole scenario will be discussed in more detail in Section 3. Figure 6 shows a two-dimensional phase plot of the merged chaotic attractor. For $\rho < 0.0037$/min, the system exhibits a hyperchaotic motion with two positive Lyapunov exponents.

Our three-population system clearly allows room for very lively dynamics. At the same time, it's worth noticing, however, that the simultaneous presence of several bacterial variants helps stabilize the system. In particular, the presence of bacteria resistant to a given virus serves to protect the population of sensitive bacteria, since the viruses are assumed also to adsorb to the surfaces of resistant cells.

3. The Chaotic Hierarchy

With the discovery of higher-order chaos in the three-population model, it becomes of interest to investigate how the behavioral complexity depends upon the number of bacterial variants. For this purpose, the external infection source ϕ is set to be constant $\phi = 0.1/(\text{min}\cdot\text{ml})$, so

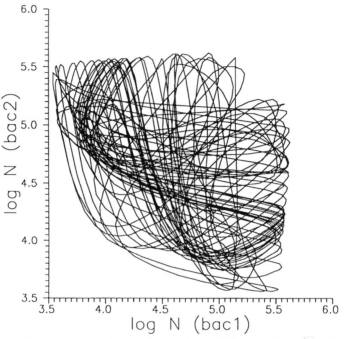

Figure 6. Two-dimensional phase plots of the chaotic attractor formed through merging of the simpler chaotic attractors of Figure 4.

that the model is fully deterministic. For simplicity we also neglect cross-infection of bacteria B_i by phage P_j (i.e., off-diagonal elements in the matrix w_{ij} are set to zero). None of these changes alter the dynamics appreciably.

In the simplest case only one population of bacteria and phages is considered: Cells feed on the substrate and multiply. At the same time, their number is reduced through washout and through successful phage attacks. Phages multiply as infected bacteria lyse and decrease in number through washout. And virus particles can also become inactivated by adsorption to the surface of already infected cells. For large dilution rates ($\rho > 0.0058/\text{min}$), both bacteria and phages arrive at a stable equilibrium state. With a plentiful supply of resources, the bacterial population successfully maintains a relatively high level of individuals. Fast dilution, on the other hand, keeps the number of viruses at a comparatively low level. The stability of the steady state implies that the system will return to the same state after a perturbation of any of the populations.

However, as ρ is decreased, the sensitivity to fluctuations grows, and at $\rho = 0.0058/\text{min}$ the steady state loses its stability. We find a Hopf bifurcation of a focus, and for $\rho < 0.0058$/min the model exhibits stable limit-cycle oscillations that increase in amplitude as ρ is decreased further. The limit cycle is characterized by bursts in the phage population, which drastically diminish the cell population. This in turn reduces the rate of viral replication, and as phages are washed out and new substrate enters the flow reactor, the bacterial population recovers. The cycle then starts anew. For sufficiently small values of ρ, the bacterial population dies out after some transient behavior. Our numerical analysis didn't show hysteresis under variation of ρ, and we conclude that the Hopf bifurcation is supercritical. The bifurcation diagram is illustrated in Figure 7.

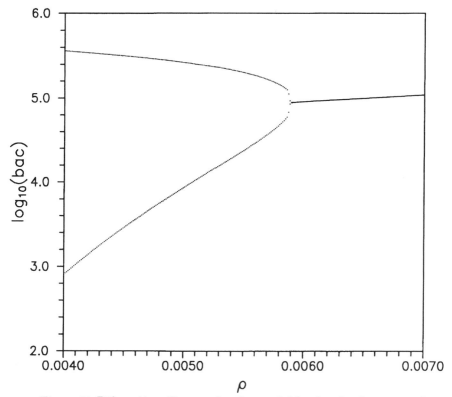

Figure 7. Bifurcation diagram for the model in the simplest case where only one population of bacteria and phages is considered. As the rate of dilution is reduced we observe a supercritical Hopf bifurcation, and below $\rho = 0.0058$/min the model exhibits self-sustained oscillations.

Let us now consider a system with two bacterial populations, each subject to attacks by specific phages. As noted above, we assume that only attacks of phage P_i on cells B_i are successful. Adsorption of phage P_i on bacteria B_j with $i \neq j$ diminishes the number of free viruses but does not produce any infected cells. Except for the initial conditions, the two predator-prey systems are identical, and for dilution rates below a critical value each subsystem is capable of generating self-sustained oscillations. These oscillations are coupled via the bacterial competition for primary resources. As the rate of dilution is reduced, the amplitude of each of the limit cycles increases and the coupling becomes stronger and stronger.

The results of this coupling between the two microbiological oscillators are revealed in the bifurcation diagram of Figure 8, which was constructed from Poincaré sections of the trajectory with the plane $B_1 = B_2$. With this technique, a self-sustained oscillation for a given value of the bifurcation parameter is represented by a single point. Starting at $\rho = 0.0050/\text{min}$, we see a limit-cycle oscillation with phase lag between the two bacterial populations. Coexisting with this limit cycle is another self-sustained oscillation (not shown in the figure) with opposite phase and amplitude relations. The initial conditions determine which of the two limit cycles the system chooses.

As ρ is decreased, the limit cycle in Figure 8 undergoes a torus bifurcation, and thereafter the model exhibits quasiperiodic solutions interrupted by phase locking—as one would expect for a system of two coupled oscillators [6]. The locked state of period 7 shows period doubling to period 14. The solution then unlocks again, and for still lower dilution rates the torus breaks down and the solution becomes chaotic. This transition occurs at approximately $\rho = 0.00448/\text{min}$, and the solution remains chaotic thereafter until the system reaches its final breakdown and bacteria and viruses die out.

As judged from the distribution of points in the bifurcation diagram, the chaotic solution changes character at $\rho \approx 0.00402/\text{min}$. To examine this phenomenon further, Figure 9a and b show two new projections of the bifurcation diagram in the interval around the critical dilution rate. Here we have plotted values for the number of infected bacteria of type 1 (I_{11}) in the Poincaré section $B_1 = B_2$ for two different sets of initial conditions. It is apparent that there are two different, coexisting chaotic solutions for $\rho > 0.00402/\text{min}$. They merge to a single chaotic solution at this value of ρ and, judging from the numerical treatment, the system possesses only a single chaotic attractor for $\rho < 0.00402/\text{min}$.

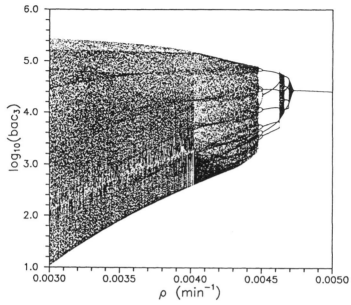

Figure 8. Bifurcation diagram for a system with two bacterial populations. With initial conditions $(S)_0 = 0.0$, $(B_1)_0 = 25,000/\text{ml}$, and $(B_2)_0 = 25,118/\text{ml}$, the simulation has been started at $\rho = 0.0050/\text{min}$ and the stationary solution followed adiabatically with a step size of $\Delta\rho = 2.0 \times 10^{-6}/\text{min}$. For each value of ρ, a transient of 10^5 min was omitted, and subsequent steady-state intersection points with the Poincaré plane $B_1 = B_2$ were plotted. Vanishing initial populations of phages and infected bacteria were assumed.

Figure 10a and b show cross sections of the two chaotic attractors coexisting for $\rho = 0.00404/\text{min}$. They are both fractionated two-tori, and can be shown to originate in the two coexisting quasiperiodic solutions found for higher values of ρ. Note how the attractors approach one another in the region of sharp foldings in the lower-left corner. Figure 10c shows the compound attractor existing for $\rho = 0.00401/\text{min}$. It is interesting to observe that this attractor retains almost all features of the two coexisting attractors right up until the merging.

In this way, the change in structure of the bifurcation diagram near $\rho = 0.00402/\text{min}$ can be attributed to a crisis associated with the collision of the coexisting chaotic attractors with their separating basin boundaries. Such a boundary crisis appears to be common in systems with two coupled self-sustained oscillators [50]. Calculating the spectrum of Lyapunov exponents for the merged attractor with $\rho = 0.0015/\text{min}$ gives $\lambda_1 = 3.02 \times 10^{-4}/\text{min}$, $\lambda_2 = 3.10 \times 10^{-6}/\text{min}$, and

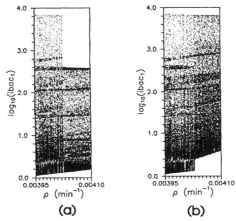

Figure 9. Detail of bifurcation diagram for the two-population model near $\rho = 0.00402/\text{min}$. At the intersection with the Poincaré plane $B_1 = B_2$, the number of infected bacteria of type 1 (I_{11}) has been plotted for two different sets of initial conditions: (a) $(S)_0 = 10.0\ \text{ng/ml}, (B_1)_0 = 20,000/\text{ml}$, and $(B_2)_0 = 10,000/\text{ml}$; (b) $(S)_0 = 0.0, (B_1)_0 = 25,000/\text{ml}$, and $(B_2)_0 = 25,118/\text{ml}$. All other state variables are initiated at zero.

$\lambda_3 = -2.06 \times 10^{-4}/\text{min}$, confirming that we have ordinary chaos with a spectrum of leading Lyapunov exponents $(+, 0, -)$. Since $\lambda_1 + \lambda_3 > 0$, the chaotic attractor is of the so-called Kaplan-Yorke type [51] with a Lyapunov dimension greater than 3.

Let us now extend the system with yet another bacterial variant, so that we are back to a model with three bacterial populations competing for the same primary resources with each attacked by a specific virus. The system is hereafter described by 16 coupled nonlinear differential equations, and as already noted, the dynamics is quite complicated. Figure 11 shows a bifurcation diagram for the three-population model. Before we discuss the diagram in more detail, it is worth noticing that it's composed of several branches with different initial conditions plotted on top of one another. In particular, we note the two periodic solutions that exist all the way from $\rho = 0.0065/\text{min}$ down to $\rho \approx 0.0031/\text{min}$. In various ranges of ρ, these periodic solutions coexist with other periodic solutions or with quasiperiodic and chaotic solutions. Not all initial conditions have been tested, however, and it is possible that the system exhibits stationary solutions other than those plotted in Figure 11.

For $\rho = 0.0065/\text{min}$ the bifurcation diagram shows the two coexisting periodic solutions with opposite directions of rotation in phase

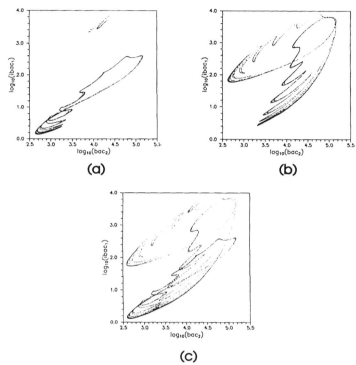

Figure 10. Poincaré sections of the chaotic attractors on both sides of the boundary crisis. (a) $\rho = 0.00404$/min, $(S)_0 = 10.0$ ng/ml, $(B_1)_0 = 20,000$/ml and $(B_2)_0 = 10,000$/ml; (b) $\rho = 0.00404$/min, $(S)_0 = 0.0$, $(B_1)_0 = 25,000$/ml and $(B_2)_0 = 25,118$/ml; (c) $\rho = 0.00401$/min, initial conditions as in (b). The chaotic attractor in (c) has developed through merging of the coexisting chaotic attractors in (a) and (b).

space, which we have seen already in Figure 2. From $\rho \approx 0.00625$/min, we also have two quasiperiodic attractors on two-tori. It is not clear where these quasiperiodic solutions come from. A possibility is that they are generated by stabilization of two unstable quasiperiodic solutions that exist already for $\rho > 0.00625$/min. For certain ranges of ρ, the quasiperiodic solutions are replaced by locked states, often with fairly high periodicity. The largest Lyapunov exponents are found to be $\lambda_1 = 4.24 \times 10^{-9}$/min, $\lambda_2 = 3.12 \times 10^{-9}$/min, and $\lambda_3 = -1.24 \times 10^{-4}$/min, indicating quasiperiodicity with a vanishing value for the two leading Lyapunov exponents.

As illustrated in Figure 12, the Poincaré section of one of the coexisting attractors obtained for $\rho = 0.005505$/min appears to consist of small line segments, many of which show a typical folded shape. For

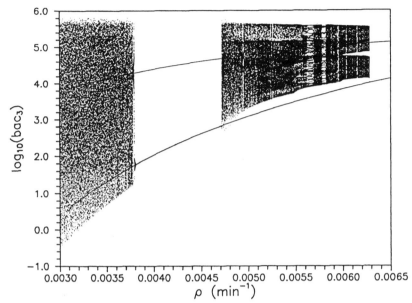

Figure 11. Bifurcation diagram for the three-population model. Several branches obtained with different initial conditions have been plotted on top of each other. For $\rho < 0.0030$/min, only hyperchaos with two positive Lyapunov exponents is found.

the corresponding aperiodic time series, the largest Lyapunov exponents are $\lambda_1 = 4.3 \times 10^{-6}$/min, $\lambda_2 = 7.5 \times 10^{-8}$/min, and $\lambda_3 = -1.9 \times 10^{-5}$/min, corresponding to the spectrum $(+, 0, -)$. Since $\lambda_1 + \lambda_3 < 0$, we have simple chaos with a fractal dimension between 2 and 3. The Lyapunov dimension is calculated to be $D_L = 2.23$.

The window in parameter space for simple chaos is quite small, however. At $\rho = 0.0055$/min, we find an attractor with a different structure (see Figure 13). The former line segments are now connected, and the attractor resembles a closed but multiply twisted ribbon. A calculation of the largest Lyapunov exponent yields $\lambda_1 = 1.86 \times 10^{-5}$/min and $\lambda_3 = -1.49 \times 10^{-5}$/min. The sum $\lambda_1 + \lambda_3$ is now greater than zero, and the attractor has experienced an increment in the integer part of its dimension. The Lyapunov dimension formula yields $D_L = 3.01$.

As ρ is decreased towards 0.0047/min, the width of the ribbon becomes larger and the attractor becomes more and more diffuse. At $\rho = 0.0047$/min, there is a crisis in which the chaotic attractor disappears as a stable solution. Depending on the initial conditions, the system now settles down into one of the coexisting periodic solutions.

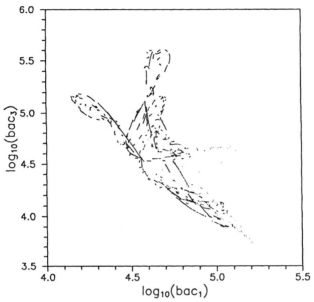

Figure 12. Poincaré section of one of the simple chaotic attractors that exist in the three-population model for $\rho = 0.005505/\min$. The cross section is taken at $B_1 = B_2$. The Lyapunov dimension is $D_L = 2.23$.

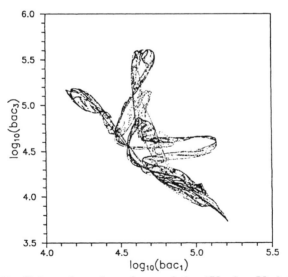

Figure 13. Poincaré section of one of the "Kaplan-Yorke" chaotic attractors existing in the three-population model for $\rho = 0.0055/\min$. The Lyapunov dimension is $D_L = 3.01$.

Self-sustained oscillations now appear to be the only solutions until for $\rho \approx 0.0037$/min, aperiodic behavior again arises, still coexisting although with the simple period-1 limit cycles. The cross section for the chaotic attractor when $\rho = 0.0035$/min is plotted in Figure 14. Even though it still contains structure and evidence of foldings, the cross section is now very diffuse. The four largest Lyapunov exponents are $\lambda_1 = 3.03 \times 10^{-4}$/min, $\lambda_2 = 1.85 \times 10^{-4}$/min, $\lambda_3 = -5.51 \times 10^{-8}$/min and $\lambda_4 = -3.41 \times 10^{-4}$/min. With two positive Lyapunov exponents we have hyperchaos (C^2), for which there are two independent directions of exponential divergence. With $\lambda_1 + \lambda_2 + \lambda_4 > 0$, the Lyapunov dimension is larger than 4. A calculation yields $D_L = 4.16$.

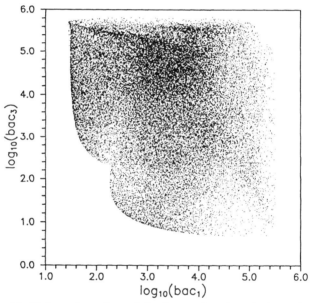

Figure 14. Poincaré section of the hyperchaotic solution which exists in the three-population model for $\rho = 0.0035$/min. The Lyapunov dimension is $D_L = 4.16$.

For $\rho < 0.0035$/min, hyperchaos appears to be the only stable solution in the system. Thus we have seen how the most complicated solution has developed from a simple limit cycle with no positive Lyapunov exponents in the one-population model, via ordinary chaos with one positive Lyapunov exponent in the two-population model, and on to hyperchaos with two positive Lyapunov exponents in the three-population model. It is interesting to note that the hyperchaotic solution is found over a significant interval of dilution rates and for most initial conditions. We do not have to specify particular parameter values carefully to obtain this type of dynamics.

As an additional numerical experiment, we have coupled four bacterial populations with their respective phages to the same substrate. Again the system starts out with simple periodic oscillations for high dilution rates, and the complexity of the dynamics gradually increases as ρ is reduced. As with three coupled populations, there are multiple coexisting solutions and the phase-space structure is very complicated. Several types of chaos can be distinguished. In view of the above results, one expects the most complex type of behavior to dominate the system for low values of ρ. Considering as an example $\rho = 0.0030/\text{min}$, we have found chaos with a Poincaré section which, in some projections, resembles a distorted cube. This is illustrated in Figure 15.

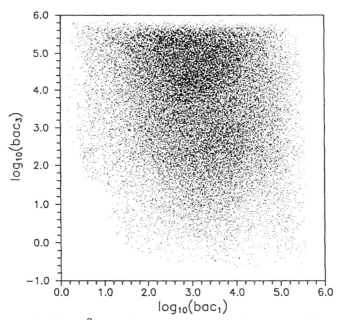

Figure 15. Hyper^2-chaotic attractor in the four-population model for $\rho = 0.0030/\text{min}$. The Lyapunov dimension $D_L = 5.99$. The basic difference when compared to simple hyper-chaos is that hyper^2-chaos requires an even higher-dimensional embedding space and looks even more like random noise.

The six largest Lyapunov exponents are $\lambda_1 = 4.2 \times 10^{-4}/\text{min}$, $\lambda_2 = 3.1 \times 10^{-4}/\text{min}$, $\lambda_3 = 1.5 \times 10^{-4}/\text{min}$, $\lambda_4 = 4.9 \times 10^{-7}/\text{min}$, $\lambda_5 = -2.2 \times 10^{-4}/\text{min}$, and $\lambda_6 = -6.7 \times 10^{-4}/\text{min}$. Three of these are positive and we thus have hyper^2-chaos (C^3) with the convention used by Klein and Baier [52]. Applying the Kaplan-Yorke estimate for the Lyapunov dimension gives $D_L = 5.99$. The hyper^2-chaotic solution is stable with respect to changes in parameters, and the basin of attraction for this mode seems to dominate phase space.

From these results it appears that the attractor dimension of an ecological system can continue to increase as more and more species are included. If we focus on low dilution rates where the most complex behavior is observed, systems with one, two, three, or four bacterial variants exhibit dynamics with zero, one, two, and three positive Lyapunov exponents, respectively. In fact, each time a new bacterial population has been added to the system, the maximum number of positive Lyapunov exponents has increased by one. Mathematically, one might expect that coupling an increasing number of nonlinear oscillators would lead to more complex dynamics. However, it seems to be common experience that this usually is not the case [53], at least for diffusive coupling.

One can argue that interactions between the species beyond those we have included in the model occur in the real world, that the significance of such interactions tends to increase with the number of species, and that they generally contribute additional stabilizing mechanisms to the ecology. This is possible, of course. On the other hand, we would like to stress that the increase in complexity that we observe in the model with the number of bacterial variants is robust and generic in the sense that it occurs for a variety of different formulations of the interspecies interactions, as well as for a wide range of parameter values. It should also be noticed that the same type of coupling was used by Levin et al. [47], and found by comparison with experiments to provide a realistic description of both the habitat and the biology of the interacting species.

To illustrate the robustness of the observed chaotic hierarchy, we have simplified the model by omitting the two delay equations for I_{i2} and I_{i3}, coupling I_{i1} directly to the phage production rate. We have also taken $\omega_{ii} = 1$ and omitted all cross couplings between populations via adsorption of phage j to cell i with $i \neq j$. We end up with a system of microbiological oscillators that are isolated, other than that the bacteria feed on the same primary resource. Performing simulations after each of these changes, we continue to find that for low-dilution-rate systems with one, two, three, or four bacterial populations produce dynamics with zero, one, two, and three positive Lyapunov exponents. This confirms the fact that coupling identical predator-prey systems to a common primary resource is the source of higher chaos in our system.

4. Spatial Inhomogeneity and Diffusive Coupling

Even though unstable dynamic phenomena may occur in ecological systems under homogeneous conditions, it is possible that the patchiness of the real world contributes significantly to the stabilization of many predator-prey interactions. By providing spatial and temporal refuges for prey, one would expect heterogeneity to widen the range of stable coexistence between species. On the other hand, one might also imagine that the introduction of spatial degrees of freedom would lead to new types of instability—including various forms of self-organizing processes, pattern formation, and waves [54–56]. Unfortunately, very little of a concrete nature is known at present about the behavior of spatially extended systems showing chaos or higher chaos in the homogeneous state.

One way to approach the problem is to divide the system into separate compartments, and consider the influence of different types of coupling between the subsystems. In a microbiological flow reactor, for instance, conditions near the walls could differ from those in the center of the reactor, leading to separate wall and liquid cultures. With insufficient stirring, the cells could also concentrate near the bottom or top of the reactor, leading to vertical gradients in cell, virus, and substrate densities. In nature, resources could be provided in a spatially inhomogeneous manner, and there could also be gradients in temperature, salinity, and other factors. Again, very little is known about the results of such compartmentation for chaotic and hyperchaotic systems.

If the individual subsystems evolve along different attractors, coupling might result in the formation of a more complex attractor, or it might lead to a much simpler dynamics through equilibration or mode locking. Pikovski [57] suggests that stabilization of diffusively coupled chaotic attractors depends upon the ratio of the Lyapunov exponent to the coupling constant. We do not know how general this result is and, in particular, we do not know what happens if the various species have widely different diffusion constants.

In an attempt to deal with these issues in a preliminary fashion, we have considered the two-chamber microbiological reactor illustrated in Figure 16. Each chamber is well stirred so that the conditions are homogeneous. The chambers are separated by a membrane through which substrate can diffuse. On the other hand, cells and viruses are retained by the membrane and forced to remain in their respective chambers. Substrate is provided to each chamber separately (at rates ρ_a and ρ_b, respectively) so that the dynamics in the two compartments

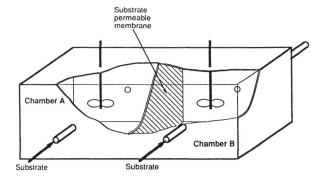

Figure 16. A two-chamber chemostat with coupling via a semipermeable membrane that allows substrate to diffuse between the two compartments. Each chamber is assumed to be well stirred, but the two dilution rates may be varied separately.

may differ. In this way, we can couple a periodic attractor in one compartment with a chaotic or hyperchaotic attractor in the other. The coupling strength can be varied by changing the diffusion parameter controlling the exchange of substrate over the membrane.

The differential equations governing the concentrations of bacteria, phages, and infected bacteria are the same as before [(Eqs. (3)–(7)], except that we now have a set of equations for each chamber. The concentrations of substrate in the two chambers are determined by the relations

$$\frac{dS_a}{dt} = \rho_a(\sigma_a - S_a) - D(S_a - S_b) - \sum_{i=1}^{3} \frac{\nu_i \gamma_i S_a B_{ia}}{\kappa_i + S_a} \qquad (14)$$

and

$$\frac{dS_b}{dt} = \rho_b(\sigma_b - S_b) + D(S_a - S_b) - \sum_{i=1}^{3} \frac{\nu_i \gamma_i S_b B_{ib}}{\kappa_i + S_b}. \qquad (15)$$

In what follows, we return to the original three-population model of Section 2. All parameters are taken to be the same as before, except that we can now vary the dilution rates ρ_a and ρ_b separately. In addition, the coupling constant D can be varied. Let us here restrict ourselves to the case where $\rho_b = 0.0060/\text{min}$, so that chamber B exhibits a periodic solution for the uncoupled system. Let us then consider how this dynamics and the dynamics of chamber A are altered as ρ_a and D change.

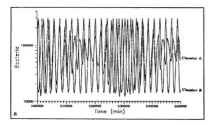
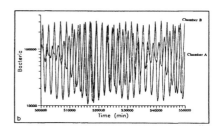

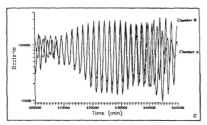
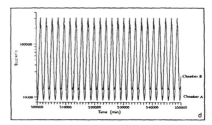

Figure 17. Temporal variations of the bacterial concentrations (variant 1) in the two chambers for $\rho_a = 0.0056$/min and $\rho_b = 0.0060$/min. For the uncoupled system these dilution rates give a quasiperiodic behavior in chamber A and a periodic oscillation in chamber B. The coupling constant is gradually increased from $D = 0.0002$/min (a), over $D = 0.0020$/min (b), and $D = 0.0040$/min (c) to $D = 0.0200$/min (d).

For $\rho_a = 0.0056$/min and $D = 0$, chamber A exhibits quasiperiodic behavior. Figure 17a–d shows the temporal variations of the bacterial populations observed in the two chambers for $D = 0.0002$/min, 0.0020/min, 0.0040/min, and 0.0200/min, respectively. Closer examination of the corresponding phase plots and the Poincaré sections reveals that for small values of D (0.0002/min), the quasiperiodic solution in chamber A modulates slightly the periodic solution in chamber B, so that both chambers display quasiperiodic behavior. When $D = 0.0004$/min and $D = 0.0008$/min, similar results are obtained. For $D = 0.0020$/min, the modulation of the bacterial population in chamber B is clearly visible in the time series.

When the coupling constant is increased to $D = 0.0040$/min, the oscillations in both chambers become more irregular. This may not be so evident from the temporal variations depicted in Figure 17c. However, the corresponding phase plots in Figure 18a and b clearly reveal a chaotic motion, and the Poincaré sections of the two attractors yield even clearer evidence for this type of behavior. A similar solution is found for $D = 0.0120$/min. Finally, for $D = 0.0200$/min, the two

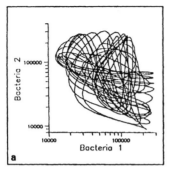 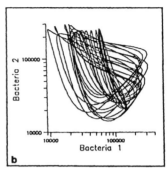

Figure 18. Phase plots for the bacterial concentrations in chambers A (a) and B (b), respectively. With dilution rates of $\rho_a = 0.0056$/min and $\rho_b = 0.0060$/min, we are coupling a quasiperiodic and a periodic solution. For a diffusion constant of $D = 0.0040$/min, this coupling leads to an overall chaotic motion.

chambers lock together to form an almost-periodic solution. The rate of diffusion is now large enough to equilibrate the resource conditions effectively in the two chambers. Details in the Poincaré sections reveal that there is still a tiny aperiodic component in each of the two attractors.

Hereafter, let us reduce the dilution rate for chamber A to $\rho_a = 0.0050$, so that this chamber in isolation exhibits a chaotic attractor. Again we shall see how this attractor is influenced by coupling to the periodic attractor in chamber B. Figure 19a and b shows the temporal variation of bacterial population 1 in the two chambers for coupling constants $D = 0.0002$/min and $D = 0.0020$/min, respectively. With low coupling, the attractor in chamber A remains chaotic, but it distorts the attractor in chamber B enough to give a slight chaotic component on top of the dominant periodic behavior. The same result is obtained for $D = 0.0010$/min. For $D = 0.0020$/min, the time plots clearly reveal that the oscillations in both chambers are irregular. The chaotic character of this variation is quite evident when examining the corresponding phase plots and Poincaré sections. It is actually possible that the behavior is hyperchaotic, but we have not yet performed a Lyapunov exponent calculation to check this conjecture.

Finally, let us reduce the dilution rate for chamber A to $\rho_a = 0.0046$/min, so that the uncoupled solution for this chamber is hyperchaotic. If we again study the effect of increasing the diffusive coupling to a periodic oscillation in chamber B, the scenario develops as follows: Beginning at a very low coupling constant of $D = 0.0002$/min,

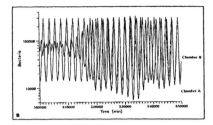 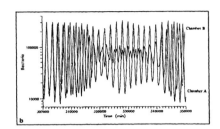

Figure 19. Temporal variation of bacterial population 1 in chambers A and B for $D = 0.0002/\text{min}$ (a) and $D = 0.0020/\text{min}$ (b). With $\rho_a = 0.0050/\text{min}$ we now couple a chaotic dynamics in chamber A with periodic oscillations in chamber B.

the hyperchaotic solution for chamber A is found to persist while the periodic oscillation in chamber B is modulated by a weak hyperchaotic component. For $D = 0.0004/\text{min}$, $0.0010/\text{min}$, and $0.0020/\text{min}$, similar results are obtained. As D is further increased, the hyperchaotic component of the motion in chamber B becomes more evident. This is illustrated, for instance, by the time and phase plots of Figure 20, which were obtained for $D = 0.0180/\text{min}$.

We conclude this preliminary investigation of diffusively coupled periodic, quasiperiodic, chaotic, and hyperchaotic attractors by noticing that in most cases the overall behavior assumes the character of the more complex of the uncoupled attractors. In a few cases, the complexity is reduced through mode locking, while in other cases our investigation indicates that the coupling leads to more complex behavior. All in all, we have not found much evidence to support the conjecture that spatial inhomogeneities and diffusion have a stabilizing influence. Stabilization is observed, however, when the supply of resources to one of the chambers is very low. It should also be noted that we have only allowed diffusion of resources. In nature, diffusion or active motion in search of food is likely to take place at several trophic levels.

5. Lysogenic Response, Mutations, and Bacteria-Virus Coevolution

We have already mentioned some of the mechanisms by which bacterial cells can defend themselves against viral attack, and we have discussed how the viruses can adapt to resistant bacteria through host-range mutations or host-controlled modifications. In the first of these processes, the adsorption specificity or the ability of the virus to utilize the repli-

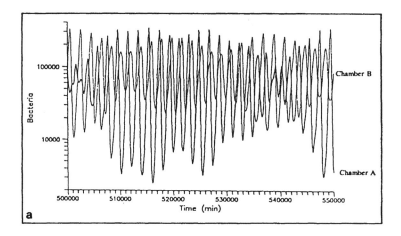

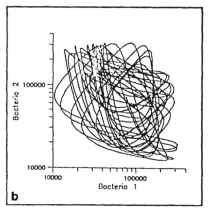

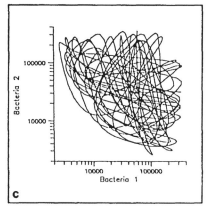

Figure 20. (a) Temporal variation of bacterial population 1 in chambers A and B for $\rho_a = 0.0046$/min and $D = 0.0180$/min. Parts (b) and (c) show the corresponding phase plots. The diffusive coupling of a hyperchaotic attractor in chamber A with a periodic attractor in chamber B leads to an overall hyperchaotic solution.

catory machinery of the cell is changed. In the second process, the viral DNA happens to escape the restriction enzymes of the cell, and the produced progeny are modified so as to be recognized as "self" by cells of the same restriction-modification type. Host-controlled modifications are distinguishable from mutations by their reversibility. If a modified virus particle completes a lytic cycle in a cell with a different restriction-modification system, it will acquire the characteristics of the new cell [45,58,59].

Instead of lysing the cells they infect, some phages are integrated into the cellular chromosome and reproduce synchronously with the cells for many generations. This is the lysogenic response discussed earlier. The inserted viral DNA is referred to as prophage. The presence of this DNA in the cell and its offsprings can be detected by the fact that every so often one of the daughter cells will lyse and liberate infectious phages. In each generation only a small fraction of the lysogenic cells lyse, and the probability that this occurs may depend on UV radiation and other external factors.

Once a cell carries a prophage, it is usually immune to infection by viruses of the same type. Often, however, this immunity is "paid for" in one way or the other, for instance through a slightly lower rate of reproduction. Then, if there are no more viruses around, nonresistant cells will outgrow resistant ones, and the immunity will again disappear. At the same time, a high density of resistant cells represents an open niche for virus mutants capable of overcoming the immunity, and sooner or later such mutants are likely to develop. This provides conditions that will select for bacteria with second-order resistance, and evolution of such bacteria sets the stage for further phage mutation. In this way, populations of bacteria and phages coevolve. Let us try to describe some of these processes by means of a simulation model.

In the first version of the model, we consider a bacterial population consisting of two variants: unmodified bacteria, which are relatively sensitive to phage attacks, and modified bacteria, for which the infection probability is many times smaller. The modified bacteria are assumed to be produced by mutation of unmodified cells, and the reverse process in which modified cells lose their resistance to phage attack is also accounted for. This is illustrated in the flow diagram of Figure 21. Except for such reverse mutations, the modification is inherited by daughter cells. The reproduction rates for the two types of bacteria are taken to be 0.050/min and 0.047/min, respectively.

As before, the culture is assumed to grow in a chemostat with a continuous supply of nutrients. With increasing bacterial populations, the growth rates are reduced due to decreasing resource availability. This decrease is described by a logistic growth correction that is assumed to depend on the total bacterial population. In addition, there is a continuous washout of bacteria by dilution. In the absence of vira, the population of unmodified bacteria will therefore outgrow and suppress the population of modified cells, and a stable equilibrium exists in which the population of unmodified bacteria is controlled by the supply of nutrients and the rate of dilution.

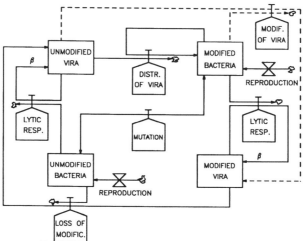

Figure 21. In its simplest form, our model of bacterium-phage evolution considers a culture consisting of two variants of the same bacterium: unmodified cells, which have the advantage of a faster reproduction rate, and modified cells, which are significantly less sensitive to phage attacks. There is a small probability that a viral DNA molecule evades the restriction system of a modified cell. This will give rise to the reproduction of modified viruses which can infect other modified cells.

From time to time a virus particle may enter the system. If the virus succeeds in infecting a bacterial cell, it will program the cell to reproduce the viral DNA. In a few cases a virus particle may succeed in penetrating the restriction system and infecting a modified bacterial cell. The new viruses produced in this way will be encoded in accordance with the modification system of the cell and can infect other modified cells. We assume that modified viruses can also infect unmodified cells. However, the progeny produced in such processes will have lost their modification.

In order to describe processes in which very few individuals are involved, the model equations are formulated in terms of discrete, stochastic transitions. At the same time, we have adopted a generic representation allowing us to extend the model to include many more populations and types of interactions. Thus, the bacterial populations are represented by a vector POP_i, with the subscript $i = 1, 2, \ldots, n$ identifying the particular variants. Similarly, the populations of phages are represented by the vector PHA_j, where the variants $j = 1, 2, \ldots, k$ may encompass a large number of different species. In the simple model $i = j = 1$ for unmodified bacteria and viruses, while $i = j = 2$ for modified bacteria and viruses, respectively. Note that POP_i and PHA_j denote total populations, not population densities as with the quantities B_i and P_j in the preceding sections.

The lytic reaction in which a phage of type j infects a cell of type i to produce new viruses of type i may be represented as

$$POP_i, PHA_j, PHA_i \rightarrow POP_i - 1, PHA_j - 1, PHA_i + \beta,$$

where, as before, β is the burst size. Assuming that the chemostat is effectively stirred so that the populations are homogeneously distributed over the available volume, the intensity of the above process may be expressed as

$$W_{\text{lytic}} = B_{ij} \times POP_i \times PHA_j. \qquad (16)$$

Here the rate constants are taken to be $B_{11} = B_{12} = B_{22} = 2 \times 10^{-7}/\text{min}$ for unrestricted lytic response, and $B_{21} = 5 \times 10^{-11}/\text{min}$ for the infection of a modified cell by an unmodified virus. This implies that a modified cell is 4,000 times as resistant as an unmodified cell toward attacks from unmodified viruses. The parameters B_{ij} may be considered the basic rate constants of the model relative to which many of the other rate constants are scaled. The assumed values for B_{ij} correspond in order of magnitude to the values assumed for α_{ij} in the previous sections if we consider a total volume of 5 μl. For larger volumes and the same population sizes, the time scales will be longer because encounters between cells and phages are less common.

The destruction of viruses by resistant bacteria is expressed by

$$POP_i, PHA_j \rightarrow POP_i, PHA_j - 1.$$

The intensity of this process is

$$W_{\text{destruction}} = C_{ij} \times POP_i \times PHA_j, \qquad (17)$$

with $C_{21} = 2 \times 10^{-7}/\text{min}$ for attacks of unmodified viruses on modified bacteria. In the other cases where the cells have no particular resistance towards the intruding viruses we have taken $C_{11} = C_{12} = C_{22} = 10^{-8}/\text{min}$, assuming, in these cases, that about 5% of the viruses are destructed. Bacterial mutation processes are expressed by

$$POP_i, POP_j \rightarrow POP_i - 1, POP_j + 1.$$

The intensity here is

$$W_{\text{mutation}} = M_{ij} \times POP_i, \qquad (18)$$

with $M_{12} \approx M_{21} = 5 \times 10^{-4}$/min. This implies that approximately 1% of all bacterial reproductions lead to a mutation. The resource-limited bacterial growth processes are described by

$$POP_i \rightarrow POP_i + 1,$$

with

$$W_{\text{growth}} = A_i \times POP_i \times \left(1 - \sum_j POP_j / POPM\right). \qquad (19)$$

With this formulation, there is no direct representation of the concentration of resources in the chemostat. As previously noted, the rate constants are taken to be $A_1 = 0.050$/min and $A_2 = 0.047$/min, respectively. In the simulations to be presented below, we have taken POPM = 3,000. Thus, due to limited resources, the total bacterial population $\sum_j POP_j$ is restricted to 3,000 cells. Because of the simultaneous removal of cells through dilution, the actual bacterial population will not attain this value. The dilution process is described by

$$POP_i, PHA_j \rightarrow POP_i - 1, POP_j - 1,$$

with the intensities

$$W_{\text{dilution}} = POP_i \times DFLOW, \qquad (20)$$

or

$$W_{\text{dilution}} = PHA_j \times DFLOW. \qquad (21)$$

DFLOW is identical to the variable ρ in the previous sections. We use different variable names to stress the significant difference in the structure and purpose of the two models.

As before, the rate of dilution is taken as a control parameter to be varied from simulation to simulation. A typical value is DFLOW = 0.02/min, corresponding to a residence time in the chemostat of 50 min. The bacterial growth rates of 0.05/min and 0.047/min correspond to doubling times of the order of 30 min. Finally, the spontaneous contamination of the chemostat by unmodified viruses is expressed by

$$PHA_j \rightarrow PHA_j + 1,$$

with

$$W_{\text{contam.}} = G_j. \qquad (22)$$

Here, $G_1 = 0.1$/min, and $G_2 = 0$.

The various processes occur with vastly different intensities that vary by orders of magnitude over time. Reproduction of bacterial cells and replication of viruses in lytic reactions thus occur at much higher rates than the processes by which bacteria and viruses are modified. To attain a reasonable dynamic range in the stochastic model, we have represented the various transition rates as Poisson processes. This implies that the probability that k transitions of a given type will take place during the time increment DT is given by

$$p_k = \frac{(W \cdot DT)^k}{k} \exp(-W \cdot DT), \qquad (23)$$

with W being one of the above specified intensities. In this way we can simulate the model with significantly larger values of DT than would otherwise be possible. For calculations involving larger populations, it would be natural to combine the stochastic description of the mutation and modification processes with a deterministic description of the growth and reproduction processes.

All our simulations with the simple 2×2 population model were performed with identical initial conditions. We start with 54 unmodified bacterial cells and no modified cells. The initial population of unmodified viruses is assumed to be $10^5 \times$ DFLOW. For DFLOW = 0.040/min, this corresponds to a viral population of 4,000 particles. In addition, there is a random contamination of the chemostat corresponding to an average inflow of 0.1 unmodified viruses per minute. The simulation period is 10,000 min, or approximately 1 week.

Figure 22 shows the results obtained for a dilution rate of DFLOW = 0.040/min. This corresponds to an average residence time for cells and viruses in the chemostat of 25 min. With this relatively high rate of dilution, the presence of viruses has little significance. As a consequence of their higher rate of reproduction, the unmodified bacteria outgrow the population of modified cells to reach a stable saturation level at approximately 1,700 cells. Through mutation, a population of approximately 200 modified cells is maintained. The initial population of unmodified phages is almost completely washed out before a significant bacterial population is established. From time to time, minor epidemics of attacks by unmodified viruses break out. These epidemics never develop into anything significant, however, and they rapidly die out again.

A somewhat lower rate of dilution allows the phages to establish a major attack on the population of unmodified bacteria. This is illustrated in Figure 23 for DFLOW = 0.030/min. The reduction of

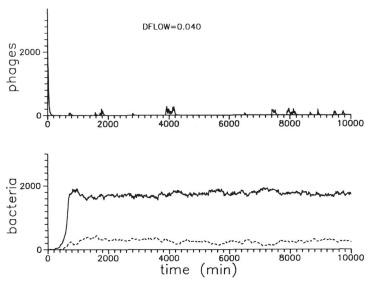

Figure 22. With a rate of dilution of DFLOW = 0.040/min, the viral populations are suppressed, and the population of unmodified cells dominates. Solid curves represent unmodified bacteria and phages; dotted curves, modified populations.

the population of unmodified bacteria brought about by the infection gives a better chance to the modified bacteria. On the other hand, the growth of this population curbs the viral infection, and the population of unmodified bacteria soon resumes its dominance. Somewhat later, a new infection occurs. The bacterial population still maintains a global stability against viral infections, however, and these infections occur as a purely stochastic phenomenon, since very few viral particles are present in the intermediate periods.

With further reduction in the rate of dilution, the phage attacks become more severe, and also more regular. Viral populations of the order of 50 to 100,000 particles are now attained in the infections. Each time, however, the growth of the modified bacteria population brought about by the infection kills off the viruses, and one can envisage that the system will develop a regular oscillatory mode with a period on the order of 3,000 min. This period is controlled mainly by the time it takes for the population of unmodified bacteria to recover after an infection, i.e., by the time it takes the bacterial population to lose the resistance developed during the infection.

If the rate of dilution is further reduced, a qualitative shift in the behavior of the model tends to occur as modified viruses now appear

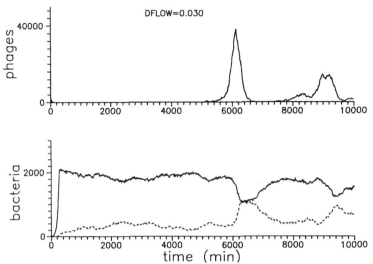

Figure 23. With a somewhat lower rate of dilution (DFLOW = 0.030/ min), the phage can establish major attacks on the population of unmodified cells. Globally, however, the bacterial population is still stable against viral infections.

in the system. Hereafter, the cells are no longer capable of killing off the phages, and a predator-prey relation develops between modified viruses and modified bacteria. The appearance of modified viruses is conditioned by the simultaneous existence of large populations of unmodified viruses and modified bacteria. In Figure 24, where DFLOW = 0.0215/min, modified viruses emerge during the second infection, i.e., at about 3,000 min. From then on the system enters a self-sustained oscillatory behavior with a period of approximately 200 min. It is interesting to note that the two bacterial populations in this model vary almost in phase. The same is true for the two viral populations. However, the populations of unmodified bacteria and viruses remain low.

If DFLOW is further reduced, modified viruses emerge even earlier. This is illustrated in Figure 25 for DFLOW = 0.015/min. At the same time, the intensity of the oscillations between modified bacteria and viruses become higher, and the oscillations become more regular. The period of these oscillations also increases slightly. With a rate of dilution as low as 0.007/min, the population of modified viruses becomes strong enough to kill off the bacterial population (see Figure 26). Thereafter, the viruses can no longer multiply, and the viral population is soon washed out of the chemostat.

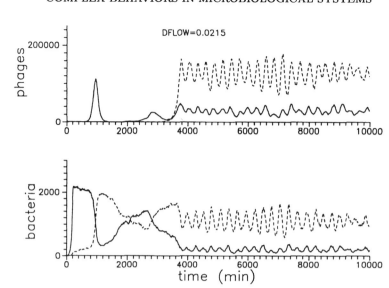

Figure 24. With DFLOW = 0.0215/min, modified viruses tend to appear in the system. Hereafter, the cells are no longer capable of killing off the phages, and an oscillatory predator-prey relation between modified viruses and modified bacteria develops.

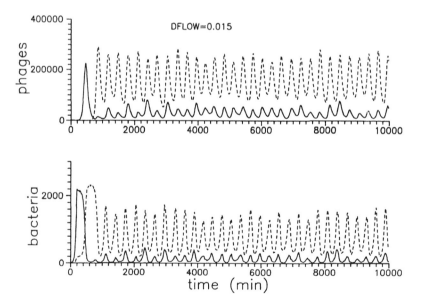

Figure 25. With DFLOW = 0.015/min, modified viruses emerge already in the first infection, and the oscillatory behavior associated with the interaction between modified viruses and modified cells becomes more intense.

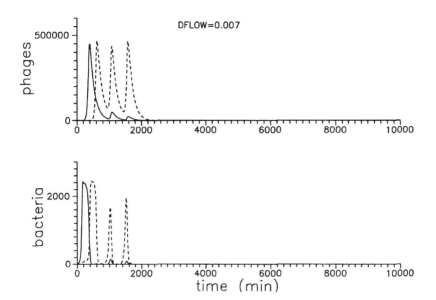

Figure 26. With DFLOW = 0.007/min, the population of modified viruses becomes strong enough to kill off the bacterial populations. Hereafter, the viruses can no longer replicate, and they, too, soon disappear.

For comparison with the above simulations with continuous dilution, Figure 27 shows the results of a simulation in which a small sample (2%) of the existing culture is taken every 400 min and allowed to regrow in an otherwise uncontaminated chemostat. This resembles the manner in which bacterial populations as applied, for instance, in cheese production in former times were subcultured for decades or maybe even centuries. In the simulation shown in Figure 27, modified viruses never emerge, and we see a modulation of the composition of the bacterial population in response to more or less random viral infections. Due to the stochasticity involved, with the same basic parameters other simulations may result in the development of modified viruses, yielding a completely different behavior of the system.

As illustrated by the above simulations, the rate of dilution is a significant control parameter that shifts the balance between viruses and bacteria. At high dilution rates, the viruses are effectively suppressed, and unmodified bacteria dominate. At lower dilution rates, the virus population becomes large enough to disturb the competition between rapidly growing, relatively sensitive cells and less rapidly reproducing, modified bacteria. In a certain range of parameters, the presence of viruses thus increases the diversity of the bacterial population [60]. If

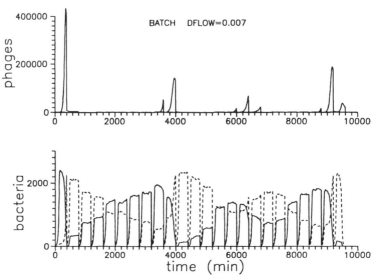

Figure 27. In this simulation, a small sample of 2% of the culture is taken out every 400 min and allowed to grow in another chemostat. Observe the modulation in the composition of the bacterial population in response to viral infections.

subculturing is performed under aseptic conditions, one often finds that the bacterial culture loses both its diversity and its viral resistance. Part of this resistance also appears to be associated with plasmides, i.e., smaller pieces of DNA existing in the bacterial cell and operating to some extend independently on the cell chromosome.

6. Modeling Bacterial Cell Division

Extensive studies have been made in order to reveal the mechanics of bacterial cell division [61]. Particular interest has been paid to the regulation of cell cycle times for *E. coli*, a bacterium that has also been investigated in many other contexts and for which more than 1,000 genes have been mapped [62]. This implies that information is available about both the nucleotide sequence and the associated protein function of this bacterium. Many of the genes involved in controlling the growth of *E. coli* have also been studied in detail. So far, however, the quantitative description has mainly concerned individual aspects of this control. What is now needed is an integrated dynamic model to test to what extent current hypotheses can account for the experimental evidence.

The cell cycle is composed of several different phases. One phase is the period it takes for the chromosome to replicate. Experimentally, one observes that replication begins at a particular site and propagates in both directions along the circular DNA molecule. For relatively fast growing cells, the time it takes for replication is nearly constant and is independent of the growth conditions. The initiation site is commonly referred to as oriC and the replication period as C-time. A typical value for the length of this period is $T_c = 40$ min.

After replication, there is a period of latency (D-time) lasting approximately 20 min before the cell divides into two halves. This period also appears to be relatively constant, and as a result, control of the cell cycle (and thereby the cell volume) under varying growth conditions rests mainly with an ability of the cell to regulate the initiation period, i.e., the period from a cell division to a new replication of the DNA molecule is initiated.

It is interesting to note that for rapidly growing cells, the time between cell divisions may be as short as 20 min. This implies that several replication processes can go on at the same time. For given growth conditions, the time between cell divisions is not constant, but there is a considerable scatter from cycle to cycle. On average, however, the regulation is capable of maintaining stable cell volumes over a variety of different growth rates, ranging from doubling times of 20 min all the way to almost no growth.

Early physiological measurements implicated a model proposed by Donachie [63] which states essentially that replication is initiated when a certain ratio between replication origins and cell mass is attained. This ratio should be constant for a range of growth conditions. At that time Donachie was unable to explain how this control took place. Later, a variety of different control mechanisms were proposed [64–66]. Gradually, an ever-increasing number of experimental results pointed to the DnaA protein as being a primary mediator of the control of initiation of control [67,68]. The concentration of this protein is regulated by specific genes (promoters) on the DNA molecule in such a way that DnaA has a negative feedback effect on its own production. Other sites, so-called DnaA-boxes, titrate the amount of DnaA produced by literally counting the molecules one by one. When this count is great enough, a replication process is initiated. The cell volume affects the control primarily by shifting the chemical balance between free DnaA and DnaA bound to boxes.

Based on these ideas we shall try to formulate a relatively detailed stochastic model that can account for some of the experimental findings concerning the control of cell cycle times in rapidly growing *E. coli*. Under conditions of slow growth, it is likely that the replication period T_c gets longer. This has not been taken into account in the model. A preliminary version of our model was presented at the 1989 International System Dynamics Conference in Stuttgart [72]. Other models have recently been developed by Margalit et al. [73] and by Mahaffy and Zyskind [74]. In particular, our model resembles the latter, except that our model is totally stochastic. This enables us to consider the statistics of cell cycle times under different growth conditions. In addition, we have performed a somewhat more detailed parameter analysis of the simulation results.

The molecular components of our model are the messenger-RNA molecules produced by the DnaA promoters, free DnaA molecules, and DnaA molecules bound to the chromosome, together with free and occupied DnaA boxes. The number of molecules of each of these components is taken as a state variable. The amount of free DnaA protein is assumed to determine when initiation of DNA replication takes place. This occurs in the model when more than 20 DnaA molecules are present in free form, while at the same time the four DnaA boxes located close to oriC are all occupied.

It is observed experimentally that there is a formation of a complex of 20 to 40 DnaA protein monomers bound to one another and bound to the four DnaA boxes close to oriC [75]. Particularly at this point, our model is clearly oversimplified. However, at present there is insufficient data to estimate the variation in binding constants as more and more DnaA molecules bind to oriC.

The number of free DnaA molecules increases as DnaA is synthesized and decreases as DnaA molecules are captured by boxes along the chromosome. Degradation of DnaA protein is assumed to be slow enough that it can be neglected. The DnaA boxes are sites on the DNA molecules with the nucleic acid sequence $\text{TTAT}_A^C\text{CA}_A^C\text{A}$ [75]. A simple estimate assuming that the nucleic acids are distributed at random along the chromosome leads to an expectation value of approximately 150 such sites. In the model we assume that there are 75 boxes distributed along the chromosome, but with a certain tendency to concentrate near oriC.

Each box allows one DnaA molecule to be captured. This binding of DnaA molecules is determined by the rates of association and dissociation for the process

$$\text{DnaA}_{\text{free}} + \text{box}_{\text{free}} \overset{K_1}{\underset{K_{-1}}{\rightleftarrows}} \text{DnaA-box}, \tag{24}$$

where K_1 and K_{-1} are the two rate constants. An estimate of K_1 can be obtained in the same way as we estimated the rate constant α_{ij} for the interaction of bacteria and viruses in Section 2. At least for the box involved in the control of the DnaA promoters, K_{-1} must be large enough to secure that the lifetime of DnaA in the bound state is much less than the cell cycle time. If not, the control will become very hazardous. It is likely that some of the other boxes are connected with the control of other cell cycle proteins and must satisfy similar conditions. Here we assume that all boxes have similar association and dissociation constants.

In the stochastic description, the intensities of the two processes are

$$W_{\text{association}} = K_1 [\text{DnaA}_{\text{free}}] \cdot [\text{box}_{\text{free}}] / V \cdot N_A \tag{25}$$

and

$$W_{\text{dissociation}} = K_{-1} [\text{DnaA-box}], \tag{26}$$

where V is the cellular volume and square brackets denote the numbers of molecules or boxes in the cell and $N_A = 6.023 \times 10^{23}/\text{mole}$ is Avogadro's number. The equilibrium constant is given by

$$K_m = \frac{K_{-1}}{K_1} = \frac{[\text{DnaA}_{\text{free}}] \cdot [\text{box}_{\text{free}}]}{V \cdot N_A \cdot [\text{DnaA-box}]}, \tag{27}$$

where the presence of V in the denominator represents the fact that for given numbers of molecules and boxes, the equilibrium will shift towards dissociation as the volume increases.

Experimental evidence suggests [70,71] that production of DnaA protein is autorepressive, and that this negative feedback control is associated with binding of DnaA molecules to boxes in the vicinity of the DnaA promoters. In the model we assume that the rate of mRNA transcription is determined by the strength of two DnaA promoters P_1 and P_2, of which P_2 is regulated in such a way that when a DnaA molecule is bound to a box between P_1 and P_2, then P_2 is deactivated.

The promoter P_1 is always active but has a relatively low rate of transcription. The intensity of mRNA transcription may then be expressed as

$$W_{\text{transcription}} = R_1 + H \cdot R_2, \qquad (28)$$

where R_1 and R_2 are the transcription rate constants for the two promoters, and the parameter H is either 1 or 0, depending on whether P_2 is active or not.

In the simulation presented below we have taken $R_1 = 0.15R_2$, so that there is only one free parameter to specify. As a base case value for the promoter strength of P_2, we have used $R_2 = 0.5$ mRNA/min. The resultant mRNA molecules are assumed to have a lifetime of 2 min, during which period they can synthesize DnaA protein. On the average, each mRNA molecule is assumed to produce one DnaA molecule.

With the initiation of the DNA replication, a series of events is set in motion. At the same time that a particular sequence is replicated, bound protein is released. And, after replication, the sequence enters a refractive period of approximately 5 min before it again becomes active. During this period, the processes of methylation, supercoiling, and folding take place. With respect to the model, this means that no DnaA protein can bind to a box for a certain period after the box has been replicated. In particular, a new initiation cannot occur as long as the boxes close to oriC are in the refractive state. However, as soon as these sequences have been methylated, the next initiation may take place before the ongoing replication is finished. This ensures that we can have cell cycle times shorter than the time it takes to copy the chromosome. Each replication process proceeds at a constant speed along the DNA molecule, so that the duration of the process corresponds to the aforementioned C-time.

The model is stochastic in the sense that the production of mRNA and the binding and dissociation of DnaA are assumed to take place in accordance with Poisson probability distributions. Since the number of molecules of each substance is relatively small, minor fluctuations can have significant consequences. So the stochastic approach is preferable to a continuous deterministic description. The model is also dynamic in the sense that we do not necessarily assume that the DnaA association and dissociation processes are fast enough to establish equilibrium conditions between free and bound protein. Between divisions the cell volume is assumed to grow exponentially over time, and each division causes the volume to be halved. The daughter cells, of course, both get a DNA molecule. The DnaA and mRNA molecules, on the other hand, are divided stochastically between the two cells in accordance with a binomial distribution. Only one daughter cell is followed after the division.

The simulation shown in Figure 28 follows the volume of a cell through a series of divisions. Despite the fluctuations, it is apparent that the volume is controlled: large values in one cycle tend to be followed by smaller volumes in the subsequent cycle. The average volumes at the times of initiation and at cell division are about $1.0 \times 10^{-15}l$ and $1.5 \times 10^{-15}l$, respectively, and both have standard deviations of 14 to 18%. For the cell age (interdivision time), the standard deviation is 25%. This agrees well with earlier experimental results, in which the volume at division was found to vary 9 to 15% and the life length, 22 to 27% [76].

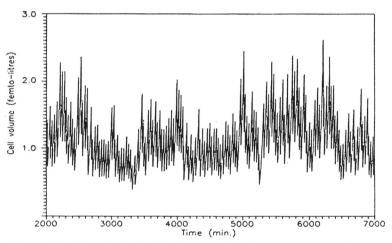

Figure 28. Simulation of the bacterial cell volume through a series of divisions under constant growth conditions. The doubling time for the cell volume is 35 min.

If simulations are performed with different promoter strengths, it turns out that the model gives stable results only within a limited range of mRNA transcription rates. When the promoter strength falls outside this range, the cell is unable to regulate its volume satisfactorily, and either very small or very large cells appear during the simulations. In order to quantify this effect, we have plotted (Figure 29a and b) the average cell volume at initiation and the standard deviation of this volume as functions of the promoter strength R_2. As the parameter in these plots, we have used the doubling time for the growth of the

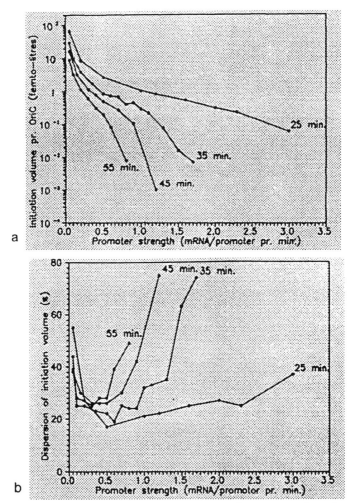

Figure 29. (a) The average initiation volume as a function of promoter strength for four different doubling times. A doubling time of 25 min represents approximately the highest growth rate that one can experience. (b) Dispersion of the initiation volume in percent of the average initiation volume for different doubling times. Outside a fairly narrow range of promoter strengths, runaway phenomena occur.

cellular volume. As noticed previously, we have restricted ourselves to relatively fast growth rates.

Inspection of the two figures reveals that the volume control is lost as R_2 becomes smaller than 0.2 or larger than 1.0 mRNA/min. The general decrease in initiation volume with increasing promoter strength is a direct consequence of the fact that the DnaA promoter should be active for a shorter time to produce the amount of DnaA required for initiation. The fact that there is a plateau on the curves for intermediate promoter strengths represents the regulatory effect of DnaA autorepression. Clearly, this effect is not as strong in the model as one could have expected. It is likely that this weakness can be remedied to some extent by introducing a growth rate dependence for the promoter strength. Experimental indication for such a dependence exists in the literature.

We have also examined why the dispersion in initiation volumes increases so dramatically as the promoter strength becomes either too small or too large. Under these conditions, the average initiation volume continues to either decrease or increase throughout the simulation without approaching a finite, stable level. The volume control fails when the promoter strength is too small, since then the two promoters together cannot produce enough DnaA protein to fill the boxes in the presence of an expanding cell volume. On the other hand, when the promoter strength is too high, the unregulated promoter alone produces more DnaA than required. The regulated promoter is never activated, and the cell divides before it has reached a sufficient volume.

Similarly, variation of the association constant for binding of DnaA protein to DnaA boxes has a marked effect on the volume control. This is illustrated in Figure 30a and b, where we have plotted the initiation volume and the dispersion of this volume as functions of the normalized association rate constant, K_1/N_A. As the parameter in these figures, we have used the equilibrium constant K_m, which is measured in moles/l. A small value of K_m implies that dissociation is relatively slow.

Inspection of Figure 30a shows that each curve has a maximum, and that the position of this maximum shifts towards higher association constants with decreasing equilibrium constant. Above the maximum, the rate constants are high enough to ensure that equilibrium conditions are maintained. In accordance with Eq. (27), the cell volume then varies inversely with K_m. Also in this range, a higher association rate constant implies that the boxes will be filled faster and that a smaller amount of DnaA needs to be produced. Consequently, the initiation volume decreases with increasing association rate constant.

On the other hand, to the left of the maxima in Figure 30a the association constant is too small to maintain equilibrium conditions between free and bound DnaA. Here the rates of the various processes become significant. In accordance with (25), a decrease of the associ-

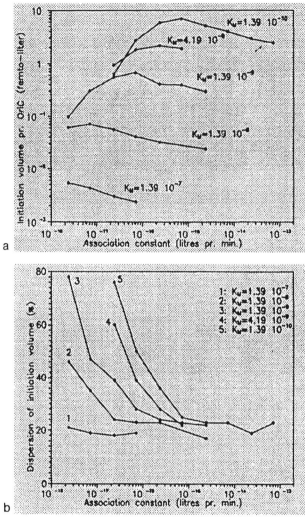

Figure 30. (a) Average initiation volume as a function of the association rate constant. The parameter is the equilibrium constant K_m in moles/l. (b) Dispersion of the initiation volume in percent of the average initiation volume as a function of the association rate constant. For small values of K_1 or K_m the association or dissociation processes become too slow to maintain equilibrium conditions.

ation constant can be compensated for by a proportionate decrease in the cell volume. Consequently, in this range stabilization can occur only at smaller and smaller cell volumes. Figure 30b illustrates the same phenomenon, namely that the stabilization of the cell volume becomes less efficient when the association and dissociation processes become too slow to respond to the changing conditions during a cell cycle. Since the ranges investigated for K_1 and K_m are biological meaningful, this example serves to show that equilibrium kinetics is not necessarily sufficient to model the processes involved in initiation control.

Figure 31 shows a simulation that follows the number of free DnaA molecules and the total number of DnaA molecules through a series of cell divisions. Most of the time there are fewer than 10 molecules of free DnaA in the cell, although this number fluctuates significantly. Peaks in free DnaA occur at irregular intervals. These represent the initiation of DNA replication and the release of DnaA molecules bound to boxes near oriC. The protein molecules are recaptured when the newly replicated area has regained its normal topology. The total amount of DnaA varies approximately in proportion with the cell volume. This implies that fluctuations in the number of free DnaA molecules can initiate replication, emphasizing the significance of the stochastic approach. However, if initiation is early in one cell cycle, it has a tendency to be correspondingly late in the next cycle.

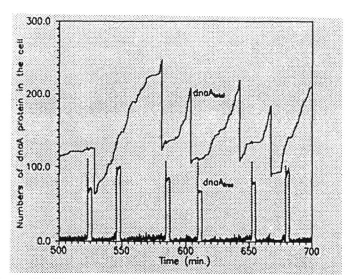

Figure 31. The total number of DnaA molecules and the number of free DnaA molecules as functions of time over a series of cell cycles.

7. Conclusion

Microbiological population systems clearly contain all the ingredients for complex dynamical behavior: delayed nonlinear feedback and instability generating multiplication processes, combined with mutation and modification processes that change the predator-prey relations and lead to coevolution and differentiation. Studies of simple models with a few species have shown that these ingredients can lead to chaos and other highly nonlinear dynamic phenomena, as well as to growth and collapse processes, such as a viral infection that dies out through the development of resistant bacteria.

However, most environments are characterized by a vast diversity of microbiological species, and to a certain extent this diversity is found in biotechnological reactors as well. The question therefore arises as to whether multispecies systems generally have simpler dynamics than those of systems with only a few species. From a mathematical point of view, one would expect the opposite to be true, since more species introduce additional degrees of freedom. On the other hand, it must be appreciated that each population within a community is likely to be involved in multiple, simultaneous interactions, and that the complexity of this network of interactions could provide a homeostatic effect. At the same time, increased diversity could provide the community with a greater capacity to withstand invasion by foreign species.

As one of the most interesting results of the present study, we have shown that the behavioral complexity of a microbiological system can continue to increase as more and more species are introduced. With one, two, three, and four bacterial populations, we find that the most complicated behavior of the predator-prey model has zero, one, two, and three positive Lyapunov exponents, respectively. This result was obtained for a model that was developed with the intention of giving a realistic account of the biological system, and that had parameter values in accordance with experimental findings. Moreover, the result was found to be generic and robust in the sense that the chaotic hierarchy persisted for a variety of different formulations of the interspecies interactions and for a wide range of parameter values.

At first it might seem somewhat disturbing that chaotic and hyperchaotic dynamics lurk in microbiological predator-prey systems. However, considering the variety of experimental and theoretical studies indicating the existence of chaos in physiological control systems, chaos should not necessarily be considered as harmful to ecological systems. It is worth bearing in mind, however, that systems with chaotic dynamics have the property that while in certain circumstances they can absorb a major blow without significant change, in other situations the slightest disturbance can redirect their motion completely.

Our model of cell division differs in two important respects from previously suggested models. In the first place, the model is dynamic, and we have shown that the association rate for binding of DnaA protein to DnaA boxes may be important. Second, the model is stochastic in all those variables where small numbers arise. This has allowed us to study the statistics of cell cycle control. With current experimental techniques it should be possible to test our predictions concerning the stability of cell volume regulation as a function of binding kinetics and promoter strengths. In particular, one can take advantage of the fact that different DnaA variants exist having different binding kinetics.

It seems likely that the cellular volume control is more stable than predicted by the model. Among the modifications of the model that could lead to increased stability, we suggest: (i) a reduced weight of the unregulated promoter, (ii) growth rate control on both promoters, and (iii) a binding kinetics that differs between boxes of different function. It is likely, for instance, that DnaA molecules have a lower affinity for sites involved in formation of the initiation complex near oriC than for other boxes, and that a more detailed representation of the formation of this complex could change our results.

Finally, it may be of interest to note that cell cycle control in virtually all eucaryotic cells (including human cells) appears to be mediated by a cell-division cycle protein called cdc2 [77], although the mechanisms of this control are doubtlessly much more complicated than those that we have described for bacterial cells. At the same time, the cell cycle control in eucaryotic cells appears to involve self-sustained chemical oscillations in the cytoplasm, and it has been proposed [78] that chaotic phenomena related to those oscillations play a role in the distribution of cycle times.

Acknowledgments

We are grateful to J. Engelbrecht, T. Atlung, and F. G. Hansen for helpful discussions of microbiological processes. A. Bjerring Nielsen, J. Sturis, B. Bodholdt, and L. Risbo are acknowledged for their contributions to initial stages of this project.

References

[1] Lorenz, E. "Deterministic Nonperiodic Flow," *J. Atm. Phys.* 20, 130–141 (1963).

[2] Ueda, Y. "Random Phenomena Resulting from Nonlinearity," *Trans. Inst. Electrical Eng.*, 98A, 167–173 (1978) (in Japanese). English translation: *Int. J. Non-linear Mech.*, 20, 481–491 (1985).

[3] Ruelle, D. "Sensitive Dependence on Initial Conditions and Turbulent Behavior in Dynamical Systems," *Ann. N.Y. Acad. Sci.*, 316, 408–416 (1979).

[4] Jensen, K. S., E. Mosekilde, and N.-H. Holstein-Rathlou. "Self-Sustained Oscillations and Chaotic Behavior in Kidney Pressure Regulation," *Proc. Solvay Institutes Discoveries 1985 Symposium: Laws of Nature and Human Conduct*, I. Prigogine and M. Sanglier (eds.), G.O.R.D.E.S., Brussels (1986).

[5] Sturis, J., K. S. Polonsky, E. Mosekilde, and E. Van Cauter. "The Mechanisms Underlying Ultradian Oscillations of Insulin and Glucose: A Computer Simulation Approach," *Am. J. Physiol.* (1991).

[6] Jensen, M. H., P. Bak, and T. Bohr. "Transitions to Chaos by Interaction of Resonances in Dissipative Systems," *Phys. Rev.*, A130, 1960–1969 (1984).

[7] Prank, K., H. Harms, Chr. Kayser, G. Brabant, L. F. Olsen, and R. D. Hesch. "The Dynamic Code: Information Transfer in Hormonal Systems," in *Complexity, Chaos, and Biological Evolution*, E. Mosekilde and Lis Mosekilde (eds.), Plenum Press, New York (1991).

[8] Goldbeter, A. (ed.) *Cell to Cell Signalling: From Experiments to Theoretical Models*, Academic Press, London (1989).

[9] Glass, L., M. R. Guevara, A. Shrier, and R. Perez. "Bifurcation and Chaos in a Periodically Stimulated Cardiac Oscillator," *Physica*, 7D, 89–101 (1983).

[10] Colding-Jørgensen, M. "Chaos in Coupled Nerve Cells," in *Complexity, Chaos, and Biological Evolution*, E. Mosekilde and Lis Mosekilde (eds.), Plenum Press, New York (1991).

[11] Berridge, M. J., P. H. Cobbold, and K. S. R. Cuthbertson. "Spatial and Temporal Aspects of Cell Signalling," *Phil. Trans. R. Soc. London*, B320, 325–343 (1988).

[12] Goldbeter, A., Y. X. Li, and G. Dupont. "Periodicity and Chaos in cAMP, Hormonal, and Ca^{2+}-Signalling," in *Complexity, Chaos, and Biological Evolution*, E. Mosekilde and Lis Mosekilde (eds.), Plenum Press, New York (1991).

[13] Glass, L. and M. C. Mackey. *From Clocks to Chaos: The Rhythms of Life*, Princeton University Press, Princeton, NJ (1988).

[14] Allen, J. C. "Factors Contributing to Chaos in Population Feedback Systems," *Ecol. Model.*, 51, 281–298 (1990).

[15] Gilpin, M. E. "Spiral Chaos in a Predator-Prey Model," *Amer. Natur.*, 113, 306–308 (1979).

[16] Inoue, M. and H. Kamifukumoto. "Scenarios Leading to Chaos in a Forced Lotka-Volterra Model," *Prog. Theor. Phys.*, 71, 930–937 (1984).

[17] Anderson, R. M. and R. M. May. "Complex Dynamical Behavior in the Interaction Between HIV and the Immune System," in *Cell to Cell Signalling*, A. Goldbeter (ed.), Academic Press, New York (1989).

[18] Grossman, Z. "Oscillatory Phenomena in a Model for Infectious Disease," *Theor. Popul. Biol.*, 18, 204–243 (1980).

[19] Aron, J. L. and I. B. Schwartz. "Seasonality and Period-Doubling Bifurcations in an Epidemic Model," *J. Theor. Biol.*, 110, 665–680 (1984).

[20] Schaffer, W. M. and M. Kot. "Nearly One-Dimensional Dynamics in an Epidemic," *J. Theor. Biol.* 112, 403–427 (1985).

[21] Kot, M., W. M. Schaffer, G. L. Truty, D. J. Graser, and L. F. Olsen. "Changing Criteria for Imposing Order," *Ecol. Med.*, 43, 75–110 (1988).

[22] Olsen, L. F., G. L. Truty, and W. M. Schaffer: "Oscillations and Chaos in Epidemics: A Nonlinear Dynamic Study of Six Childhood Diseases in Copenhagen, Denmark," *Theor. Popul. Biol.*, 33, 344–370 (1988).

[23] MacArthur, R. "Fluctuations of Animal Populations, and a Measure of Community Stability," *Ecology*, 36, 533–536 (1955).

[24] Bull, A. T. and J. H. Slater. *Microbial Interactions and Communities*, Vol. 1, Academic Press, New York (1982).

[25] Hairston, N. G., J. D. Allan, R. K. Colwell, D. J. Futuyma, J. Howell, M. D. Lubin, J. Mathias, and J. H. Vandermeer. " The Relationship Between Species Diversity and Stability: An Experimental Approach with Protozoa and Bacteria," *Ecology*, 49, 1091–1101 (1968).

[26] Waagner Nielsen, E., J. Josephsen and F. Kvist Vogensen. "Lactic Starters: Improvement of Bacteriophage Resistance and Application of DNA-Technology," *Danish Journal of Agronomy*, Selected Research Reviews, pp. 35–45, (1987).

[27] Hugenholtz, J. "Population Dynamics of Mixed Starter Cultures," *Neth. Milk Dairy J.*, 40, 129–140 (1986).

[28] Stadhouders, J. and G. J. M. Leenders. "Spontaneously Developed Mixed-Strain Cheese Starters. Their Behaviour Towards Phages

and Their Use in the Dutch Cheese Industry," *Neth. Milk Diary J.,* 38, 157–181 (1984).

[29] Kristensen, H., L. Risbo, E. Mosekilde, and J. Engelbrecht. "Complex Dynamics in Bacterium-Phage Interactions," *Proc. 1990 European Simulation Multiconference,* Erlangen-Nuremberg, June 10–13, 1990, pp. 636–641, B. Schmidt (ed.), Simulation Councils Inc., San Diego, CA.

[30] Nielsen, A. B., H. Stranddorf, and E. Mosekilde. "Complex Dynamics, Hyperchaos and Coupling in a Microbiological Model," *Proc. 1991 European Simulation Multiconference,* Copenhagen, June 17–19, 1991, pp. 751–756. E. Mosekilde (ed.), Simulation Councils Inc., San Diego, CA.

[31] Wolf, A., J. B. Swift, H. L. Swinney, and J. A. Vastano. "Determining Lyapunov Exponents from a Time Series," *Physica D,* 16D, 285–317 (1985).

[32] Packard, N. H., J. P. Crutchfield, J. D. Farmer, and R. S. Shaw. "Geometry from a Time Series," *Phys. Rev. Lett.,* 45, 712 (1980).

[33] Grassberger, P. and I. Procaccia. "Measuring the Strangeness of Strange Attractors," *Physica,* 9D, 189–208 (1983).

[34] Rössler, O. E. "An Equation for Hyperchaos," *Phys. Lett.,* A71, 155–157 (1979).

[35] Matsumoto, T., L. O. Chua, and K. Kobayashi. "Hyperchaos: Laboratory Experiment and Numerical Confirmation," *IEEE Trans. Circuits Syst.,* CAS–33, 1143–1147 (1986).

[36] Peinke, J., B. Röhricht, A. Mühlbach, J. Parisi, Ch. Nöldeke, and R. P. Huebener. "Hyperchaos in Post-Breakdown Regime of P-Germanium," *Z. Naturforsch.,* 40a, 562–566 (1985).

[37] Killory, H., O. E. Rössler, and J. L. Hudson. "Higher Chaos in a Four-Variable Chemical Reaction Model," *Phys. Lett. A,* 122, 341–345 (1987).

[38] Dmitriev, A. S., Yu. V. Gulyaev, V. Ya. Kislov, and A. J. Panas. "Mode Pulling and Competition in a System with Chaotic Dynamics," *Phys. Lett. A,* 128, 172–176 (1988).

[39] Thomsen, J. S., E. Mosekilde, and J. D. Sterman. "Hyperchaotic Phenomena in Dynamic Decision Making," *J. Syst. Anal. Model. Simulation,* 9, 137–156 (1992).

[40] Rössler, O. E. "The Chaotic Hierarchy," *Z. Naturforsch.,* 38a, 788–801 (1983).

[41] Baier, G., J. S. Thomsen, and E. Mosekilde. "Chaotic Hierarchy in a Model of Competing Populations," submitted for publication in *J. Theor. Biol.*

[42] Bruckner, E., W. Ebeling and A. Scharnhorst. "Stochastic Dynamics of Instabilities in Evolutionary Systems," *Syst. Dynam. Rev.*, 5, 176–191 (1989).

[43] Schwartz, M. "The Adsorption of Coliphage Lanbda to Its Host: Effect of Variations in the Surface Density of Receptor and in Phage-Receptor Affinity," *J. Mol. Biol.*, 103, 521–536 (1976).

[44] Chandrasekhar, S. "Stochastic Problems in Physics and Astronomy," *Rev. Modern Phys.*, 15, 1–89 (1943).

[45] Arber, W. "Host-Controlled Modification of Bacteriophage," *Annu. Rev. Microbiol.*, 19, 365–377 (1965).

[46] Monod, J. "La Technique de culture continue: théorie et applications," *Ann. Inst. Pasteur*, 79, 390–410 (1950).

[47] Levin, B. R., F. M. Stewart, and L. Chao. "Resource-Limited Growth, Competition, and Predation: A Model and Experimental Studies with Bacteria and Bacteriophage," *Amer. Natur.*, 111, 3–25 (1977).

[48] Crawford, J. D. and E. Knobloch: "Classification and Unfolding of Degenerate Hopf-Bifurcations with O(2) Symmetry: No Distinguished Parameter," *Physica D*, 31, 1–48 (1988).

[49] Sturis, J. and E. Mosekilde. "Bifurcation Sequence in a Simple Model of Migratory Dynamics," *Syst. Dynam. Rev.*, 4, 208–217 (1988).

[50] Grebogi, C., E. Ott and J. A. Yorke. "Basin Boundary Metamorphosis: Changes in Accessible Boundary Orbits," *Physica*, 24D, 243–262 (1987).

[51] Kaplan, J. L. and J. A. Yorke. "Chaotic Behavior and Multidimensional Difference Equations," in *Lecture Notes in Mathematics*, Vol. 730, H.-O. Peitgen and H.-O. Walther (eds.), Springer, Berlin (1979), pp. 204–227.

[52] Klein, M. and G. Baier. "Hierarchies of Dynamical Systems," in *A Chaotic Hierarchy*, G. Baier and M. Klein (eds.); World Scientific, Singapore (1991).

[53] Badola, P., V. R. Kumar, and B. D. Kulkarni. "Effects of Coupling Nonlinear Systems with Complex Dynamics," *Phys. Lett. A*, 155, 365–372 (1991).

[54] Li, Y.-X., D.-F. Ding, and J.-H. Xu. "Chaos and Other Temporal Selforganization Patterns in Coupled Enzyme-Catalyzed Systems," *Commun. Theor. Phys.*, 3, 629–638 (1984).

[55] Li, Y.-X. and A. Goldbeter. "Oscillatory Isozymes as the Simplest Model for Coupled Biochemical Oscillators," *J. Theor. Biol.*, 138, 149–174 (1989).

[56] Sevcikova, H. and M. Marek. "Wave Patterns in an Excitable Reaction-Diffusion System," *Physica D*, 49, 114–124 (1991).

[57] Pikovsky, A. S. "On the Interaction of Strange Attractors," *Z. Phys. B—Condensed Matter*, 55, 149–154 (1984).

[58] Krüger, D. H. and T. A. Bickle. "Bacterio-Phage Survival: Multiple Mechanisms for Avoiding the Deoxyribonucleic Acid Restriction System of Their Hosts," *Microbiol. Rev.*, 47, 345–360 (1983).

[59] Bode, V. C. and A. D. Kaiser. "Changes in Structure and Activity of DNA in a Superinfected Immune Bacterium," *J. Mol. Biol.*, 14, 399–417 (1965).

[60] Levin, B. R. "Frequency-Dependent Selection in Bacterial Populations," *Phil. Trans. R. Soc. London B*, 319, 459–472 (1988).

[61] Ingraham, J. L., O. Maaløe, and F. C. Neidhardt. *Growth of the Bacterial Cell*, Sinauer Associates, Sunderland, MA (1983).

[62] Bachmann, B. J. "Linkage Map of *Escherichia coli* K-12. Edition 7," *Microbiol. Rev.*, 47, 180–230 (1983).

[63] Donachie, W. D. "Relationship Between Cell Size and Time of Initiation of DNA Replication," *Nature*, 219, 1077–1079 (1968).

[64] Wickner, S. H. "DNA Replication Proteins of *Escherichia coli*," *Annu. Rev. Biochem.*, 47, 1163–1191 (1978).

[65] Tomizawa, J.-I. and G. Selzer. "Initiation of DNA Synthesis in Escherichia coli," *Annu. Rev. Biochem.*, 48, 999–1034 (1979).

[66] Nossal, N. G. "Prokaryotic DNA Replication System," *Ann. Rev. Biochem.*, 53, 581–615 (1983).

[67] Zyskind, J. W., L. T. Deen, and D. W. Smith. "Temporal Sequence of Events During the Initiation Process in *Escherichia coli* DNA Replication: Roles of the dnaA and dnaC Gene Products and RNA Polymerase," *J. Bacteriol.*, 129, 1466–1475 (1977).

[68] Kellenberger-Gujer, A., D. Podhajska, and L. Caro. "A Cold Sensitive dnaA Mutant of *E. coli* which Overinitiates Chromosome Replication at Low Temperature," *Mol. Gen. Genet.*, 162, 9–16 (1978).

[69] Hansen, F. G. and K. V. Rasmussen. "Regulation of the dnaA Product in *Escherichia coli*," *Mol. Gen. Genet.*, 155, 219–225 (1977).

[70] Atlung, T., E. S. Clausen, and F. G. Hansen: "Autoregulation at dnaA-Gene of *Escherichia coli* K-12," *Mol. Gen. Genet.*, 200, 442–450 (1985).

[71] Braun, R. E., K. O'Day, and A. Wright. "Autoregulation of the DNA Replication Gene dnaA in *E. coli* K-12," *Cell*, 40, 159–169 (1985).

[72] Bodholdt, B., B. B. Christensen, J. Engelbrecht, E. Mosekilde and J. Sturis. "Modeling Control of DNA-Replication in Bacterial Cells," in *Computer Based Management of Complex Systems*, P. M. Milling and E. O. K. Zahn (eds.), Springer-Verlag, Berlin (1989), pp. 255–262.

[73] Margalit, H., R. F. Rosenberger, and N. B. Grover. "Initiation of DNA-Replication in Bacteria: Analysis of an Autorepressor Control Model," *J. Theor Biol.*, 111, 183–199 (1984).

[74] Mahaffy, J. M. and J. W. Zyskind. "A Model for the Initiation of Replication in *Escherichia coli*," *J. Theor. Biol.*, 140, 453–477 (1989).

[75] Fuller, R. S., B. E. Funnell, and A. Kornberg. "The dnaA Protein Complex with the *E. coli* Chromosomal Replication Origin (oriC) and other DNA-sites," *Cell*, 38, 889–900 (1984).

[76] Koch, A. L. "Does the Initiation of Chromosome Replication Regulate Cell Division?" *Adv. Microbiol. Physiol.*, 16, 49–98 (1977).

[77] Murray, A. W. and M. W. Kirschner: "What Controls the Cell Cycle," *Sci. Amer.*, March 1991, 34–41.

[78] Mackey, M. C., M. Santavy, and P. Selepova. "A Mitotic Oscillator with a Strange Attractor and Distributions of Cell Cycle Times," in *Nonlinear Oscillations in Biology and Chemistry*, H. Othmer (ed.), Springer-Verlag, Berlin (1986), pp. 34–45.

CHAPTER 8

Chaotic Dynamics of Linguistic-like Processes at the Syntactic and Semantic Levels: In Pursuit of a Multifractal Attractor

JOHN S. NICOLIS AND ANASTASSIS A. KATSIKAS

Department of Electrical Engineering
University of Patras
Patras, Greece

1. Syntactic Level

The relation between symbolic language (be it the genetic code, natural language, or computer language) and the dynamics of thinking, e.g., information processing, has always been—and to great extent remains—enigmatic. Historically, the first scientist to postulated the existence of a bilateral relationship (a functional feedback loop) between language and information processing was the linguist Benjamin Lee Whorf. In the early 1950s, Whorf argued essentially that a poor linguistic repertoire drastically restricts the number of categories into which the user of such a language splits the external word. For several years this conjecture remained a linguistic curiosity; some structural anthropologists (working with aborigines) took it seriously, though, and they found that in a number of cases the Whorfian hypothesis suggesting a relativity of categories apparently holds true. For example, in the aborigine dialect "the Bassa" in Liberia, the seven conventional discrete categories into which most Indo-European languages divide the visible part of the continuous electromagnetic spectrum are replaced by just two categories—so those people can recognize and *name* only

Cooperation and Conflict in General Evolutionary Processes, edited by John L. Casti and Anders Karlqvist.
ISBN 0-471-59487-3 ©1994 John Wiley & Sons, Inc.

two "colors," although physiologically speaking they do not suffer from color blindness [2,3].

Such examples are encouraging in the following sense: They suggest that a dynamical modeling of the language—which abounds in experimental material from cognitive psychology—can reveal even in the worst case the tip of the cerebral iceberg, i.e., the software of biological information processing. What *minimal* tools do we need from nonlinear dynamics, in general, and chaotic dynamics, in particular, to construct the macroparameters characterizing a language at the syntactical level? It's important to understand at the outset that the very need for a *symbolic representation* of a dynamical phenomenon arises from sheer parsimony: namely, from the necessity to classify phenomena in ways both compact and general enough (e.g., universal) so as to allow for argumentation by analogy or for a unified description of whole sets of groups of phenomena similar or isomorphic to the specific one under observation. This immediately brings to mind the time-honored methods of classical statistical mechanics, which starts from a given probability density function at the microscopic level and takes its moments and cross moments, thereby forming (a few) *collective properties* that represent the needed macroparameters or "order parameters" of the phenomenon under consideration.

In quite a number of interesting cases, though, the probability density functions involved are highly inhomogeneous or fractal. This implies either the impossibility of forming just a few representative collective properties or the breakdown altogether of the law of large numbers. Let us examine briefly how such a state of affairs may arise.

Consider a dynamical system whose full description at the "microscopic level" involves the time evolution of a master equation with respect to the pertinent coarse-grained probability density function $P(x;t)$, where x is the N-dimensional (discrete or continuous) variable in state space. The probability of finding the system at the point x in state space at a time t will increase due to the transitions from other points x' and will decrease due to transitions leaving x. So the master equation takes the classical form

$$\frac{dP(x;t)}{dt} = \sum_{x'} \omega(x,x')P(x';t) - P(x;t)\sum_{x'\neq x} \omega(x',x), \qquad (1)$$

where $\omega(x,x')$ is the transitional probability rate for jumping from $x' \rightarrow x$ and $\omega(x',x)$ is the transitional probability rate for jumping from x to x'. The asymptotically stable probability density function $P(x;t)$ (if it exists at all) is in many cases multihumped, so that the mean value, far from being the most probable moment, is the least probable. Note also that in quite a few cases the transition probability rates are nonlinear functions of x. Multiplying both members of Eq. (1) by $x, x^2, \ldots$ and integrating or summing up in the interval of x, we get a set of phenomenological higher-level macroscopic equations with respect to the various moments $\langle x \rangle, \langle \delta x^2 \rangle$ of the probability density function $P(x;t)$: namely,

$$
\begin{aligned}
\frac{d}{dt}\langle x \rangle &= f_1\{\langle x \rangle, \langle \delta x^1 \rangle, \ldots\}, \\
\frac{d}{dt}\langle \delta x^2 \rangle &= f_2\{\langle x \rangle, \langle \delta x^2 \rangle, \ldots\}.
\end{aligned}
\tag{2}
$$

Due to the nonlinear character of the operators f_i (as a result of the nonlinearity of the transitional probabilities), we get an infinite number of coupled nonlinear differential equations with respect to the collective properties of the microscopic dynamics (moments and cross moments). There are instances where those moments may become comparable (e.g., in the vicinity of a bifurcation). Under such circumstances, the descriptions at successive hierarchical levels may get mixed up. In many examples of physical systems, hierarchical systems are "almost decomposable" in the sense that one can study the dynamical behavior at a given level, taking what is just "underneath" as a boundary condition and what takes place "above" as a constant. Such an operation can be carried out whenever the rate constants from level to level differ by several orders of magnitude (lower levels being characterized by larger rate constants or smaller relaxation times).

In symbolic or linguistic systems, however, this decoupling of levels of description cannot be easily carried out, due to the existence of evolutionary feedforward-feedback loops mediating the dynamics between successive levels. Specifically, in linguistic systems one makes use of two hierarchical levels simultaneously, where statements and metastatements are interlocked. We will consider this issue later. That having been said, the (impatient) reader might at this point deduce that the idea of a dynamical theory of symbolic processes is a hopeless one. The time-honored method of forming abstractions by taking averages is not the only way to proceed. In fact, it is the *wrong* approach in the case where *few* details of a pattern, *few* keywords in a linguistic scheme, or *few* variables of a dynamical system may represent the lion's share of the full description of the system. In such cases, taking averages would amount to "washing away" the pertinent and crucial features of the pattern under investigation—thereby blurring it completely. Effective reduction of the number of degrees of freedom of a nonlinear dynamical system can be accomplished in two far more effective ways: First, by recognizing a most remarkable property of phase transition processes, namely, that in the vicinity of a bifurcation point, the dynamics of the system is entirely determined by a very small number of key variables— in terms of which all others can be expressed and, secondly, by pointing out that in the case of dissipative chaos, reduction in the numbers of degrees of freedom of the state space occurs when asymptotically stable regimes (attractors) exist. In such cases, the dynamical system starting from a set of initial conditions of dimension N sooner or later gets "self-imprisoned" in a compact subset of the state space (the attractor), which is characterized by a *much lower dimension.* The aim of this paper is to suggest a minimal complexity chaotic model that hopefully captures or, rather, mimics the salient features of a linguistic scheme, be it natural language, genetic language, or artificial texts, at both the syntactical and the semantic levels.

What are these salient, *sine qua non,* characteristics of a linguistic scheme at the syntactical and the semantic levels? Let us start from the bottom of Figure 1 (although the linguistic levels are connected with a "bootstrap" rather than a strict hierarchical fashion). At the syntactical level, the sole aim of a language is the grammatically admissible formation of texts (sequences of words separated by the pause) with *enough Markovian memory* to allow the recipient to devise a code for detecting and correcting (single or multiple) errors. Another prerequisite is that a linguistically admissible text must display polarity (palindromes should be discouraged). For example, language is read only from left to right, or top to bottom, and so forth. Meaning has no place at the syntactical level.

At the semantic level, on the other hand, meaning should emerge through the unambiguous partitioning of a set of undifferentiated raw (environmental) stimuli impinging on the user's receptors. This partitioning should map onto multiple coexisting categories—memories or "abstracts." For instance, attracting (and compressing) subsets. Also, some mechanism should be invoked for the intermittent jumping of the cognitive processor from attractor to attractor—thereby providing a metalanguage of communication among the coexisting categories-attractors. A common characteristic at both the syntactic and semantic levels is the high selectivity of a language. Let us elaborate on this a bit more.

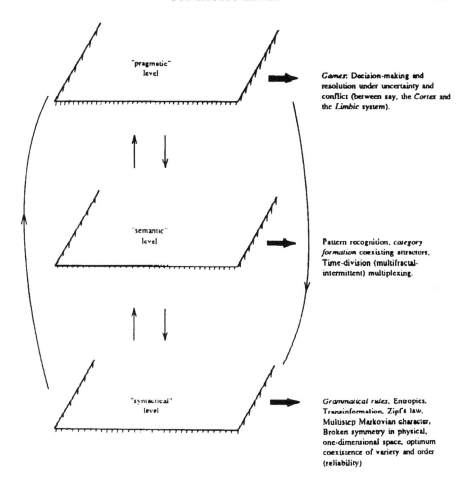

Figure 1. A layout of the linguistic dynamical processes at *three* hierarchical levels: the syntactical, the semantic, and the pragmatic. The exact way these levels are interconnected remains unknown.

When we read a text, say of a natural language possessing Λ symbols (including the pause), we are always witnessing the appearance of an extremely small subset of all *a priori* possible sentences of length ψ, which is Λ^ψ. If it were not so, the utterance, say in English, of a sentence of a modest length of 300 symbols, would be expected with a probability 27^{-300}, a sort of miracle! Likewise, if all polypeptide chains in the genetic language were grammatically admissible, the appearance of a given protein of a modest length, say, for example, ~ 300 amino acids, would be expected with an *a priori* probability 20^{-300}, again close to a miraculous event! What happens in practice is that for some reason—which we intend to justify in our modeling—only a small percentage of all possible combinations Λ^ψ are realizable, hence grammatically admissible. Both natural and genetic texts also display polarity.

At the semantic level, on the other hand, a similar filtering process seems to take place when a (human) processor scans a pattern (see Figure 2). Contrary to the methods employed by artificial intelligence, far from "pixelizing" a pattern and paying equal attention to all possible features, we instead disregard at the very outset the overwhelming majority of all features, so that the processor directs its intermittent attention to a small subset of details—in particular, the "sharp" ones—characterized by small radii of curvature. Finally, there is the matter of *self-consistency,* a characteristic of any linguistic scheme that exists, so to speak, at the interface between the syntactical and the semantic levels.

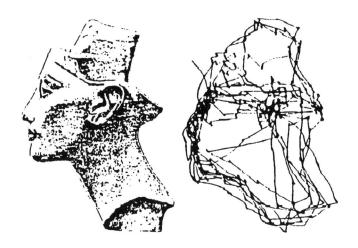

Figure 2. Scanning a bust of Nefertiti.

Let us give an elementary example of how the self-consistency of any self-referential linguistic statement whose proof on solely logical grounds may well be undecidable (see Figures 3 and 4), but which could be dealt with if dynamics enter the problem, e.g., if this sentence is isomorphically mapped onto a digital feedback loop and the asymptotically stable regimes (attractors) of the loop are investigated. Then for those subsets of initial conditions for which attractors exist, the sentence is self-consistent; otherwise, its logical closure remains inconclusive.

To illustrate, consider the sentence: "This sentence has X letters." Take the initial condition X_1 = "one" and verify the truthfulness of the ensuing statement. One gets $F(X_1) = 25$. Proceed to the second iteration, as it were, by taking now X_2 = "twenty-five." You then get $F(X_2) = 32$. Next, you take X_3 = "thirty-two" and you come up with $F(X_3) = 31$. Finally, in the fourth iteration X_4 = "thirty-one" and, lo and behold, you get $F(X_4) = 31$. That is, you have landed on the attractor. There is another *basin of attraction* that is a subset of numbers leading to the second attractor, "thirty-three" (Figure 5), and beyond that there exists, of course, an uncharted set of initial conditions (Xs) for which the consistency of "This sentence has X letters" remains undecided.

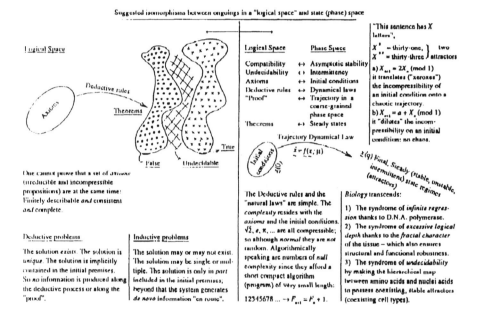

Figure 3. Suggested isomorphisms between "logical space" and state (phase) space.

– The Model –

Now we intend to use the simplest possible chaotic models (e.g., a nonlinear, one-dimensional map $X_{n+1} = F(X_n; \psi)$ with a single control parameter) as the generator of a linguistic text, thereby modeling the ongoings at the syntactical level. The text generated (see below) will be compared via a set of key macroparameters with texts referring to

T can be proved both true and false in the system

INCONSISTENCY

T can be proved neither true nor false in the system

INCOMPLETENESS

Figure 4. Mapping between Gödel's numbers and theorems. Each symbol and logical operation of Gödel's system is mapped onto a prime number. Any statement in this system can thus be represented by the product of prime numbers associated with each of its constituent elements. The motivation is that any whole number can be decomposed *uniquely* into a product of its prime divisors (e.g., $2 \times 3 \times 11 = 66$ is the *only* way to factor 66 into primes).

So any given number corresponds uniquely to a particular logical proposition composed of strings of symbols corresponding to its prime divisors: Every "theorem" simply corresponds to a number—its "Gödel number."

Hence, each *metamathematical* statement has a *unique* Gödel number, so there exists a complete correspondence between arithmetic itself and statements *about* arithmetic. Contrast the sentence: *"The theorem possessing Gödel number X is undecidable."*

We may find the Gödel number of the above proposition and use it as the value of X above. The resulting statement with X substituted into it is then a *theorem stating its own undecidability.* So undecidability is a "syndrome" of all systems that contain arithmetic—since the latter allows for setting up the correspondence with meta-statements via the prime number decomposition. Hence, the deliberation(s) of our very minds keeps part of itself beyond our own grasp.

natural, genetic, and musical language in order to verify or disprove the hypothesis that even simple chaotic systems, balanced precariously between variety (disorder) and reliability (order), can generate impec-

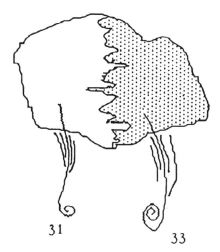

Figure 5. Sketch of the basin of attraction for the attractors "31" and "33" for the consistency of the sentence "This sentence has X letters."

cable grammatical texts, i.e., sequences of symbols that respect, as we said before, the *sine qua non* salient features of any conceivable language.

Our dynamical model—the generator of the text—will be essentially the logistic map $x_{n+1} = rx_n(1 - x_n)$, both in the fully chaotic regime $(r = 4)$ and at the first accumulation point $(r \sim 3.57\ldots)$, where the first multifractal (Feigenbaum) attractor shows up. In the first case $(r = 4)$, the invariant measure is a smooth continuous (hyperbolic) function having only two singular points at $x = 0$ and $x = 1$. It supports the attractor in the "solid" interval from 0 to 1.

In the second case, the invariant measure is essentially a highly inhomogeneous set of δ-functions, and it supports the attractor in an asymmetric Cantor set with fractal dimension $D_0 = 0.538$. This attractor, however, is highly inhomogeneous (multifractal), so a whole spectrum of dimensions is necessary in order to characterize it. In this case, our interest centers on the highly inhomogeneous shape of the invariant measure.

If we impose a finite partition of subintervals ("symbols") on the attractor, few symbols and, consequently, few words (strings of symbols between two pauses) will claim importance from the point of view of *a priori* probabilities in a long text that we are going to generate by iterating the map, thereby creating a succession of strings of symbols (words) or a long artificial text. We hope that by comparing the macroparameters (see below) for the statistics of such texts in the two cases ($r = 4, r = 3.57\ldots$), we will be able to discriminate among generators creating languages with fewer key words ($r = 3.57\ldots$) versus more homogeneous ones ($r = 4$). An intermediate case involves values of the control parameters around "band merging and mixing," such as the value $r = 3.7$, where the invariant measure is still very "spiky" but nevertheless occupies a "solid" part of the unit interval.

– Why Dissipative Chaos? –

Suppose you release a small animal (e.g., a kitten) on the floor. Even if it is satiated, even if it does not look after a partner, even if it does not run away from an enemy, the healthy animal is still going to happily occupy itself in a ceaseless game-like exploratory activity. What is the survival value, if any, underlying such seemingly purposeless behavior? Very simply, the animal is "disturbing the universe" by its positive Lyapunov exponents! But let's get serious. *Any* act of information processing involves two separate phases: an expanding phase and a contracting phase, which may be executed by the organism involved either in succession or (usually) in unison by means of a recursive feedback loop.

1. During the *expanding phase,* the organism-processor via its motor activity makes a subset of alternatives manifest themselves both in number and *a priori* probabilities by disturbing the environment.

2. During the *contracting phase,* the processor, via its sensors, contracts (compresses) the basin of attraction created earlier onto one (out of many) coexisting attractor-categories.

It is imperative then for a biological processor to possess coexisting chaotic strange attractors in order to comply with both of these requirements (steady states and limit cycles provide only for the contracting phase and are useful only as classifiers provided that the processor has already been given, or has created, the set of alternative responses, i.e., reactions to "raw stimuli" from the environment).

Most people tend to take *a priori* probabilities for granted—literally as given *a priori*. Perhaps few realize that behind an *a priori* probability lies much action, much trial and error, and most important, a convergent process. The very definition of *a priori* probability presumes that $P_i = \lim_{N \to \infty} N_i/N$, where N is the total number of trials (stimuli) and N_i is the number of trials under scrutiny. It is not enough just to allow $N \to \infty$. The limit must exist. This requirement of convergence—which lends dynamical overtones to the definition of probability—is more clearly manifested in the way a Markov chain of N states and a matrix of transition probabilities P_{ij} converges to its *a priori* steady states u_i. This convergence takes place via a cascade of iterations given by the linear map

$$u_{i_{t+1}} = \sum_{j=1}^{N} P_{ji} u_{j_t}. \tag{3}$$

The necessary and sufficient condition of convergence to steady-state attractors u_i is that there must be an integer ψ such that the matrix $[P^{\psi}]_{ij}$ possesses at least one column with all elements different from 0. Let us provide a counterexample, consisting of a system for which the definition of an *a priori* probability does not hold.

Think of an elevator moving between two floors, call them 1 and 2. Someone approaches the nontransparent door of floor 1. Does he have the right to claim that the elevator will be on either floor with equal probability? Not at all! Indeed, the frequency of the elevator visiting the two floors is given by the expression $\sum_i = [1 + (-1)^i]/2$, $i = 0, 1, 2, \ldots$. This sum clearly does *not* converge but oscillates between 0 and 1. As our observer approaches the nontransparent door, he has to consult a counter. If, for example, the number is even, he pushes the button, while if the number is odd, he can confidently push the door. But in order to do anything he has to get some auxiliary information, so there is no *a priori* probability to think about.

We have considered this simple example just to show that a chaotic strange attractor provides, among other things, the *deus ex machina* that via its positive Lyapunov exponents creates *variety* or *entropy* by *amplifying* initial uncertainties in directions in state space and provides order (information) by *constraining* initial uncertainties along directions characterized by the negative Lyapunov exponents. More generally, for linguistic processes we can claim that the subtle interplay between unpredictability (variety) and reliability (order) that underlies any linguistic scheme is typical of the behavior of a class of low-dimensional nonlinear dissipative dynamical systems (models):

$$\dot{x} = F(x; \mu), \quad \nabla \cdot F < 0. \tag{4}$$

In these systems, the ensuing unpredictability, due essentially to the nonlinear character of the underlying process, is manifested either

1. Through a sensitive dependence on control parameter(s) μ, giving rise to instabilities, bifurcations, broken symmetries, and multiplicity of solutions (scenarios) beyond some instability point μ_c.

2. Through a sensitive dependence on initial conditions $x(0)$, giving the local (uniform or nonuniform multifractal) exponential separation of nearby trajectories, or the chaotic behavior of the variables involved.

In compensation, there exist *three* mechanisms for moderating this unpredictability and establishing some order. Namely,

1. The drastic reduction of the number of degrees of freedom in the vicinity of a bifurcation and the emergence of essentially only a few dominant "order parameters" in a reduced state-space description (the "center manifold"). These order parameters may subsequently interact in a nonlinear fashion, giving rise to low-dimensional dissipative chaos.

2. The existence of (multiple) attractors possessing invariant measures in the dynamical system governed by the interplay among the order parameters. (Further reduction of the number of pertinent variables may take place in the case of homogeneous attractors whose invariant measures are highly irregular as will be seen later).

3. A (possible) renormalization or fractal scale-invariant syndrome in the evolution of the system, as a result of which a study of the statistics of the time series involved in a restricted window of the variable and the parameters may give results which hold invariant for many other windows or scales of variables and parameters. In short, information is produced not only via cascading bifurcations giving rise to broken symmetry, but also via successive iterations giving rise to ever-increasing resolution.

Beyond a certain resolution interval, ever-present microscopic fluctuations are no longer smeared out but get amplified, passing from lower to higher levels. It is these fluctuations that essentially account for new information generated by the evolving nonlinear dissipative system.

For an attractor simulating the dual aspect of a cognitive system (or a "processor"), there exist *two* basic requirements: large dynamical storage capacity *and* good compressibility. Stable steady states and stable limit cycles of information dimension zero and one, respectively, are very poor as *dynamical* information storage units (they cannot improvise from within, as it were; their algorithm is too inflexible, with too few degrees of freedom. Nevertheless, they are ideal as information compressing gadgets or "photographic" categorizers.

Strange chaotic attractors, on the other hand, via a harmonious combination of expanding and converging trajectories in state space, e.g., by possessing positive (λ_+) and negative (λ_-) Lyapunov exponents can, in principle, satisfy both requirements. They may possess a considerable information dimension, as well as higher-order dimensions, making them suitable for dynamical and selective information storage, while being "attractors" $(\sum |\lambda_-| > \sum \lambda_+)$, they serve also as information compressors.

– Why a Multifractal Attractor? –

A classical strange attractor in one dimension *is* a self-similar Cantor set, although not necessarily a symmetrical one (see below). While such a set is characterized by a *single* scaling exponent, the fractal or Hausdorff dimension D_0, *multifractals* are described by two sets of scaling exponents, one for the supporting fractal "dust" on the interval and one for the probabilities of the associated invariant measure. Let us elaborate on this point.

We take the unit interval from which a number of open subintervals are removed, leaving nonoverlapping or disjoint line segments of length r_i. To each line segment r_i, we assign a weighting factor or probability P_i. Now we recursively iterate [or rather allow the dynamical system $\dot{x} = F(x)$ to do it] this process of removal and redistribution of probabilities in a scale-invariant fashion until the transients die out. We then end up with a generalized Cantor set having probabilities associated with individual *points* of the remaining dust. The set of these final probabilities stands for the frequency with which the underlying dynamical system visits the different points of the attractor—a Cantor dust—and constitutes the invariant measure of the attractor. The whole edifice then forms a multifractal attractor on a fractal support.

Let us now see how we can properly introduce the *interplay* between the *two* scaling exponents mentioned above. Following [11] and [12], we start with the definition of the Hausdorff dimension of a set. It is given by the limit as $r \to 0$ of the expression $-\log N / \log r$, where N is the minimal number of pieces of diameter r that completely cover the set, i.e.,

$$D \sim \lim_{r \to 0} \frac{\log N}{\log \frac{1}{r}}, \tag{5}$$

or

$$\lim_{r \to 0} Nr^D = \text{const.} \tag{6}$$

Thus, D is the exponent that keeps the product Nr^D finite and non-zero as the resolution increases, i.e., as $r \to 0$. Consider now N_G, the *initial* number of equal-length pieces of the recursive generator on the interval and r_G, the common length of the segments of that generator. After ψ iterations, these numbers become N_G^ψ and r_G^ψ, respectively, which means that instead of the above relation, we can now write

$$\lim_{\psi \to \infty} (N_G r_G^D)^\psi = C^{te}, \tag{7}$$

or

$$N_G \, r_G^D = 1. \tag{8}$$

Hence,

$$D = \frac{\log N_G}{\log(1/r_G)}. \tag{9}$$

This means that for a strictly self-similar process, taking the limit is not really necessary. For generating segments of different lengths r_i, we will now have

$$\lim_{\psi \to \infty} \left(\sum_{i=1}^{N} r_i^D \right)^\psi = c^{te}, \tag{10}$$

or

$$\sum_{i=1}^{N} r_i^D = 1. \tag{11}$$

Thus far we have been considering a nonsymmetric Cantor set. For a dynamic that creates a multifractal process with line segments r_i and visiting probabilities P_i, we now introduce *two* exponents: τ for the supporting intervals r_i, and q for the probabilities P_i. So we now have to consider the expression $\lim_{\psi \to \infty} \left(\sum_{i=1}^{N} P_i^q r_i^\tau \right)^\psi$ and look for values of q and τ for which this expression remains finite and, in particular,

$$\sum_{i=1}^{N} P_i^q r_i^\tau = 1. \tag{12}$$

For $q = 0$ (a simple fractal set), τ of course corresponds to D. For an easy way of discovering a relationship between q and τ, let us try

the classical symmetrical Cantor set ($N = 2, r_i = \frac{1}{3}$) and take $P_i = \frac{1}{2}$.
Equation (12) then becomes

$$2^{-q}3^{-\tau} + 2^{-q}3^{-\tau} = 1, \tag{13}$$

or

$$\tau = (1 - q)\frac{\log 2}{\log 3}, \tag{14}$$

or

$$\tau = (1 - q)D, \tag{15}$$

(since $D = \log 2/\log 3$ is the Hausdorff dimension of the set). From
Eq. (15) we see that for $q = 0, \tau$ indeed corresponds to D—but it *may*
differ from the Hausdorff dimension for $q \neq 0$. We may then envision a
whole hierarchy of generalized dimensions D_q, $q = 0, \pm1, \pm2, \ldots, \pm\infty$,
which as we are going to see immediately, provide a measure of the
inhomogeneity of the distribution of points on the attractor. This inhomogeneity can in turn be related to the existence of a *spectrum of*
singularities of the invariant measure itself.

We divide the attractor into $n(r)$ boxes of linear dimension r and
denote by $P_i(r)$ the probability that the trajectory on the strange attractor visits the box i, i.e.,

$$P_i(r) = \lim_{N \to \infty} N_i/N, \tag{16}$$

where N is the total number of points on the attractor and N_i is the
number of points in the ith cell. Consider now the $\sum_i^{n(r)} P_i^q$ for $q > 1$.
It stands for the total probability that q points of the attractor are
within one box. The generalized dimension D_q, which is related to this
qth power of P_i, is defined to be

$$D_q = \lim_{r \to 0} \frac{1}{q - 1} \frac{\log \left[\sum_{i=1}^{n(r)} P_i^q(r)\right]}{\log r}. \tag{17}$$

This measures *correlations* among different points on the attractor,
thereby characterizing the attractor's degree of inhomogeneity or its
"multifractality"—that is, the degree with which the attractor prefers
to visit certain areas (or subsets) in state space at the expense of others
that are visited less frequently. For $q = 0$, we obtain from Eq. (17)

$$D_0 = \lim_{r \to 0} \frac{\log \sum_{i=1}^{n(r)} 1}{\log \tau} = \lim_{r \to 0} \frac{\log n(r)}{\log \tau}, \tag{18}$$

which is just the Hausdorff dimension of the attractor. An attractor characterized only by its fractal dimension D_0 then appears perfectly homogeneous, since it visits all boxes quite democratically. After applying L'Hospital's rule when $q = 1$, Eq. (17) gives

$$D_1 = \lim_{r \to 0} \frac{\sum_{i=1}^{n(r)} P_i(r) \log_2 P_i(r)}{\log_2 r}, \tag{19}$$

the so-called information dimension, since $- \sum_{i=0}^{n(r)} P_i(r) \log P_i(r)$ is the information gained if we know $P_i(r)$ and learn that the trajectory is in a specific cell.

Equation (19) tells us how this information scales as $r \to 0$. Obviously, $D_1 \leq D_0$ and $D_{\max} = D_0$. For $q = 2$, Eq. (17) yields the (two-point or dyadic) correlation dimension

$$D_2 = \lim_{r \to 0} \frac{\log \sum_{i=1}^{n(r)} P_i^2(r)}{\log_2 r}. \tag{20}$$

We can see easily that $D_2 < D_0$. Note also that for a self-similar fractal with equal probabilities $P_i = 1/n(r)$, Eq. 17 always yields

$$D_q = \frac{1}{q-1} \frac{\log[n(1/n)^q]}{\log r} = \frac{\log n(r)}{\log \left(\frac{1}{r}\right)} = D_0, \tag{21}$$

as expected. The (coarse-grained) probabilities $P_i(r)$ entering Eq. 17 are related to the invariant density measure $\bar{\mu}(x)$ of the attractor via

$$P_i(r) = \int_{|\underline{x}_i - \underline{x}| \leq r} \bar{\mu}(\underline{x}) \, d\underline{x}, \tag{22}$$

where $\underline{x}_i$ denotes the center of the box i; $\bar{\mu}(\underline{x})$ is seldom a smooth function of x. On a fractal support materialized by a "Cantor dust" $\bar{\mu}(\underline{x})$ is, strictly speaking, a highly inhomogeneous set of δ-functions, which for practical reasons may in some cases be smoothed out. Even for the logistic map $x_{n+1} = \epsilon x_n(1 - x_n)$ in the fully chaotic regime ($\epsilon = 4$), where $\bar{\mu}(\underline{x})$ is continuous and the support of the attractor is the "solid" interval from 0 to 1, the invariant measure possesses *two* hyperbolic singularities: at $x = 0$ and $x = 1$. If $P_i(r)$ *diverges* as $r \to 0$ $[P_i(r \to 0) \sim r^{a_i}]$, the invariant measure at $\underline{x}_i$ has a singularity whose strength is characterized by the exponent a_i (in the hyperbolic case $a_i = -1/2$). Since different points $\underline{x}_i$ on the attractor can have different strengths a_i, it is useful to introduce the function $f(a)$, which measures the Hausdorff dimensionality of the subset of points x_i on the attractor that have the *same* exponent a. The quantity $f(a)$ then characterizes the static distribution of points on the attractor having singularities of strength a.

This concept of a *distribution* $f(a)$ of fractal dimensions associated with subsets of singularities of strength a of the invariant measure begs the question: What is the relationship between D_q and $f(a)$? We are asking this question since both the functions $f(a)$ and $r(q) = (1 - q)D$ describe essentially the *same* aspects of a multifractal. The answer (the proof is irrelevant for the purpose of this paper) is that $f(a)$ and D_q are connected via a negative Legendre transformation. To uncover an explicit relationship for D_q as a function of r_i and p_i, we again take up the expression $\lim_{\xi \to \infty} \left(\sum_{i=1}^{N} P_i^q r_i^\tau \right)^\xi$ and introduce N constant bin sizes $r_i = r$, which for $\xi \to \infty$ does not affect the values of r and q for which the above limit converges. Thus we obtain

$$r = r(q) = - \lim_{r \to \infty} \frac{\log \left[\sum_{i=1}^{N} P_i^q \right]}{\log r}, \tag{23}$$

and, with Eq. (17), we get

$$r(q) - (1 - q)D_q. \tag{24}$$

For a self-similar fractal, the generalized dimensions D_q are then obtained directly from the P_i and the r_i. Using $\sum_{i=1}^{N} P_i^q r_i^\tau$ and Eq. (24), we obtain

$$\sum_{i=1}^{N} P_i^q r_i^{(1-q)D_q} = 1. \tag{25}$$

Now we provide a nontrivial application. Using Eq. 25, we will give an *analytical* calculation $D_q = f(q)$ for the case of a "cardinal" multifractal attractor—which will be used later as a linguistic generator. The multifractal attractor is the one associated with the "notorious" logistic map $x_{n+1} = \epsilon x_n (1 - x_n)$ at the threshold of chaos, namely, at the first accumulation point $\epsilon_\infty = 3.5699456 \ldots$. This is where the first unstable periodic trajectory appears, so that the attractor is no longer a stable limit cycle of period 2^ξ (as for $\epsilon < \epsilon_\infty$) but an uncountable set of points forming a Cantor set.

Let us have a look at the bifurcation diagram (x versus ϵ) of the logistic map (see Figure 6). For each value of ϵ, the figure shows the set

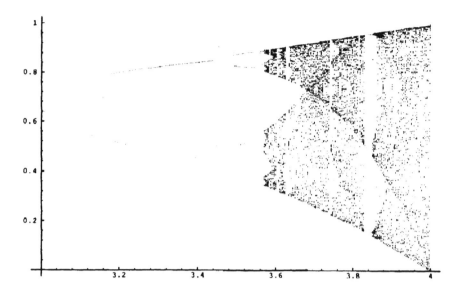

Figure 6. The bifurcation diagram of the logistic map.

of points in the unit interval that are the corresponding attractor. To be sure, this is just the *support* of the attractor. What is missing from the picture is the associated invariant measure, which should perhaps be displayed on a vertical plane perpendicular to the page and parallel to the x-axis.

Here we are particularly interested in the topology of the support of the Feigenbaum attractor at the first accumulation point ϵ_∞. Figure 6 gives some of the intermediate attracting periodic sets along the cascade of period-doubling bifurcations, which in the limit give rise to the strange attractor. What is the recurrent or iterative philosophy behind the formation of this attractor? What are the basic segments r_i from which this process gets started? To put it succinctly, what are the *scaling laws* underlying this period-doubling cascade?

Between the two Feigenbaum constants we are interested in the one associated with the scaling of the variable x along the cascade. Specifically, there is a scaling coefficient $\mu = 2.50290787\ldots$ that gives (see Figure 6) the "density-packing index" of the attracting points x^* between two successive generations of bifurcations. This means that the distance in the unit interval along the x-axis between the bifurcating point $x_0 = 0.5$ is μ times smaller than the distance between its "parent" and the point 0.5. The distances d_n of the points in a 2^n-cycle that are closest to $x = 0.5$ have various ratios d_n/d_{n+1}, which tend to μ as

$n \to \infty$. So the set is not exactly self-similar. Furthermore, its two halves are asymmetric. The left half of the bottom line is the line above it, compressed by a factor $\mu \sim 2.5$, while the right half of the bottom line is the line above it, compressed by a factor $\mu^2 = 2.5^2$. What then is the procedure for calculating the spectrum of the generalized dimensions D_q of the Feigenbaum attractor? Again we use the master equation, Eq. (25).

In this example we start with $N = 2$ main pieces, r_1 and r_2, both with the same initial weight (probability) $p_1 = p_2 = \frac{1}{2}$, where $r_1 \sim \frac{1}{\mu}$ and $r_2 \sim \frac{1}{\mu^2}$ (i.e., $r_1 \sim 0.408$ and $r_2 = r_1^2$). Then for $q = 0$ we obtain

$$r_1^{D_0} + r_2^{D_0} = 1. \tag{26}$$

From Eq. (26) we find that

$$r_1^{D_0} = \frac{\sqrt{5} - 1}{2} = 0.618. \tag{27}$$

Thus

$$D_0 \cong \frac{\log 0.618}{\log 0.408} \sim 0.537. \tag{28}$$

This is the Hausdorff dimension of the Feigenbaum attractor.

Let us calculate the information dimension D_1 for $q = 1$. This will be calculated from the expression $r(q) = (1 - q)D_q$, namely,

$$D_1 = -\left.\frac{d\tau}{dq}\right|_{q=1} \tag{29}$$

More precisely, the identity

$$\sum_{t=1}^{N} P_i r_i^\tau = 1 \tag{30}$$

leads to

$$D_1 = \frac{\sum_{i=1}^{N} P_i \log p_i}{\sum_{i=1}^{N} P_i \log r_i}. \tag{31}$$

For $P_i = 1/N$,

$$D_1 = \frac{N \log N}{\sum_{i=1}^{N} \log \frac{1}{r_i}}. \tag{32}$$

In our particular case, $N = 2, r_i \cong 0.408, r_2 = r_1^2$, so $D_1 = 0.515$. For $q \neq 0$, $q \neq 1$ so we come to the general expressions for the generalized dimension(s) of the Feigenbaum attractor as a function of q. From Eq. 25 and $p_1 = p_2 = 1/2$, we obtain

$$r_1^\tau + r_1^{2\tau} = 2^q, \tag{33}$$

which gives

$$D_q = \frac{\log\left[\frac{1}{2}\left(\sqrt{1 + 2^{q+2}} - 1\right)\right]}{(1 - q)\log r_1}. \tag{34}$$

The plot of D_q versus q is shown in Figure 7, where q varies from $-\infty$ to $+\infty$. The corresponding values of D_q range from

$$D_{-\infty} = \frac{\log 2}{\log \frac{1}{r_1}} \tag{35}$$

down to

$$D_{+\infty} = \frac{\log 2}{\log \frac{1}{r_2}}. \tag{36}$$

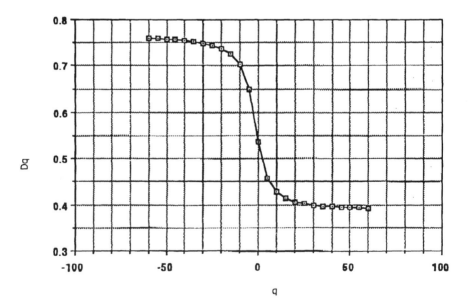

Figure 7. Plot of D_q vs. q.

In the general case, as $q \to +\infty$ only the *highest* probability p_{max} in the sum of the expressions in Eq. (17) counts. On the other hand, for $q \to -\infty$ the smallest probability p_{min} dominates the sum. Let us recall that the sites with low probability on the attractor are visited infrequently. The trajectory spends most of its time in high probability subsets. It is expected that this high selectivity characterizing the topology of strange attractors, in general, and multifractal attractors, in particular, will somehow be manifested in the characteristic macroparameters of the artificial texts to be generated, if such attractors are used as primitive linguistic generators. Comparison of these macroparameters (to be specified in subsequent sections) with the corresponding macroparameters from natural, genetic, and even musical "texts," shows to what extent simple low-dimensional hardware can mimic the salient features of the complex dynamic edifice we call language at both the syntactical and the semantic levels.

– The Invariant Measure of Maps and Flows –

The invariant measure $\overline{\mu}(x)$ determining the density of the iterates of a (unimodular) map $x_{n+1} = F(x_n)$ over the unit interval is formally defined to be

$$\overline{\mu}(x) = \lim_{n \to \infty} \frac{1}{n} \sum_{i=0}^{n} \delta\{x - F^i(x_0)\}. \tag{37}$$

The system is called *ergodic* if $\overline{\mu}(x)$ does not depend on x_0. In this case, Eq. (37) allows us to write time averages over any function as averages over the invariant measure.

The evolution equation for the invariant measure in our simple one-dimensional model can be defined as follows: An initial point x_0 evolves to $F(x_0)$ after one iteration. This means that a δ-function distribution $\delta(x - x_0)$ evolves after one time step to $\delta\{x - F(x_0)\}$, which can in turn be written as

$$\delta\{x - F(x_0)\} = \int_0^1 \delta\{x - F(y)\}\delta\{y - x_0\} \, dy. \tag{38}$$

Generalizing this to the evolution of an arbitrary density $\mu_n(x)$ at time n, we obtain the Frobenius-Perron equation

$$\mu_{n+1}(x) = \int_0^1 \mu_n(y)\delta[x - F(y)] \, dy. \tag{39}$$

Equation (39) governs the time evolution of $\mu_n(x)$. The invariant measure is then defined to be $\bar{\mu}(x) = \lim_{n\to\infty} \mu_n(x)$, or

$$\bar{\mu}(x) = \int_0^1 \bar{\mu}(y)\delta[x - F(y)]\,dy. \tag{40}$$

When the map concerned is in the periodic regime (before the first accumulation point r_∞), the invariant measure is just a set of δ-functions. As we will see, at the first accumulation point r_∞ the invariant measure becomes fractal (continuous, but nowhere differentiable). In the fully chaotic regime ($r = 4$ for the logistic map), the invariant measure is everywhere smooth (see Figure 8).

The simplest $\bar{\mu}(x)$ is associated with the symmetrical triangular "tent map" given by

$$F(x) = \begin{cases} 2x & \text{for } x \leq 1/2, \\ 2(1 - x) & \text{for } x > 1/2. \end{cases} \tag{41}$$

Then Eq. 40 becomes

$$\bar{\mu}(x) + \frac{1}{2}\left[\bar{\mu}\left(\frac{x}{2}\right) + \bar{\mu}\left(1 - \frac{x}{2}\right)\right], \tag{42}$$

which has the obvious normalized solution $\bar{\mu}(x) = 1$. This solution is unique as we can immediately see by starting from an arbitrary initial condition $\bar{\mu}_0(x)$ and iterating n times with Eq. (39). This yields

$$\mu_n(x) = \frac{1}{2^n}\sum_{j=1}^{2^{n-1}}\left[\mu_0\left(\frac{j-1}{2^{n-1}} + \frac{x}{2^n}\right) + \mu_0\left(\frac{j}{2^{n-1}} - \frac{x}{2^n}\right)\right], \tag{43}$$

which converges to

$$\bar{\mu}(x) = \lim_{n\to\infty}\mu_n(x) = \frac{1}{2}\left[\int_0^1 \mu_0(x)\,dx + \int_0^1 \mu_0(x)\,dx\right] = 1. \tag{44}$$

From this result, we can immediately deduce the invariant measure of the logistic map $x_{n+1} = 4x_n(1 - x_n)$ in the fully chaotic regime ($r = 4$). By making the change of variables $x_n = \frac{1}{2}[1 - \cos(2\pi y_n)]$ and substituting, we get

$$\frac{1}{2}[1 - \cos(2\pi y_{n+1})] = [1 - \cos(2\pi y_n)][1 + \cos(2\pi y_n)]$$
$$= \frac{1}{2}[1 - \cos(4\pi y_n)], \tag{45}$$

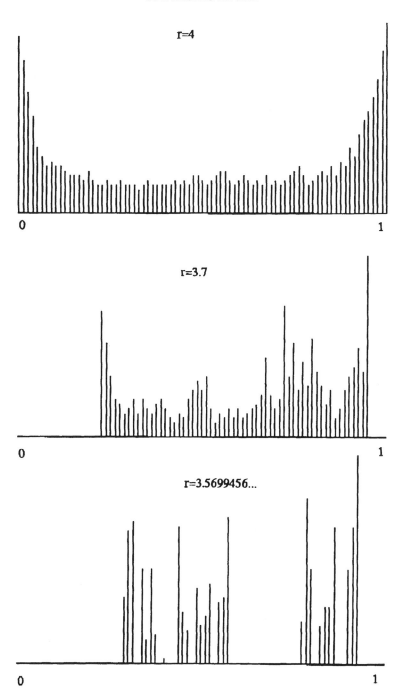

Figure 8. The invariant measure for the logistic map for r_∞, $r = 3.7$, and $r = 4$.

from which we get $y_{n+1} = 2y_n \pmod 1$ or $y_n = 2^n y_0 \pmod 1$. This implies for the logistic map,

$$x_n = \frac{1}{2}[1 - \cos(2\pi 2^n y_0)], \tag{46}$$

where $y_0 = \frac{1}{2\pi} \arccos(1-2x_0)$. Since x_0 taken at random is an irrational number with probability 1, the sequence $\{x_n\}$ gives rise to an aperiodic (chaotic) series.

The invariant measure can now be calculated as

$$\begin{aligned}
\bar{\mu}(x) &= \lim_{n\to\infty} \frac{1}{n} \sum_{n=0}^{n-1} \delta(x - x_n) \\
&= \lim_{n\to\infty} \frac{1}{n} \sum_{n=1}^{n-1} \delta x - \left[\frac{1}{2}[1 - \cos(2\pi y_n)]\right].
\end{aligned} \tag{47}$$

Since $\mu(y) = 1$, we find that

$$\bar{\mu}(x) = \int_0^1 \bar{\mu}(y)\delta\left\{x - \frac{1}{2}[1 - \cos(2\pi y_n)]\right\} dy \tag{48}$$

or

$$\bar{\mu}(x) = \frac{2}{\frac{d}{dx}\left\{\frac{1}{2}[1 - \cos(2\pi y)]\right\}} = \frac{1}{\pi}\frac{1}{\sqrt{x(1-x)}}, \tag{49}$$

i.e. a (smooth) hyperbolic probability density function with two singularities at $x = 0$ and $x = 1$. The information produced by the logistic map in this fully chaotic regime is then

$$I = \frac{1}{\pi} \int_0^1 \frac{\log[4(1-2x)]}{\sqrt{x(1-x)}} dx = 1 \text{ bit/iteration.} \tag{50}$$

– Chaotic Dynamics and Markov Partitions –

At this point, the reader should perhaps be reminded that the reason for reviewing the above properties of simple chaotic maps is the hope of being able to use these kinds of dynamics for the construction of primitive linguistic schemes displaying a mixture of order (simple grammatical rules) and variety (unpredictability and information). To this end, it is now natural to impose on the unit interval of a simple unimodal map (e.g., the logistic map) an arbitrary partition of subintervals—each standing for a symbol. (One of those symbols may be taken as the blank space used to separate strings of symbols into words.)

However, some caution must be exercised. In order to construct an alphabet that remains invariant along the flow (or the cascade of iterations producing the text), care should be taken that the chosen partition be *Markovian:* namely, that the partition be on nonoverlapping subintervals. This means that the transition points defining the boundaries between adjacent subintervals should remain invariant under the dynamics. So we are seeking a master equation whose time evolution describes the transient trajectory of a *coarse-grained* initial measure $P_0(x)$ towards (hopefully) an asymptotically stable regime that will be identified as the *coarse-grained* invariant measure $\bar{\mu}(x)$. Here, however, we run into a small difficulty: So far, the only equation we have describing the evolution of an arbitrary $P_0(x)$ towards the invariant measure $\bar{\mu}(x)$ is the *fine-grained* one—namely, the Frobenius-Perron equation. But the Frobenius-Perron equation fails to define a physically meaningful stochastic process on account of the singular character of the pertinent transitions probabilities $\delta\{x - F(y)\}$. G. Nicolis and C. Nicolis have recently shown [5] that coarse graining can lead to an exact image of dynamics in the form of a regular Markov process obeying a master equation.

In order to construct the grammatical rules governing the (essentially stochastic, that is, chaotic) transition among the members of our alphabet, we proceed as follows: Let $p_n(x)$ be the fine-grained probability density function on the interval at some transient time n. Now define properly the states—"letters"—of the underlying process by introducing the probability vector

$$P_n = \left\{ \int_{M_1} p_n(x)\, dx, \dots, \int_{M_\Lambda} p_n(x)\, dx \right\}, \tag{51}$$

where Λ is the number of subintervals on the unit interval along x_n. Assuming a "coarse-grained" initial condition $p_0(x)$ taking a *constant* value ("letter") in each of the cells of the partition, we may deduce [5] from the (fine-grained) Frobenius-Perron equation a (coarse-grained) master equation

$$P_n = W P_{n-1}. \tag{52}$$

The transition probability matrix W_{ij} gives the probability of jumping in one step (iteration) from the element B_i of the partition to the

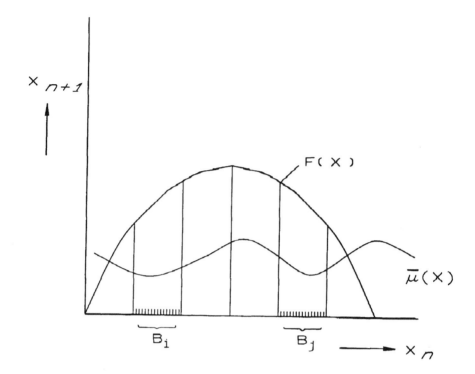

Figure 9. Partitioning the unit interval into Λ subintervals ("letters") B_i.

element B_j. The quantity $\overline{\mu}(x) = \lim_{n \to \infty} P_n$ now gives the coarse-grained invariant measure. In general, and without making further appeal to Eq. (52), we may calculate the $\Lambda \times \Lambda$ elements of the transition matrix $W^\xi(B_j|B_i)$ of moving from element B_i to element B_j in ξ steps $\xi = 1, 2, \ldots$ as follows (see Figure 9): We subdivide each interval B_i into say 100 points, and perform $\approx 10{,}000$ iterations on each of them to make sure that the transients have died out. The 100 points will by then have spread to other subintervals. A fraction of them will be located within the partition B_j. After the transients have died out, the common area of $F^\xi(B_i)$ and B_j expresses the number of elements of B_i reaching B_j after ξ iterations. So, in normalized form,

$$W^\xi(B_j|B_i) = \frac{\overline{\mu}\{F^\xi(B_i) \cap B_j\}}{\overline{\mu}\{B_j\}}, \tag{53}$$

or, what is equivalent for the case of invertible maps,

$$W^\xi(B_j|B_i) = \frac{\overline{\mu}\{F^{-\xi}(B_j) \cap B_i\}}{\overline{\mu}\{B_i\}}. \tag{54}$$

Here $\mu\{\cdot\} = \int_c \bar{\mu}(x)\,dx$, where c is the pertinent interval.

The entropy of the chosen partition, or the average amount of information needed to locate the system in state space (the so-called "monitor parameter"), is now given by the Shannon entropy (the entropy of the transmitter)

$$S = -\sum_{i=1}^{\Lambda} \bar{\mu}(B_i) \log_2 \bar{\mu}(B_i) \text{ bits},\tag{55}$$

where $\bar{\mu}(B_i) = \int_{B_i} \bar{\mu}(x)\,dx$. The Λ values may be calculated from the W_{ij} elements from the $(\Lambda - 1)$ equations of the linear system

$$\bar{\mu}(B_i) = \sum_{j=1}^{\Lambda} \bar{\mu}(B_j) W_{ji}\tag{56}$$

and the normalization condition

$$\sum_{i=1}^{\Lambda} \bar{\mu}(B_i) = 1.\tag{57}$$

The average amount of information created by the linguistic system per transition per unit time (the so-called "predictive factor") is given by the Kolmogorov-Sinai entropy for the chosen partition, namely,

$$S_K = \sum_{i=1}^{\Lambda}\sum_{j=1}^{\Lambda} \bar{\mu}(B_i) W_{ij} \log_2 W_{ij} \text{ bits}.\tag{58}$$

However, the macroparameter characterizing the degree of "grammatical coherence" of the created Markovian chain is the "mutual information" or "transinformation"

$$I(\xi) = \sum_{i=1}^{\Lambda}\sum_{j=1}^{\Lambda} \bar{\mu}\{F^{\xi}(B_i) \cap B_j\} \log_2 \frac{\bar{\mu}\{F^{\xi}(B_i) \cap B_j\}}{\bar{\mu}\{F^{\xi}(B_i)\}\bar{\mu}\{B_j\}} \text{ bits}.\tag{59}$$

This stands for the information stored in a symbol of the sequence about what is going to emerge ξ iterations (or ξ time units) later. The quantity $I(\xi)$ gives the information transferred between two steps ξ time units apart. The quantity $I(\xi)$ may decrease monotonically or it may not, as will be seen in our examples. In the latter case, we may infer the existence of *more than one time scale* in the sequence of symbols, and thereby see some sort of "regeneration" or "persistence" of memory.

Now Eq. (59) can be deduced straightforwardly from the Shannon expression of information transfer $I(x-y)$ between a transmitting set X of Λ messages x_i with *a priori* probabilities $p(x_i)$ and a set Y of Λ *replicas* y_i at the receiving end, separated by a noisy channel with a $\Lambda \times \Lambda$ error matrix $P(x_i/y_j)$. We can express this mathematically as

$$I(x \to y) = \sum_{i=1}^{\Lambda} \sum_{j=1}^{\Lambda} P(x_i, y_j) \log_2 \frac{P(x_i, y_j)}{P(x_i)P(y_j)} \text{ bits.} \qquad (60)$$

The transformation from Eq. (60) to Eq. (59) can be calculated if we assume that the transmitter is a dynamical system proceeding in time under the action of a map $P(x_i|y_i)$ and y_i is simply the (uncertain) value of x_i, η time units later.

Another important parameter is the sequence entropy S_L, which contains information about the richness of the language. It is given by

$$S_L = -\sum_{i=1}^{N} P(C_L) \log_2 P(C_L), \qquad (61)$$

where $P(C_L)$ is the probability of the formation of words C_L with length L. From Eq. (61) we can also express the block entropy

$$S_I = S_{L+1} - S_L, \qquad (62)$$

where S_{L+1}, S_L are the Shannon entropies for words of length $L+1$ and L, respectively. S_I expresses the information change when the text is increased by one more symbol.

– An Alternative Way of Producing Information-Rich Symbolic Sequences: The Quest for Spatial Asymmetry –

So far our efforts at generating Markovian strings have used (coarse-grained) iterated maps possessing a smooth invariant measure. We have not looked for any particular relevance to the linguistic schemes; we are still in the primitive stage of searching for *sine-qua-non* (just necessary but by no means sufficient) conditions producing a happy mixture of variety and reliability.

In this section, we look at a second possibility: To generate Markovian strings of higher order displaying *explicitly* a broken symmetry in one dimension [39]. The idea is that information-rich, one-dimensional strings manifesting an asymmetry in physical space (e.g., by not being palindromes) can, in principle, be produced by an underlying chaotic dissipative flow (characterized by time irreversibility). The necessity of broken symmetry in the modeling of the strings arises from natural languages (where it is simply an empirical fact), and even more important from the (universal) character of the genetic language. In fact, much of what we call genetic information is carried in biological (self-reproducing) structures by a one-dimensional aperiodic and asymmetric structure, the DNA.

Although almost perfectly reproducible from generation to generation, the codon sequence of DNA is fundamentally unpredictable or incompressible (to a human reader—but not apparently to an enzyme), in the sense that its global structure cannot be inferred from knowledge of a part of it, however large. On the other hand, the extremely large number of random sequences makes it very unlikely to select on an *a priori* basis the particular set of sequences that is going to play the "seed" in an observed morphogenetic phenomenon. For example, the number of random polypeptides around 100 amino acids long that have to be scanned in order to select a specific protein is $\sim 20^{100} \sim 10^{130}$. On these grounds, the spontaneous appearance of a given protein with probability 10^{-130} would rightly be considered a miracle. In practice, though, the number of observed combinations of polypeptide chains does not appear to exceed $\sim 10^{30}$. What is needed therefore is a dynamical process capable of producing with *high probability* extremely complex (information-rich) aperiodic sequences of states. The instability of motion associated with dissipative chaos allows the model to explore a *self-limiting,* yet sufficiently large (selected with probability *one*) set of states in phase space. So the selection of a particular sequence out of a very large number of *a priori* equiprobable ones simply does *not* arise. The chaotic strange attractor acts as a *selection* mechanism, rejecting beforehand the overwhelming majority of random sequences and keeping only those *compatible* with the underlying deterministic dynamics.

But the genetic code is characterized not only by a particular, albeit unpredictable sequence of codons along the DNA strand. It is also constrained by the fact that this sequence is "read" by the enzymes in only one direction, corresponding in fact to the sequence $5' - 3'$ of the nucleotide string. S. Luria [40] has provided evidence that *synthetic* biomolecules lacking *polarity* are unable to act as efficient messengers. So this preferred direction of reading allows the introduction into the modeling procedure of attractors enjoying asymptotic stability (irreversibility) and hence *reproducibility.*

Finally, it is closer to biological reality to consider the emergence of each symbol nucleotide as a nonlinear phenomenon taking place far from *equilibrium* and beyond a certain arbitrary threshold. Consequently, and following [39], we will consider a dissipative chaotic flow whose state variables $x_1, x_2, x_3, \ldots$ display sustained (aperiodic) oscillations. We assume that whenever a variable crosses a certain (arbitrary) level with positive slope, the corresponding symbol $x_1, x_2, x_3, \ldots$ *falls out* and gets "typed" onto a one-dimensional tape. In this manner we obtain a one-dimensional string of threshold-value digits generated by the dynamics. One can see immediately that if the process is a zero-order Markov chain (e.g., a Bernoulli-shift sequence), any string of such a process constitutes a palindrome, namely,

$$P(x_1, x_2, \ldots, x_n) = \prod_{i=1}^{n} P(x_i) = P(x_n, x_{n-1}, \ldots, x_1). \qquad (63)$$

A well-known example of a chaotic attractor giving rise to a Bernoulli process is the symmetric tent map (or the logistic map for $r = 4$), when the probability of being in the left or right half of the unit interval is considered. Consider next a *first-order* Markov chain $x_1, x_2, \ldots, x_n, \ldots$. The probability of the appearance of a given finite-length word is $P(x_1, x_2, \ldots, x_n) = u_1(x_1)P(x_2|x_1)\ldots P(x_n|x_{n-1})$, where $u_1(x_1)$ is the *a priori* probability of x_1. Let us evaluate the probability that this Markov chain generates the reverse sequence. For *two* symbols (two-dimensional systems, therefore no chaos!), it is immediately seen that any sequence x_1, x_2 is palindromic:

$$\begin{aligned} P(x_1, x_2) &= u_1(x_1)P(x_2|x_1), \\ P(x_2, x_1) &= u_2(x_2)P(x_1|x_2). \end{aligned} \qquad (64)$$

However, since $u_1(x_1) = u_1(x_1)P(x_1|x_1) + u_2(x_1|x_2)$ and $P(x_1|x_1) + P(x_2|x_1) = 1$, we obtain from Bayes' Rule

$$P(x_2|x_1)u_1(x_1) = P(x_1|x_2)u_2(x_2) = P(x_1, x_2) = P(x_2, x_1). \qquad (65)$$

So one must go to at least *three* dimensions in order to get conditions of unidirectionality:

$$P(x_1, x_2, x_3) \neq P(x_3, x_2, x_1). \qquad (66)$$

Indeed, we have

$$P(x_1, x_2, x_3) = u_1(x_1)P(x_2|x_1)P(x_3|x_2), \tag{67}$$

while

$$P(x_3, x_2, x_1) = u_3(x_3)P(x_2|x_3)P(x_1|x_2). \tag{68}$$

On the other hand, there exist classes of multidimensional systems for which palindromes exist—the most obvious case is a system satisfying the principle of detailed balance: $P_{ij} = P_{ji}$. A broader class of such systems involves transition probability matrices with cyclical symmetry. So a dynamical system that can be mapped to a first-order Markov process involving at least three symbols generates asymmetric sequences, and thus can be used for coding biological (and therefore linguistic) information. In order to examine cases allowing for the generation of higher order Markov sequences, Nicolis et al. [39] took up the case of Rössler's attractor,

$$\frac{dx_1}{dt} = -x_2 - x_3,$$

$$\frac{dx_2}{dt} = x_1 + ax_2, \tag{69}$$

$$\frac{dx_3}{dt} = bx_1 - cx_3 + x_1x_3.$$

For the parameter values $a = 0.38$, $b = 0.3$, and $c = 4.5$, this system has the chaotic strange attractor shown in Figure 10.

Taking the arbitrary thresholds $x_{1C} = x_{2C} = x_{3C} = 3$ and starting from $x_{10} = x_{20} = x_{30} = 1$, one generates via the threshold-crossing mechanism suggested above a sequence of symbols that includes *only* the three-element groups ("hypersymbols") $r_1 = x_3x_2x_1$, $r_2 = x_3x_1x_2x_1$, $r_3 = x_3x_1$. This fact suggests the existence of strong correlations among the three basic variables or the existence of *sharp* transition rules—which implies a higher-order Markov chain. One can in principle investigate the statistical properties of the sequences produced as follows: By integrating Eq. (62) and letting the transients die out, the system is on the attractor. One then generates sequences with a statistically significant number of symbols and hypersymbols. Then the number of recorded singlets (x_1, x_2, x_3), doublets $(x_1x_1, x_1x_2, \dots)$, and triplets $(x_1x_2x_3, \dots)$ are counted and the conditional probabilities

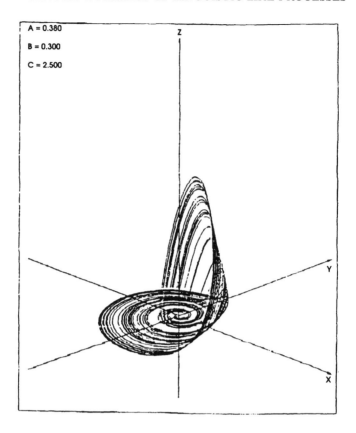

Figure 10. The Rössler attractor.

are deduced from

$$P(x_j|x_i) = \frac{P(x_i, x_j)}{P(x_i)}, \tag{70}$$

$$P(x_k|x_i, x_j) = \frac{P(x_i, x_j, x_k)}{P(x_i, x_j)}, \tag{71}$$

where x_i, x_j, x_k, etc. take the values x_1, x_2, x_3. The absence of the doublet $x_1 x_1$ in the recorded (ergodic) sequence from Eq. (69) indicates that the sequence is not completely random. Also, the absence of the triplet $x_2 x_1 x_2$ excludes the possibility that the sequence refers to a single-memory Markov chain, since in that case one could write $P(x_2 x_1 x_2) = u_2(x_2) P(x_1|x_2) P(x_2|x_1)$, where in the right-hand side the terms are all positive. One suspects therefore that we have a higher-order Markov process—where the memory of a symbol goes several steps into the past. In [39] it is pointed out that good agreement with the numerical frequencies of septuplets of symbols is reached by assuming a fifth-order Markov chain

$$P(A,B,C,D,E,F,G) = P(A,B,C,D,E,F)P(G|B,C,D,E,F)$$
$$= P(A,B,C,D,E) \tag{72}$$
$$P(F|A,B,C,D,E) = P(G/B,C,D,E,F),$$

where $P(G|B,C,D,E,F)$ stands for the numerically computed conditional probability of the last symbol, given B,C,D,E,F. Another important detail emerging from the numerical calculations is that from the 3^7 possible sequences of length seven that can be constructed from the symbols x_1, x_2, x_3, only 21 are actually realized by the dynamics of the Rössler attractor. This shows the strong filtering action performed by this attractor.

– Zipf's Law –

In this section we deal with a particular statistic that we *know* characterizes (some) natural languages: the so-called Zipf's law, first established empirically by Zipf in 1949 for the English language [6]. Here we present a chaotic dynamics model that creates Markovian strings of symbols, as well as strings of words, and discuss its possible connection with Zipf's law. Specifically, we have already seen that starting from a deterministic dynamical system operating in the chaotic regime, one may generate strings of symbols obeying well-defined Markov statistics (as well as *long-term memory, non-Markovian ones*). Inasmuch as randomness ensures variety and information generation while order ensures reliability, it's reasonable to expect that chaotic dynamics should be relevant to biological information processing, in general, and to linguistic dynamics, in particular. In this section, we show that by combining the strings of symbols in *words* interrupted by the pause (blank space), we get statistics characterized by an inverse power law similar to Zipf's law of experimental linguistics.

Power law statistical distributions have been investigated rather extensively during the last ten years or so in connection with self-similar (fractal) processes in physics, biology, cognitive psychology and the social sciences [7,8]. In a recent paper [9], inverse power laws were even attributed to the distribution of natural constants—which is not that surprising if one realizes that natural constants express *invariances,* either in scale or in translational and rotational transformations. On the other hand, Zipf's law,

$$p(x) \sim \frac{0.1}{x}, \tag{73}$$

relating the probability $p(x)$ of the appearance of a word with its rank order x in a long *written* text, belongs to the (still uncharted) domain of mathematical or dynamical linguistics. As it holds in Eq. (73), it is applicable to the first 8,727 common words, since

$$\sum_{x=1}^{8,727} \frac{0.1}{x} = 1. \tag{74}$$

It can be generalized to the form

$$P(x) \sim \frac{A}{x^\lambda}, \tag{75}$$

with A and λ being empirically determined parameters. In this form, Zipf's law can be made to hold for the full repertoire of a given natural language.

We are suggesting here that for people working in the field of chaotic dynamics as applied to biological information processing, i.e., for those involved in partitioning and abstracting ("compressing") a large set of raw stimuli—long sequences of symbols—onto a much more restricted set of attractors [1,4,10], it is tempting to search for isomorphisms between the statistical properties of natural (and also genetic and artificial) languages (e.g., the syntax) and $1/f$ spectra related to *intermittent* processes. Under certain conditions, such processes characterize either the ongoings within a simple given strange inhomogeneous or *multifractal* attractor, or they characterize the irregular and sequential transitions among *coexisting* strange attractors—both stable and unstable (repellors)—with subbasins separated by fractal boundaries.

Any natural language may be considered as a partition that the biological processor imposes on the external world. Biological processors are constrained to form concepts or cognize patterns inasmuch as they form attractors (eigenfunctions) of the linguistic operator involved. In such a view, the information dimension of each attractor is meant to stand roughly (i.e., plus or minus a constant) for the length of the minimal algorithm or code necessary to dynamically generate the category represented by such an attractor [10]. Basically, this means that every time the specific attractor is visited by the dynamics of the system (which may permute the coexisting attractors on a time-division-multiplexing basis), a synonymous word is produced, i.e., an element of its own basin of attraction. In this way we can envisage the dynamical formation of language at the *semantic* level. But we are still operating at the syntactical level.

Coming back then to Eq. (75), we will try to deduce this empirical law of Zipf's by means of a variational formulation, e.g., by employing the maximum entropy principle [7] as follows: We search for the probability density function $p(x)$ characterizing a word of rank order x that maximizes the *a priori* uncertainty

$$S(x) = -\int p(x) \log p(x)\, dx, \quad \int p(x)\, dx = 1, \tag{76}$$

or the maximum information per word conveyed by a language using such a syntax under a given constraint. Such a constraint, given in the form $\int \xi(x) p(x)\, dx = c^{te}$, is associated with a certain average cost. If we choose $\xi(x) \sim \ln(x)$, we end up with the functional

$$H(p) = -\int p(x) \log p(x)\, dx - \lambda_1 \int p(x)\, dx - \lambda_2 \int \log p(x)\, dx. \tag{77}$$

The distribution $p(x)$ maximizing $S(x)$ will then be deduced from the condition $\partial H(p)/\partial p$, which leads to

$$p(x) \sim \exp(-\lambda_1 - \lambda_2 \log x) = \frac{A}{x^{\lambda_2}}, \tag{78}$$

where A is a normalization constant over the x interval. It is obvious from Eq. (78) that $p(x)$ obeys the scaling relation

$$p(\alpha x)\, d(\alpha x) = \alpha^{1-\lambda_2} p(x)\, dx, \tag{79}$$

which implies a lack of fundamental scale in the process underlying $p(x)$. If the rank order of x with a probability density function $p(x)$ is known in a given interval, that interval can be extended; the scaling implies that the fluctuations of a random variable x are generated at each scale in a statistically identical (self-similar) fashion. For N words, Zipf's law written in the normalized form becomes

$$p(s) \sim \frac{x^{-1}}{\sum_x x^{-1}}. \tag{80}$$

This gives rise to an entropy

$$S = -\sum_x p(x) \log_2 p(x) \cong \frac{1}{2} \log_2 N + \log_2(\log_2 N) - \gamma \log_2 e, \tag{81}$$

where $\gamma = 0.5772\ldots$ is Euler's constant.

The crux of the matter is now that there is a physical meaning behind this constraint: $\eta(x) \cong \ln x$. This comes about by considering that in a large sample of text, long words are used less frequently than short words. So the rank order x_L of a word of length L belonging to a language having an alphabet of K symbols (including the pause) can be taken to be the sum of *all* words of length less than or equal to L, namely,

$$X_L = \sum_{i=0}^{L} K^i = \frac{K^{L+1} - 1}{K - 1}. \tag{82}$$

Consequently, our cost function $C = \int \log_2 xp(x)\, dx$ can be written as

$$\sum^{L} \log N_L \frac{X_L^{-1}}{\sum^{L} X_L^{-1}} = \sum_{L=1}^{L_0} \left[\{(L+1)\log K - \log K\} \frac{K^{-L}}{\sum_{L=1}^{L_0} K^{-L}} \right]$$

$$\cong \sum^{L} \left[(L \log K) \frac{K-1}{1 - K^{-L_0}} \frac{1}{K^L} \right] = \frac{(K-1)\log K}{1 - K^{-L_0}} \sum_{L=1}^{L_0} LK^{-L}. \tag{83}$$

Taking into account the equality

$$\sum_{L=1}^{L_0} LK^{-L} = -K \frac{d}{dK} \left(\sum_{L=1}^{L_0} K^{-L} \right), \tag{84}$$

we get

$$C \sim \frac{(K-1)\log K}{1 - K^{-L_0}} K \frac{d}{dK} \left[\frac{1 - K^{-L_0}}{K - 1} \right]$$

$$= \frac{\log K}{1 - K^{-L_0}} \frac{K}{K - 1} \{(1 + L_0)K^{-L_0} - L_0 K^{-L_0+1} - 1\} \tag{85}$$

or

$$K^{-L_0} \cong \left(\frac{K}{K - 1} + \frac{C}{\log K} \right) \frac{1}{L_0}. \tag{86}$$

For *small* costs, we have

$$L_0 K^{-L_0} \cong \frac{K}{K - 1}. \tag{87}$$

Therefore,

$$X_L = K^L \frac{K}{K - 1} - \frac{1}{K - 1} \cong L_0 K^{L - L_0} - \frac{1}{K - 1}, \tag{88}$$

or, taking the logarithm of both sides, we end up with the relation

$$L \cong L_0 - \log_K L_0 + L' \log_2 X_L \cong L_0 + L' \log_2 X_L. \tag{89}$$

This means that the length of a word is roughly proportional to the logarithm of its rank order. The cost C_L of a word is in turn proportional to its length L, allowing us to write $C_L \sim C_0 + C' \log_2 X_L$. Since the length of a word is also proportional to the *time* a certain device (natural or artificial) needs in order to read, write, or produce that word, the significance of the constraint used above in our variational principle becomes clear: By some process of selection, those languages that have survived are the ones that were able to convey the maximum possible information for a given cost. That cost equals the *average time* needed to produce the words of the language.

– Numerical Results and Simulations –

In this section we will present some results about Zipf's law, transinformation, block entropy, and Markov orders. Let us first try to derive Zipf's law from purely "chaotic" dynamical arguments. These arguments have already been touched upon in the preceding sections. Specifically, we consider a dissipative recurrent dynamical system $\underline{x}_{n+1} = \underline{F}(\underline{x}; \xi)$ possessing a smooth invariant probability density $\overline{\mu}(x)$.

We partition the state space into N nonoverlapping cells (symbols) B_i, such that the boundaries between cells are preserved by the dynamics, thereby ensuring the invariance of the alphabet on the cascade of iterations.

The shift process induced by the recurrence scheme given earlier on such a Markov partition $\underline{B}$ generates sequences of N symbols. As mentioned above, the probability distribution of these strings obeys a master equation giving rise to an irreversible approach to a stationary state (the invariant measure):

$$P_{n+1}(i) = \sum_{j=1}^{N} W_{ij} P_n(j), \tag{90}$$

where

$$W_{ij}^t = \frac{\overline{\mu} F^t(B_j) \cap B_i}{\mu B_i} \tag{91}$$

and $\overline{\mu}(B_i) = \lim_{n \to \infty} P_n(B_i)$.

• *The Partitions*

We believe that the logistic map is a simple chaotic model with enough complexity to serve as a text generator. For that reason, we first consider this map in the fully chaotic region ($r = 4$) and use the following Markovian partitions:

1. A partition with *three* cells, the so-called *partition 3A,* whose cells are separated by the points of the period-2 unstable orbits $x_1 = 0.345$ and $x_2 = 0.905$. These points define a three-box partition. The resulting three states, α, β, and γ, which can also be viewed as the letters of the alphabet. These letters are continuously transformed one to another by the dynamics as a first-order Markov chain, whose conditional probability matrix is given by the earlier general formula as

$$W_{ij} = \begin{bmatrix} 1/2 & 1/2 & 0 \\ 0 & 1/2 & 1/2 \\ 1 & 0 & 0 \end{bmatrix}.$$

Taking into account the invariant measure $\bar{\mu}(x)$, we will show that starting from the above alphabet, one can generate sequences of words having well-defined statistical properties.

We start by choosing one of the symbols of the alphabet to be the pause (blank space). As the dynamics unfolds in the shift space, the remaining $N - 1$ symbols are organized into words B_L of varying lengths L, interrupted by the pause. We want to investigate the probability $P(B_L)$ of the formation of such words. Choosing first β to be the pause, the language is limited to words involving a single nontrivial letter α, since the role of γ is trivial ($W_{\gamma\alpha} = 1$). Using the explicit form of the above transition probability matrix, we obtain

$$P(B_L) = W_{\alpha\alpha}^{L-1} = \left(\frac{1}{2}\right)^{L-1}, \quad L \geq 2. \tag{92}$$

If, on the other hand, we use γ as the pause, a richer repertoire emerges involving two nontrivial symbols, α and β. In this case, we obtain

$$P(B_L) = \sum_{m=1}^{L-1} W_{\alpha\alpha}^m W_{\alpha\beta} W_{\beta\beta}^{L-1-m} = \left(\frac{1}{2}\right)^L (L - 1). \tag{93}$$

Although Eqs. (92) and (93) differ significantly for small integer values of L, they tend asymptotically to the same relation for long words, i.e., as $L \to \infty$.

2. We now use a modified model of the logistic map,

$$x_{n+1} = 4x_n(1 - x_n/4), \tag{94}$$

with

$$\mu(x) = \frac{1}{\pi}\left[\frac{x}{4}\left(1 - \frac{x}{4}\right)\right]^{-1/2}. \tag{95}$$

In the *partition 4A*, the interval is divided into four *equal and nonoverlapping* subintervals, and in each subinterval a symbol is mapped. For the above partition, elementary geometry shows that the transitional probability matrix W_{ij} among the four "letters" takes the form

$$W_{ij} = \begin{bmatrix} P_{11} & P_{12} & P_{13} & 0 \\ 0 & 0 & 0 & 1 \\ 0 & 0 & 0 & 1 \\ P_{41} & P_{42} & P_{43} & 0 \end{bmatrix},$$

where $P_{11} = P_{41} = 2 - \sqrt{3}, P_{12} = P_{42} = \sqrt{3} - \sqrt{2}$, and $P_{13} = P_{43} = \sqrt{2} - 1$. Two cases were examined for this nonlinear map. In the first case, the pause symbol for the alphabet is "D" with "D" $= (3,4)$. In the second case, the pause symbol is taken to be "C," with "C" $= (2, 3)$. For the geometrical construction of this partition, we use the midpoint $(X = 0.5)$, the fixed point $(X^* = 0.75)$, and its inverse image (e.g., $X^{*-1} = 0.25$).

3. The *partition 3* of the logistic map in the interval $[0, 1]$ with $r = 4$ is constructed using the fixed point of the map and its pre-image.

4. In the same fashion, a *partition 4* is constructed by the fixed point of the above map, together with its pre-image and pre-pre-image. The construction of *partition 3* and *partition 4* is shown in Figures 11 and 12.

5. A partition with 12 symbols, *partition 12*, is generated in the same way. For partition 12, the invariant measure gives the unique equilibrium distribution of the 12-state Markov process [15]:

$$p_{eq}(k) = (0.26, 0.13, 0.13, 0.26, 0.26, 0.52, 0.52,$$
$$0.26, 0.26, 0.13, 0.13, 0.26)/\pi.$$

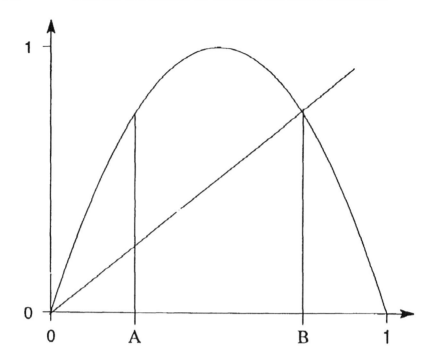

Figure 11. Geometrical construction of *partition 3*.

The transition probability matrix W_{ij} now takes the form

$$
\begin{bmatrix}
P_{11} & P_{12} & P_{13} & 0 & 0 & 0 & 0 & 0 & 0 & 0 & 0 & 0 \\
0 & 0 & 0 & 1 & 0 & 0 & 0 & 0 & 0 & 0 & 0 & 0 \\
0 & 0 & 0 & 0 & 1 & 0 & 0 & 0 & 0 & 0 & 0 & 0 \\
0 & 0 & 0 & 0 & 0 & 1 & 0 & 0 & 0 & 0 & 0 & 0 \\
0 & 0 & 0 & 0 & 0 & 0 & 1 & 0 & 0 & 0 & 0 & 0 \\
0 & 0 & 0 & 0 & 0 & 0 & 0 & P_{58} & P_{59} & P_{510} & P_{511} & P_{512} \\
0 & 0 & 0 & 0 & 0 & 0 & 0 & P_{68} & P_{69} & P_{610} & P_{611} & P_{612} \\
0 & 0 & 0 & 0 & 0 & 0 & 1 & 0 & 0 & 0 & 0 & 0 \\
0 & 0 & 0 & 0 & 0 & 1 & 0 & 0 & 0 & 0 & 0 & 0 \\
0 & 0 & 0 & 0 & 1 & 0 & 0 & 0 & 0 & 0 & 0 & 0 \\
0 & 0 & 0 & 1 & 0 & 0 & 0 & 0 & 0 & 0 & 0 & 0 \\
P_{121} & P_{122} & P_{123} & 0 & 0 & 0 & 0 & 0 & 0 & 0 & 0 & 0
\end{bmatrix}.
$$

The rules of the grammar of the above partitions arise from the construction of the transition probability matrix W_{ij} [14]. From the construction of the partition, we can see that this partition is generic and is Markovian, since each cell is mapped onto another

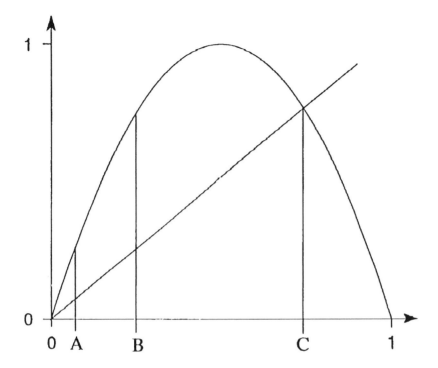

Figure 12. Geometrical construction of *partition 4*.

cell or a union of cells. Each cell, no matter how small, participates in this N-state Markov process. So it is possible to construct a family of partitions with as many cells as the number of letters (~ 27) in the alphabet of any modern natural language we want to examine.

6. Next we examine *partition 12* for the cases of $r = r_\infty$ and $r = 3.7$. We must say here that *partition 12* is not the same in these three cases because the main element for the construction of the partition is the fixed point x^*, which is calculated each time by the expression $x^* = 1/r$, i.e., the fixed point depends on r.

7. Finally we examine a dissipative flow, i.e., we use the Rössler attractor, and examine the Zipf statistics for the three "hypersymbols" $\alpha = x_3 x_2 x_1$, $\beta = x_3 x_1 x_2 x_1$, $\gamma = x_3 x_1$. This hypersymbol sequence is random, indicating that the compression achieved in the hypersymbols has removed much of the redundancy in the original text.

● *Zipf 's Law*

Let us now see the behavior of the above partitions in Zipf's law. First we examine *partition 3A* [57]. Figure 13 depicts the dependence of the logarithm of $P(B_L)$ versus L or versus B_L, where B_L is the rank order obtained by iterating the partition and then monitoring the frequency of appearance of strings of symbols of given length between two pauses. The full lines stand for Eqs. (92) and (93). We note that the agreement between simulation and theory is quite satisfactory. Let us again recall that the rank order X_L of a word of length L is the sum of *all* words of length equal to or less than L:

$$X_L = 1 + \sum_{i=1}^{L} K^i, \tag{96}$$

where K is the number of symbols in the alphabet excluding the pause. We have seen that for long words in which $X_L \gg 1/(K-1)$, one can write

$$X_L \sim K^{L-L_0}, \tag{97}$$

where L_0 is defined by

$$\frac{K}{K-1} = K^{-L_0}. \tag{98}$$

If we now apply Eq. (97) to the first example of the logistic map, we have $K = 2$, i.e., $L_0 = 1$, and so $X_L \sim 2^{L+2}$. Consequently, from Eqs. (92) and (93) we obtain for large values of L,

$$P(X_L) \sim \frac{1}{X_L}, \tag{99}$$

which is exactly Zipf's law. Implicit in the above syllogism is the fact that the dynamical system generates the successive symbols at *regular* time intervals.

Zipf's law for *partition 4A* is examined for two cases, as shown Figures 14 and 15. We can see that in both cases there is a similarity to Zipf's law. Also, for *partition 3* and *partition 4*, Figures 16 and 17 show a Zipf-like behavior. In the case of *partition 12*, we present results from simulations of Zipf's law with 11 possible positions of the "pause" in the interval $[0, 1]$. The symbol "α" is excluded because for this partition the element $P_{\alpha\alpha} \neq 0$, as we see from the matrix W_{ij} (see Figure 18).

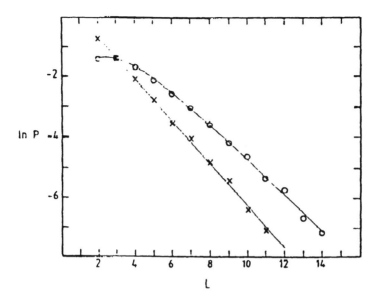

Figure 13. Probability of words of length log P plotted against L. The words are generated by the shift process induced by the dynamics of the partition 3A. The crosses and circles stand for the numerically computed probability using, respectively, letter β and letter γ as the pause. Solid lines represent the analytically deduced laws.

We also show *partition 12* for r_∞ and $r = 3.7$ in Figures 19 and 20. Here we must say that because of the probability density function of the logistic map at r_∞, the map does not visit all the cells of the partition but only a few of them. The same thing happens when $r = 3.7$, but now the system visits more cells. Zipf's law was checked for the above partitions with different number of words for the same partition (25,000, 42,000, and 80,000 iterations, i.e., "letters" for *partition 12* at r_∞ and 20,000, 40,000, and 80,000 iterations for *partition 12* with $r = 3.7$). We can see from Figures 21 and 22 that after a given number of words, the Zipf's-like law for a given partition is independent of the number of words examined. A general comparison of all the artificial languages for Zipf's-like law is presented in Figure 23.

Next we look at natural languages. The primary reason is that we want to show how simple chaotic dynamics can in principle capture the characteristic trend of a very complicated phenomenon like a natural language. We use texts from contemporary Greek and English prose. Because of the different structure of the grammatical rules in Greek and in English, we have made some changes in the sample Greek

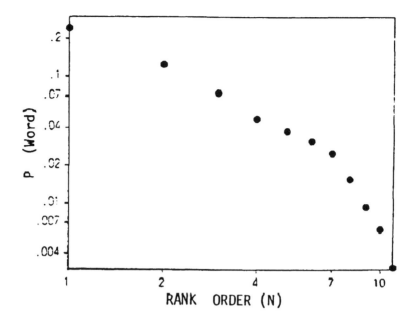

Figure 14. Zipf's-like law behavior for the *partition 4A*. Pause symbol is "C."

text. Specifically, we treated all the different declinations of nouns and adjectives in each case as a single word. But the difference is rather minor. For Greek, the text under study consisted of 50 pages from the book *Antilegomena* by L. M. Panayiotopoulos, having a total of 8,253 words and 2,615 different words. Out of these 2,615 different words, there exist 51 with distinct probabilities [13]. As we can see, the lion's share of the total probability is associated with ∼ 50 words.

For English, we use passages in the book *Dreams of Reason* by H. J. Pagels with 7,214 words in total and 1,947 different words. The comparative results are presented in Figure 24. By way of comparison, the languages produced by the logistic map with 4 letters, e.g., by *partition 4A* and the *partition 4* for genetic languages, are also used. For the genetic language, two strings are examined (nucleotide sequences): length $l_1 = 4,854$ letters or 1,618 words for the first sequence (RNA polymerase III), and length $l_2 = 8,109$ letters or 2,073 words for the second (embryonic cDNA). We note that in these two cases we have an alphabet with only 4 letters; A, C, G, T, and only 64 possible words, each with 3 letters. The plot of a Zipf-like law for these two strings is presented in Figures 25 and 26.

At this point, we must admit that even a monkey-typed language

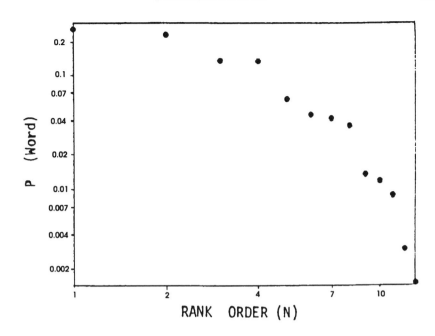

Figure 15. Zipf's law behavior for the *partition 4A*. Pause symbol is "D."

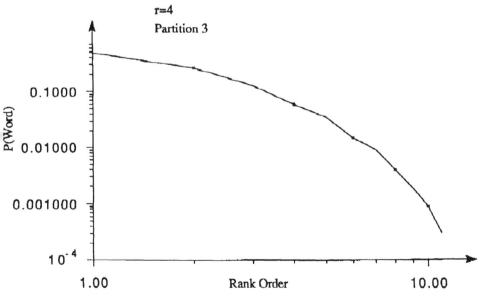

Figure 16. Zipf's law behavior for the *partition 3*.

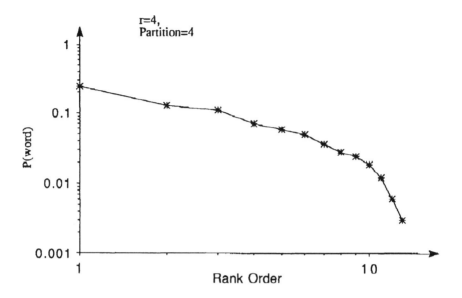

Figure 17. Zipf's law behavior for the *partition 4A*.

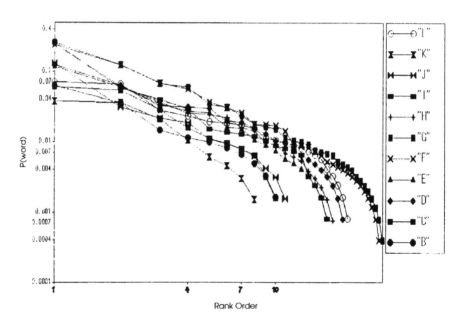

Figure 18. Comparison of Zipf's-like law behavior for the *partition 12*, $r = 4$. Because of the symmetry of the W_{ij}, we can observe similar behavior for "mirror" letters of the partition.

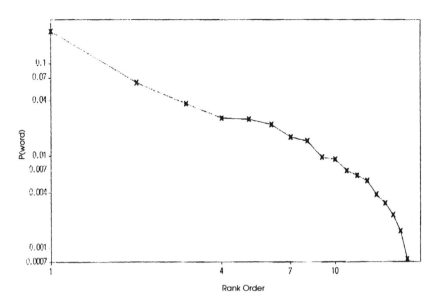

Figure 19. Zipf's-like law behavior for the *partition 12*, $r = r_\infty$. Pause symbol is "G."

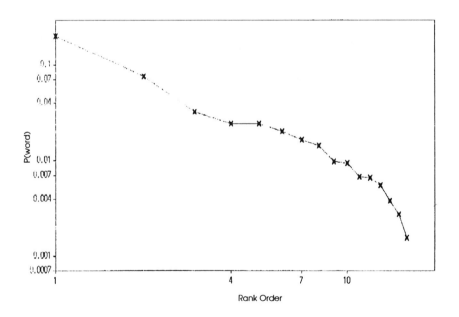

Figure 20. Zipf's-like law behavior for the *partition 12*, $r = 3.7$. Pause symbol is "G."

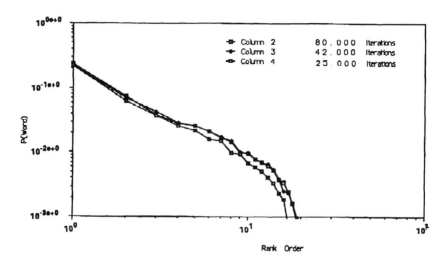

Figure 21. Comparison Zipf's-like law behavior for the *partition 12*, $r = r_\infty$ for three different numbers of iterations. Pause symbol is "F."

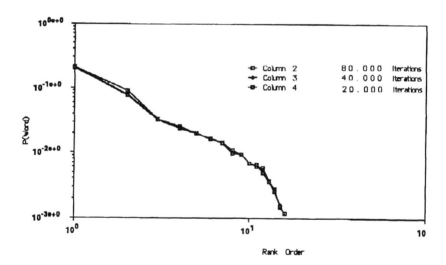

Figure 22. Comparison Zipf's-like law behavior for the *partition 12*, $r = 3.7$ for three different numbers of interactions. Pause symbol is "F."

mimics Zipf's law. But in this case the number of words produced is uncountable, forming a Cantor set. Therefore, the monkey language is useless, since for any language the number of words may be infinite but should be countable, hence lexicographically manageable [11]. In the case of the Rössler attractor, Figure 27 depicts the numerically

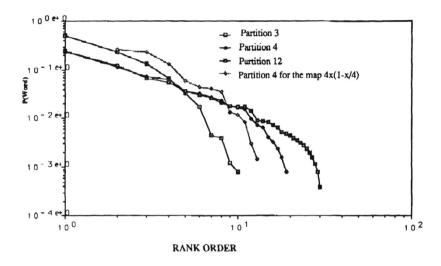

Figure 23. Comparison of Zipf's-like law behavior for artificial languages.

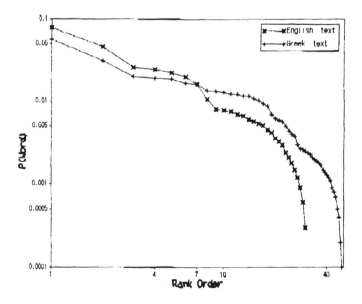

Figure 24. Comparison of Zipf's-like law behavior for natural languages.

computed $\ln P(B_L)$ versus L (or versus $\ln B_L$) for the hypersymbol sequence—using γ as the pause. The dependence is now more complicated than in the case of the logistic map, but the general trend is more or less the same.

We have shown earlier that a dynamical system generates the suc-

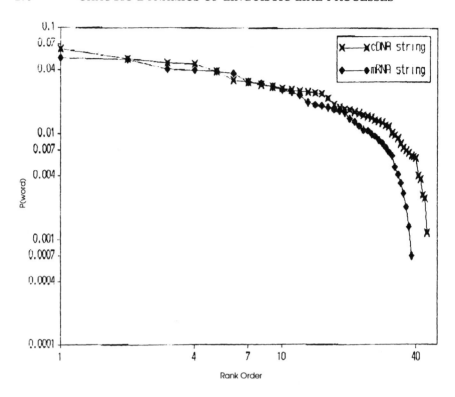

Figure 25. Comparison of Zipf's-like law behavior for genetic languages.

cessive symbols at *regular* time intervals. But in dissipative *flows*—such as the Rössler attractor—this is not the case. Perhaps this inhomogeneity in time is partly responsible for the "jerks" seen in Figure 27. The mean time for the formation of a word increases with its length. However, there is a large dispersion around the mean, leading to crossovers in the times of formation observed in a given realization of particular words of different length. This overlapping may be the source of the observed deviation from Zipf's law. In order to grade the three processors corresponding to Eqs. (92) and (93) and the Rössler attractor, we introduce the *entropy* of the process as

$$B_L S_I = -\sum P(B_L) \log P(B_L). \tag{100}$$

This quantity depends on the partition [without being confused, of course, with the entropy of the partition $-\sum_i^{2,3} \overline{\mu}(B_i) \log_2 \overline{\mu}(B_i)$]. In other words, it depends on the algorithm generating the stochastic process $\{B_L\}$ from the original dynamical system. The computation

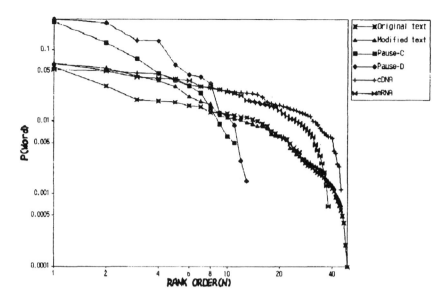

Figure 26. Comparison of Zipf's-like law behavior for natural, genetic, and artificial languages.

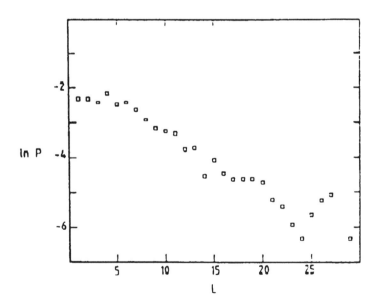

Figure 27. Probability of words of length $\log P$ plotted against L. The words are generated by Rössler's model in the hypersymbol space, γ used as the pause.

of S_I from Eqs. (92) and (93) and the numerical values from Figure 27 gives successively $S_I = 2\ln 2 \sim 1.39$, $S_I = 1.88$, and $S_I = 2.87$ for Rössler's model. This trend highlights the increasing variety of the repertoire of the corresponding languages.

• *The Transinformation*

Transinformation [see Eq. (59)] is a generalization of the correlation function applicable to symbolic (not necessary numerical) sequences, like linguistic texts or musical pieces. Transinformation is the natural tool to detect correlations in symbolic sequences. The function $I(t)$ may decrease monotonically, but we shall see that it also possible for $I(t)$ *not* to decrease monotonically. For every one of the partitions discussed above, the transinformation was examined. Let us first consider the artificial texts.

Figure 28 presents a comparison of $I(t)$ for *partition 3–4A* and *partition 12* for $r = 4$. In Figure 29, a comparison of the transinformation for natural languages is presented. We see from the fluctuations of $I(t)$ that there must be a strong correlation between the letters of the examined languages. A comparison of $I(t)$ for biological languages is presented in Figure 30. We can see that in this case too there is a correlation between the "letters," but this occurs because genetic strings are Bernoulli shifts; therefore, there are a lot of palindromes, giving rise to fluctuations in Eq. (59). In the above comparisons, we note a nonmonotonic decay. This suggests the existence of more than one time scale in the sequence of symbols, indicating a regeneration or persistence of memory. Let us now look at the results from *partition 12*.

For a musical language, we examine the "Partita 1 for solo violin" from J. S. Bach. We can regard music as a language with 12 different letters: the 8 notes and the 4 dieses. The length of the text is about 12,000 letters. In Figure 31, a comparison of $I(t)$ for *partition 12* when $r = 4$ and the musical language is presented. Now consider the case when $r = r_\infty$. We note the nonmonotonic decay of $I(t)$ in this case, too, just as with in the natural and biological languages (see Figure 32). This result tends to confirm our hypothesis that for the transinformation of natural, genetic, and musical texts, a multifractal process is probably involved. Finally, $I(t)$ is studied for *partition 12* for the case $r = 3.7$. Here we observe (see Figure 33) a monotonic decay.

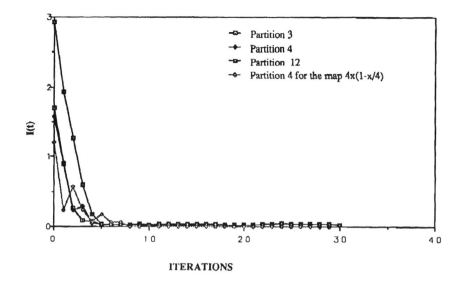

ITERATIONS

Figure 28. Comparison of the transinformation $I(t)$ for partition 3, partition 4, partition 4A, and partition 12 for the logistic map in the fully chaotic regime.

• *The Block Entropy*

We now calculate the block entropy for the above partitions. For comparison with partition 4, partition 4A, and partition 12, *random sequences* were created, each with the same number of symbols. To be more specific, first a random sequence with four symbols (A, B, C, D) was created (i.e., a zero-order Markov chain) and was compared with partition 4, partition 4a, and DNA and RNA strings. The results are presented in Figure 34. We can see that the S_I of the random sequence and of DNA and RNA strings decreases step by step, in contrast to the S_I of the two partitions, where the difference $S_{L+1} - S_L$ seems to converge. In addition, a random sequence with 12 symbols was created in order to compare it with partition 12 for $r = 4$ and with the musical language. The results are presented in Figure 35. Again the S_I of the random sequence seems to be decreasing at each step, while the S_I of partition 12 and that of the musical language again seen to converge to a stable limit.

• *The Markovian Order*

We now look at the Markovian order of these artificial languages. For partition 4A we display a χ^2 test that was devised to verify the order

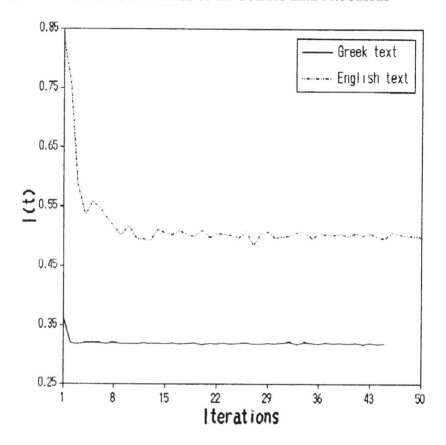

Figure 29. Comparison of the transinformation $I(t)$ for natural languages.

of the Markov process of the artificial text. Using 10,000 iterations, we may generate $\sim 1,500$ words. The results are recorded starting at $t = 500$, in order to eliminate all transients. The number of observed singlets, doublets, or triplets is counted. The conditional probabilities are then deduced from relations like

$$P(B|A) = \frac{P(AB)}{P(A),} \tag{101}$$

$$P(C|AB) = \frac{P(ABC)}{P(AB).} \tag{102}$$

The fact that doublets like DD or CC never occur shows that our sequence is not completely random. Good agreement with numerically computed frequencies of septuplets is reached by assuming a fifth-order Markov chain

$$P(A_1 \ldots A_7) = P(A_7|P(A_2 \ldots A_6) \cdot P(A_6|A_1 \ldots A_5), \tag{103}$$

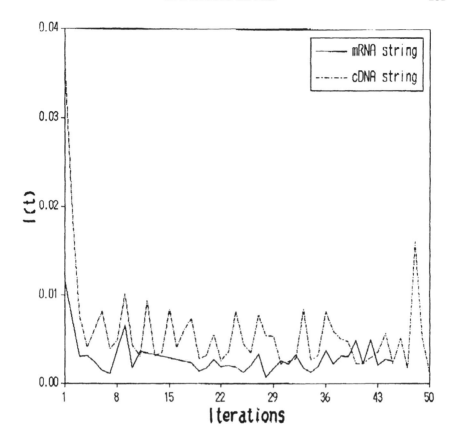

Figure 30. Comparison of the transinformation $I(t)$ for genetic languages.

where $P(A_7|A_2 \ldots A_6)$ represents the numerically computed conditional probability of the last symbol given by the quintuplet $A_2, A_3 \ldots A_6$. A χ^2 test has been devised to verify the order of the Markovian character of the artificial text,

$$
\begin{bmatrix}
1 & 1-0 & 9086 \\
2 & 2-1 & 2793 \\
3 & 3-2 & 503 \\
4 & 4-3 & 123.1 \\
5 & 5-4 & 36.5 \\
6 & 6-5 & 0.629
\end{bmatrix}.
$$

The first column indicates the degrees of freedom, the second the compared orders, and the third the numerical value of χ^2. We see that out of 4^7 (16,384) possible words, only 253 appear in the text. This corroborates the hypothesis that the simple chaotic generator we have been

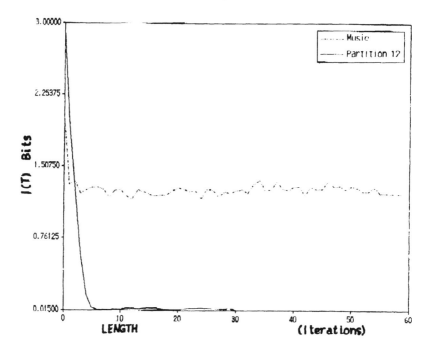

Figure 31. Comparison of the transinformation $I(t)$ for partition 12, $r = 4$, and musical language.

using acts essentially as a linguistic filter, admitting the emergence of a language with variety but also good reliability—a mixture, in short, of innovation and order.

We also check the extent to which the language created by our chaotic generator displays a broken symmetry in one-dimensional physical space, thereby avoiding palindromes. We found that among $\sim 1,500$ words, there were only seven palindromes [13]. It is of interest to see the role of the *control parameter* of the dynamical system in affecting both the interval length and the degree of the Markovian memory of the generated text. We again consider the logistic map $x_{n+1} = rx_n(1 - x_n)$ in the unit interval, and impose partitions 3 and 4. Using the χ^2 test, we calculate the Markovian order k of the generated test for partitions 3 and 4 (see Figures 36 and 37). In Figures 36 and 37, the lower value of the control parameter is not the same. Specifically, for partition 3 we have $r_{\min} = 3.9$, because after this critical value our dynamical system visits all the cells of the partition. For the same reason, in partition 4 we take $r_{\min} = 3.928$. For natural languages, experiments by Shannon show that English is a Markov chain with order between 5 and 7.

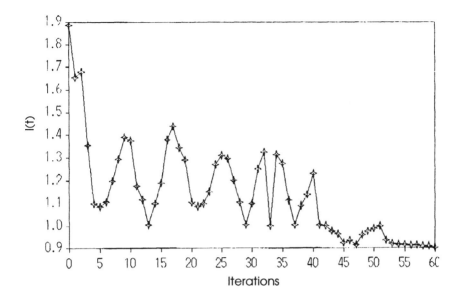

Figure 32. Calculation of the transinformation $I(t)$ for partition 12 for the Feigenbaum attractor $(r = r_\infty)$.

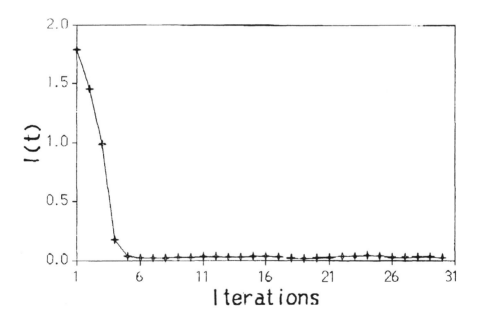

Figure 33. Calculation of the transinformation $I(t)$ for partition 12 for $r = 3.7$.

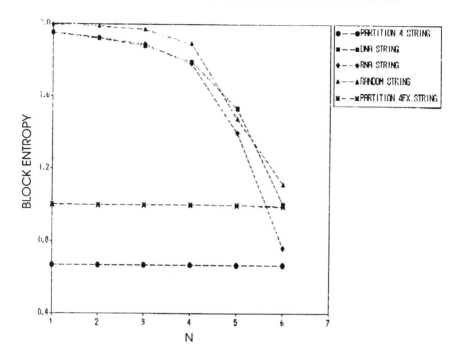

Figure 34. Comparison of block entropy for partition 4, partition 4A, the DNA sequence, RNA sequence, and the random sequence with four symbols.

A cognitive processor recognizes the external world in a coarse-grained fashion. For single-attractor systems, this implies the quantization of state space for flows or the shift space for maps. For dissipative dynamical systems with multiple *coexisting attractors,* the *categories* under consideration *are* the coexisting attractors and the algorithm of jumps from one to the other forms of the metalanguage at the *semantic level.* In this section, we have considered two particular forms of coarse graining: partitioning of the state space into cells, or boxes, satisfying the Markovian condition, and monitoring the crossings by the state variables of appropriately defined threshold values. Deterministic chaos induces stochastic processes of varying complexity into the discrete state space. We have identified mechanisms leading to a statistical distribution of words of a given rank, satisfactorily described by Zipf's law, especially for maps. We have also identified the reasons restricting the applicability of this law in dynamical flows.

– Fractals, Multifractals, and Their Connection to Chaos –

We have seen already that in the case of a chaotic motion the most

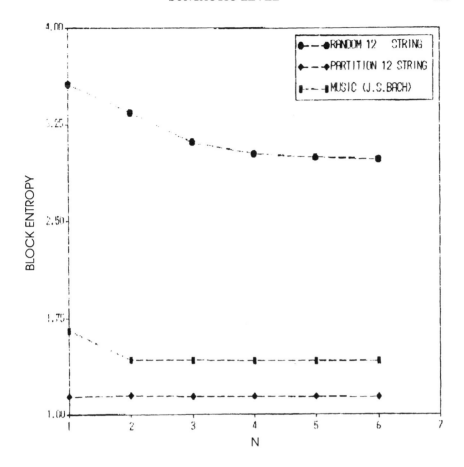

Figure 35. Comparison of block entropy for partition 12, the musical sequence, and the random sequence with 12 symbols.

important distribution on strange attractors is the invariant measure $\bar{\mu}(\underline{x})$, which describes how often a given part of the attractor is visited by the chaotic trajectory in the long-term limit. This invariant measure may be more or less smooth. In the case where $\bar{\mu}(\underline{x})$ is a fractal itself, the attractor is called multifractal or highly inhomogeneous. Let us elaborate on this point.

One of the crucial features of a chaotic system, stemming from its exponential sensitivity to initial conditions, is its unpredictability: One observes nearby trajectories diverging (on the average) exponentially in time. However, there exist time variations (fluctuations) of this "chaoticity." Consequently, the *mean* exponential growth of uncertainty in the initial state does not exhaust the typical behaviors. So

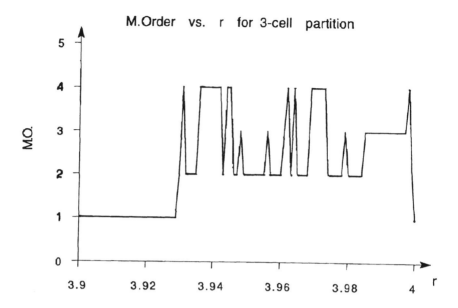

Figure 36. Calculation of the Markovian order versus the control parameter r for partition 3 of the logistic map.

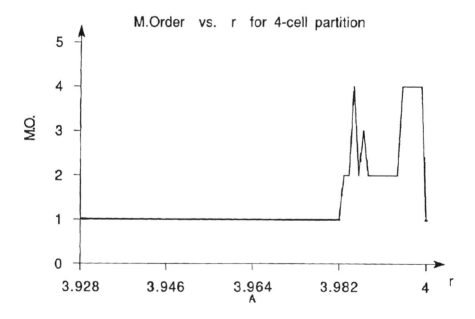

Figure 37. Calculation of the Markovian order versus the control parameter r for partition 4 of the logistic map.

one can observe a regular motion in state space for a long time, interrupted suddenly by randomly distributed bursts of strong chaoticity.

This phenomenon, called *temporal intermittency,* plays an important role in pattern recognition, as we shall see in next section. A very simple example of intermittency is seen in the one-dimensional map $x_n = f_\epsilon(x_{n-1})$, where for $\epsilon = 0, f_\epsilon(x)$ has a tangent contact with the bisector line $x_n = x_{n+1}$. For small ϵ, the state x_n spends a large number of iterations ($\sim \epsilon^{1/2}$) in the regular or laminar phase, i.e., near the bisectrix. Nevertheless, x_n is finally expelled and a chaotic burst follows, to be followed in turn by a "re-injection" onto the laminar channel, and so forth. This behavior is also displayed by chaotic flows like the Lorenz attractor at $r \geq r_c \sim 166.07$, which has regular oscillations interrupted by randomly distributed bursts that become more and more frequent as r increases.

Here we use the term "intermittency" in a somewhat broader sense than that used in the transition to stochasticity through a tangent bifurcation. For us, every *inhomogeneity* in the degree of chaoticity of the dynamical system is included in the class of intermittent behaviors. As we have already seen, quantities like the Lyapunov exponent (which is the first moment of the instantaneous deviation between two nearby trajectories) and the Kolmogorov entropy cannot give a complete characterization of intermittency. Rather, they constitute global indices of the exponential divergence of nearby trajectories and cannot measure the *fluctuations* of the "chaoticity degree" along a given trajectory. We have shown how to achieve the goal of describing the *inhomogeneity* of chaotic attractors. We have proposed *new* indices like the generalized *dimensions* or *higher moments* of the instantaneous deviation between nearby trajectories.

The intermittency then appears as a *manifestation of the multifractality* with respect to the time dilations in the trajectory's space. In short, the main idea consists in the characterization of the scaling structure of an object (dynamical or "frozen" trajectory) by means of a set of indices—the generalized dimensions D_q. For any microscopic compact or homogeneous object like the cube or the sphere, the surface-to-volume ratio is small, since the ratio is inversely proportional to the linear size of the body. However, there exist highly inhomogeneous objects with large surface-to-volume ratios. They play an important role, for example, in many biological tissues (and the concomitant biological function) involved in respiration, digestion, and information generation, transfer, storage, and recall. For example, the need for rapid gas exchange justifies the existence of the large surface-to-volume ratio seen in the human lung, where the area of the respiratory surface is on the order of ~ 100 m^2 while the enclosed volume is a few liters.

Perhaps the most general way to define a fractal system is to state that in such systems the surface or volume area depends on the measurement *resolution* over several orders of magnitude, and the surface or volume dependence on the resolution converges very slowly—if at all. Furthermore, over several orders of magnitude the observed surface or volume dependence on the resolution (grid size) follows a power law behavior with a noninteger exponent smaller than the Euclidean dimension of the space the object is embedded within. Fractals have no smallest or largest scale, displaying self-similarity over many orders of magnification—no matter how much one blows them up, they always reveal *new* information.

What is the *meaning* of a "fractional" dimension? The fractal trajectory is simply avoiding the full potential for exploration afforded by the embedding dimension of the phase space. In the plane, for example, one type of fractal trajectory has a dimension ~ 1.58 instead of 2. Thus nonergodic behavior can be considered isomorphic to syntactical rules or grammatical constraints, whereas a fully ergodic trajectory would simply correspond to a nongrammatical sentence.

Let us now consider a probability distribution on the unit interval constructed by the following rule: At the first step, the middle third of the interval is made more (or less) probable than the outer two thirds. Let the probability of each of these outer pieces be P_1, while the probability of the middle third is $P_2 = 1 - 2P_1$, where $P_2 > P_1$. At the next step, each piece is again divided into thirds and the probability is redistributed within each of these nine pieces so that the ratios within each third are the same as those of the distribution made at the first stage. This procedure is repeated *ad infinitum*, leading to the distribution becoming more and more inhomogeneous. The density of the asymptotic distribution (which may be taken as *the invariant measure* of some chaotic map or flow) obtained after an infinite number of steps turns out to be *discontinuous* everywhere. Studying the properties of such a distribution requires a grid of finite size ϵ.

Let us subdivide the unit interval into n subintervals (boxes) of size $\epsilon = (\frac{1}{3})^n$, with $n \gg 1$. The probability of box m (m is an integer from 0 to n) can take on one of the values

$$P_m = P_1^m P_2^{n-m}. \qquad (104)$$

Since the number of boxes 3^n is much greater than m, the above distribution of box probabilities is *degenerate*, i.e., there are $\binom{n}{m}2^m$ boxes having the same measure P_m. The first question is: Which boxes give the bulk of the total probability when the grid is refined, i.e., when $\epsilon \to 0$? Although the most probable box is that in the middle of the unit interval, which has measure $P_0 = P_2^n$, its contribution becomes negligible as $n \to \infty$ since $P_2 < 1$. On the other hand, the rarest boxes ($P_n = P_1^n$) are numerous, but the total measure of those boxes is also negligible, since $(2P_1)^n \to 0$ as $n \to \infty$. Therefore, we conclude that for large n a few columns very close to some medium "height" give the main contribution. Specifically, for $n \to \infty$, there exists a single index $m = m_1(n)$ between 0 and n such that only boxes with measure P_{m_1} contribute to the total probability $N_{m_1} \cdot P_{m_1} \to 1$. Thus, *none* of the other boxes are important contributors to the total measure—which is concentrated almost exclusively on the above "hot spots," in which the index m_1 can be calculated in the following way [41].

In evaluating the total probability $N_m P_m$, we use Stirling's formula $\ln \xi! \cong \xi(\ln \xi - 1)$ for $\xi \gg 1$. Thus we obtain

$$\ln(N_m P_m) = n \ln(n) - m \ln(m) - (n - m)\ln(n - m) + $$
$$m \ln 2 + m \ln P_1 + (n - m)\ln P_2. \quad (105)$$

This expression takes its maximum at $m_1 = 2P_1^n$. Hence,

$$\ln(N_{m_i}) = -n[2P_1 \ln P_1 - P_2 \ln P_2]. \quad (106)$$

The number N_{m_i} of boxes on which the bulk of the total invariant measure is concentrated increases exponentially with n. Since the resolution is $\epsilon = 1/3$, N_{m_1} may be written as $N_{m_1} = e^{-f_1}$, where

$$f_1 = \frac{2P_1 \ln P_1 + P_2 \ln P_2}{\ln(1/3)}. \quad (107)$$

These "main" boxes cover a fractal subset of the unit interval, where f_1 is the fractal dimension of this subset.

In order to test the degree of inhomogeneity (or multifractality) of the above invariant measure, we look at the qth power of the box probabilities, i.e., the qth-order coherence of the set. The total amount of these box probabilities is

$$\sum_m N_m P_m^q = (2P_1^q + P_2^1)^m. \quad (108)$$

Here again columns of a certain height in the coarse-grained $\bar{\mu}(x)$ dominate this sum. For fixed q, the dominating boxes are those with probability P_{m_q}, where

$$m_q = n\frac{2P_1^q}{2P_1^q + P_2^q}. \tag{109}$$

Again this number increases exponentially with n, so we can write $N_{m_q} = \varepsilon^{-f_q}$. Thus, a subset of fractal dimension F_q gives the dominant contribution to the sum of the qth power of the box probabilities. For $q \neq 1$, this fractal subset *differs* from that contributing to the total measure. In our specific example

$$f_q = \left[\frac{2P_1^q \ln P_1^q + P_2^q \ln P_2^q}{2P_1^q + P_2^q} - \ln(2P_1^q + P_2^q)\right]\frac{1}{\ln(1/3)}. \tag{110}$$

By changing the "degree of coherence" q, boxes with higher (or lower) probabilities (that is, fractal regions with more or less dense occupations) provide the main contributions. So the contribution of the sum of different powers of the box probabilities is dominated by *different* fractal subsets on the unit interval. The spectrum f_q of their fractal dimensions provides the characteristic function of this *highly inhomogeneous* (or multifractal) set.

The box probability P_{m_q} can be written in the form $P_{m_q} = \epsilon^{a_q}$, where a_q is the qth-order singularity exponent of the invariant measure. Then from Eqs. (104) and (109), the expression for the singularity exponent becomes

$$f_q = \frac{2P_1^q \ln P_1 + P_2^q \ln P_2}{2P_1^q + P_2^q} \times \frac{1}{\ln(1/3)}. \tag{111}$$

The expression for the spectrum $f(a_i)$ of the fractal dimensions of the singularities a_i can be deduced from f_q and a_q by eliminating the variable q. Then the spectrum of the dimensions D_q can be computed from $f(a)$.

The question now arises: How does the above analysis fit in to the context of our main theme (the role of fractal or multifractal attractors in information processing)? Although the answer to this question belongs mainly to the next section, it is possible here to make some introductory remarks on the use of *multifractals* in modeling chaotic dynamics in perception and cognition. We start by mentioning some well-known facts from cognitive psychology.

When we recognize a pattern or read a book, the number of key features or key words or key concepts we scan and memorize is an extremely small percentage of the total information received; *very few* details claim the lion's share in pattern recognition—something that cartoon and caricature artists (as well as composers of music) know very well.

On the other hand, from the epoch-making discoveries of Hubel and Wiesel in neurophysiology, we can justify in the above respect (optical) pattern recognition: Hypercolumn neurons in the occipital cortex seem to fire selectively to details associated with angles, edges, or, in general, stimuli having a *small radius of curvature,* e.g., eyes, lips, and ears. Our work on Zipf's law indicates that the first 20 to 40 keywords in a natural or artificial language essentially monopolize a long text. Such examples lead us to suggest [44] that perhaps such a "scandalous" selectivity is an *inhomogeneous* (that is, a *multifractal*), *intermittent* mechanism of perception and cognition taking place in the processing apparatus of the cerebral cortex. Specifically, the thalamocortical pacemaker (whose one-dimensional trace is the recorded EEG from the scalp) is supposed to play the role of the dynamical agent that *forms* and *permutes* the neuronal mosaic attractors in the cortex (that is, it scans the memory) on a time-division-multiplexing basis. Such a dynamics of intermittent jumps from attractor to attractor constitutes something much more complicated than a purely random noise affair. In 1981, it was suggested [42] that a reliable information processor should behave as a chaotic attractor *itself.* That is, the process of moving from attractor to attractor (from cognitive category to cognitive category) should take the form of a *metalanguage* characterized by a subtle interplay between variety and reliability—properties that a chaotic attractor can in principle have—as we have been advocating.

Since then a number of experimental investigations [43] seem to justify the above assertion. The crux of the matter is that cognitive psychology appears to suggest that of all available attractors categories in the cortex, few seem to be visited systematically by the pace-making processor—itself a strange attractor. We have here an intermittent process among "hot" and "cold" spots—attractors (categories) that are visited repeatedly and almost exclusively—that help build the mind's *Weltanschauung,* and attractor categories that are visited sporadically and whose existence may seem quite academic.

Should we infer and subsequently test from the available EEG data the possible *multifractality* of the thalamocortical pacemaker? Note that this processor may display considerable homogeneity in periods of drowsiness, epilepsy, deep sleep, or even REM sleep, but once in the regime of performing mental tasks or perceptual and cognitive phases (e.g., the Rorschach test) it might display a highly *inhomogeneous* or multifractal nature. The very technique via which one may infer the possible existence and principal features of a (low-dimensional) attractor [such as the embedding dimension, as well as the spectra of D_q and $f(a)$] behind a recorded one-dimensional trace of it—like the EEG—will be presented in following sections. In the case of attractors displaying a fractal invariant measure, such as the Feigenbaum attractor, e.g., the logistic map around the first accumulation point, the essential new thinking is that the different symbols are participating in the total measure in a very undemocratic way. Very few hot spots (features, symbols) monopolize the dynamics of the pattern recognition—it is as if a *deus ex machina* provided this parsimonious mechanism of information selection, which saves time and energy from the processor during the execution of mental task or the recognition of a face from very few sharp edges of its own attractor.

The final justification for the existence of this selective mechanism will come when we prove that indeed edges and other details of small radii of curvature in a pattern are loaded with an excessive amount of information as compared with the smoother ones.

2. Semantic Level

– Introduction –

The semantic linguistic level has to do with the way stimuli—words—are partitioned (compressed) into abstract *categories* (coexisting, multiple attractors) and the way those attractors establish a metalanguage of communication among themselves. Such communication takes place via an irregular sequence of jumps from attractor-category to attractor-category. In this case of unstable attractors (repellors), no external excitation (noise) is necessary. The intermittent behavior takes place spontaneously. In the case of stable attractors, we need external excitation.

In our theory, the source of this excitation is provided in the mammalian brain by the thalamocortical pacemaker, whose one-dimensional trace is the routinely recorded EEG. This oscillator is itself a chaotic attractor, so the process of moving among the individual attractor-categories takes place in a more or less orderly fashion that nonetheless leaves room for a considerable amount of variety. Our conjecture is that under the regime of pattern recognition, or more generally, the execution of mental tasks, the thalamocortical pacemaker becomes highly inhomogeneous or multifractal.

– Chaotic Dynamics in Biological Information Processing: A Heuristic Outline –

A fixed, but otherwise random, environmental time series impinging on the input of a certain biological processor passes through practically undetected with high probability. A very small percentage of environmental stimuli, though, are captured by the processor's nonlinear dissipative operator as initial conditions, i.e., as solutions of the processor's dissipative dynamics. The processor in such cases is instrumental in compressing or abstracting those stimuli, thereby causing the external world to collapse from a previous regime of a "pure state" of suspended animation onto a set of stable eigenfunctions or categories—chaotic strange attractors. The characteristics of this cognitive set depend on the operator involved and the hierarchical level where the abstraction takes place. Here we model the physics of such a cognitive process and the role that the thalamocortical pacemaker of the (human) brain plays in both stimulating the individual attractors and permuting them on a time-division-multiplexing basis. A synthesis of the Markovian process taking place within each individual attractor-memory, and the Markovian or semi-Markovian process involving the intermittent jumping among the different attractor-memories is discussed.

– Mechanistic (Syntactical) vs. Biological (Semantic) Processing –

As the interface between the artificial and the biological gets more and more blurred, it becomes increasingly important to devise a definition of life—for legal and moral purposes, at least. For the time being, this does not seem feasible. A necessary condition exists, however: a biological system is self-replicating. This property implies the ability to simulate patterns and, in turn, the ability to compress data and come up with abstract algorithms. In this section we will examine how this need for compressibility of environmental stimuli may be dynamically implemented by a biological processor. We have to face the problem of information transaction. Hence, it is appropriate to start with a few comments about the relevance of classical information theory for such a task.

Shannon's information theory is a branch of combinatorics. Predictably, then, in the first ten years or so after the invention of this theory by Shannon, many physicists and biologists busied themselves with calculating the information content of, say, long macromolecular chains (DNA, in particular) merely by estimating the number of "complexions" or configurations of a given structure and taking the logarithm. As may be surmised, the entire enterprise was misguided from the outset and the results were quite meaningless, i.e., operationally useless. No wonder information theory amounts to thermostatics with no evolutionary outlook at all. It essentially boils down to a variational principle that aims at striking a compromise between two conflicting requirements: maximizing the speed of transfer and minimizing the reproduction error of a message (stationary time series or vector). The message, once it is selected from a known set at the transmitting end, travels towards the receiver (that has a copy of this set of messages) through a noisy channel that corrupts it, i.e., it rotates the message in state space. The variational principle then amounts to devising, under certain constraints, an optimum code or a one-to-one mapping between the members of the transmitting set in the initial state space and a state space of much larger dimension.

In the best case, such an inflation of the state space generates the quasiorthogonality of the mapped vectors, thus ensuring some immunity from single- or multiple-channel errors. Provided that the coding operation is linear, the inverse operation takes place (reduction of dimensionality) at the receiving end. Obviously, in such a formalism of one-to-one mapping—which serves admirably the need for reliable message copying—there is no room for simulation, abstraction or memory-forming processes. The Shannon information processor acts simply as a linear transformer, and the coding-decoding sequence amounts essentially to diagonalizing a matrix of conditional probabilities.

To better understand the difference between the above paradigm and the one we are about to suggest (that of an hierarchical dissipative mapping), we examine how two important issues, namely the storage of information and the generation of information, can be treated by a mechanistic and a biological processor, respectively. By "mechanistic," we mean a processor unable to make abstractions and, moreover, one that once faced with a self-referential scheme (like any linguistic scheme) suffers a deadlock instead of searching for self-consistency.

First let us choose a word, say "virus," and ask: How much information is stored in this word? A mechanistic processor would proceed as follows: The processor would read, say, 1,000 pages from the *Encyclopedia Britannica*—just to collect enough data for the information needed to generate a 27×27 transition probability matrix between the symbols $i \rightarrow j$ of the English alphabet—and then would proceed to calculate the probability of the appearance of the word "virus" as

$$P = P_v P_{iv} P_{ri} P_{ur} P_{su} P_{-s},$$ (112)

where P_v is the *a priori* probability for v. The *a priori* probability of the 27 symbols are calculated from a system of 26 linear equations:

$$P_i = \sum_{j=1}^{27} P_{ij} P_j, \tag{113}$$

and the condition

$$\sum_{i=1}^{27} P_i = 1. \tag{114}$$

The entropy, or information, stored in the above words would be $S = I = \log_2 (1/P)$ bits.

It is important to say that we oversimplify our calculation by accepting a single-step Markov chain. In a natural language (such as English), the chain involved seems to be on the order of five to seven memory steps. However, our main argument is still valid.

A biological processor, on the other hand, would proceed very differently. Instead of calculating the entropy of the word, it would try to estimate its *complexity* by the length of the minimal program needed for its reproduction. To this end, the processor would play the game of Twenty Questions, claiming that the amount of information stored in "virus" is approximately the number of Yes-No answers (or the minimal number of judicious dichotomies) the processor is capable of. It should be borne in mind, though, that a prerequisite for playing such a game is the existence of *names* or compressed versions of alternatives in the cognitive repertoire of the processor. To make this particular point more graphic, consider the task of calculating the information content in a pack of cards.

There are $52! \sim 10^{68}$ configurations of such a system, so one would might naively assume that in about $\log_2 10^{68}$ Yes-No questions it would be possible to settle the issue of how the deck is arranged. That is, the deck contains this many bits of information. But this is an illusion: to proceed to the next question means that we are in possession of the names, or the compressed versions, of the (many!) different alternatives. Otherwise, we cannot form categories and, hence, make any decisions. Since, however, the overwhelming majority of the card configurations are incompressible (for a proof, see Appendix A), such a task simply cannot be carried out. So the issue remains unsettled.

Thus, although a mechanistic processor would conclude that a pack of cards contains ~ 220 bits of information, his biological counterpart would claim that the task is operationally impossible and therefore meaningless.

Now let us see how our two friends M(echanistic) and B(iological) would respond to a second challenge, the estimation of information generation physically prompted by a cascade of broken symmetries. This is one way of producing information; the other is via cascading iterations of increasing resolution. Beyond a critical point, microscopic noise is no longer smeared out but declares its presence from microscopic to macroscopic levels. In many cases, when a dynamical system becomes destabilized via a shifting of one or more control parameters, the emerging branches in the plane of the variable(s)-control parameter(s) are symmetric, i.e., whenever a certain branch is selected by a random choice (microscopic fluctuations), the "catoptric" pattern is *a priori* equally probable and tunneling to that catoptric branch afterwards is allowed with finite probability—which nonetheless decreases rapidly with the scale of the system.

Under such circumstances we speak of "repealed" rather than "broken" symmetry. More precisely, the answer to the question "repealed or broken symmetry?" seems to depend on the observational time window. It is possible, though, to imagine a set of asymmetric initial and/or boundary conditions through which all but one branch of the bifurcation diagram get suppressed—thereby allowing for broken symmetry. Physical agents of such a transfer of asymmetry from the environment to the evolving dynamical system might be a gravitational or a polarized electromagnetic field [30]. At a more subtle level, even in the absence of an external or an *ad hoc* asymmetry, one may invoke the weak nuclear interaction, whose influence is not only manifested in an asymmetry at the atomic level, but also in the preferred chirality of biological macromolecules [31]. Be that as it may, genuine broken symmetry is possible. Hence, genuine information generation can be attributed to physical causes. How would such information generation be appraised and measured by a mechanical or a biological processor? We argue that the mechanical processor would act as follows:

• Formulate the master equations for the probability density function of the dynamical system involved, well before and after the instability causing the symmetry breaking

$$\dot{P}_I = f_I(\cdot), \quad \dot{P}_{II} = f_{II}(\cdot),$$

respectively. (In between, the probability density function is multi-humped.)

• Find the asymptotically stable solutions (if any) of the above master equation

$$f_I(\cdot) = 0 \rightarrow P_{I\infty}.$$

• Calculate the information production as the entropy difference, namely,

$$I = S_{II} - S_I = -\sum P_{II\infty} \log_2 P_{II\infty} + \sum P_{I\infty} \log_2 P_{I\infty},$$

where the sums are taken over the sets of the discrete states that the system could assume—before and after the instability.

The biological processor, on the other hand, would measure the information generation by the increase in the length of the minimal description (complexity) needed for the replication of the pattern—before and after the instability.

For example, consider a sphere. The information necessary for its replication is its radius. Now make a hole on the surface of the sphere at some point. To replicate this pattern, you need "en plus" the coordinates ϕ, θ and the radius of the hole (if it's circular). One may think that the complexity of a pattern increases monotonically as the cascade of bifurcations leading to broken symmetry goes on. However, this does not seem to be the case. Again, consider the case of a perfect, solid homogeneous sphere in which you start drilling small holes. Initially, the complexity increases. But beyond a certain point (if you drill *too* many holes) the previous solid sphere turns into a more or less homogeneous and isotropic spherical cloud of dust. So we infer that the curve of complexity versus the number of broken symmetries is (single?) humped. The maximum complexity is to be found after a finite number of "mutilations" of the initially perfect symmetric pattern.

We have deliberated at some length in this introduction on the different (and equally legitimate) *modi operandi* of a mechanistic and a biological processor just to show that the M-processor works on the syntactical level, whereas the B-processor works at both the syntactic and the semantic levels, and presumes the ability of the system to display hierarchical discrimination and perform abstractions—which leads to a compressed version (memory) of any of the alternatives involved. In this case, the length of the minimal program is just the information dimension of the attractor to whose basin the initial message belongs.

– Where Is Information Stored? –

Although the overwhelming majority of environmental stimuli are incompressible by a given biological processor, a small subset of them (a basin) are, in principle, compressible up to a given number of bits [28]. Thus for the members of such a subset, it is meaningful to ask the above question. In our opinion the answer has emerged from a very simple and very original experiment performed some years ago [29] by Glass and Perez.

These authors took a blank sheet of paper and photocopied it using an imperfect machine. Eventually, some black dots appeared here and there. They repeated the process on the same sheet (up to ~ 20 times) until they obtained a rather homogeneous set of dots—a perfectly random (noisy) and informationless pattern, it seems. Then they made a transparency of this sheet and superimposed this transparency on the original— thereby forming a linear map of the plane to itself.

Playing with three available parameters of this map (the scale a, along the x-axis, the scale b along the y-axis, and the angle of rotation Θ), they were able to see patterns (circles, spirals, rays emerging from a center of hyperbolic trajectories) consisting of dynamical flows (eigenfunctions) in the neighborhood of the singularities as deduced from the eigenvalues of the map—i.e., depending on a, b, and Θ.

Thus, we conclude that information is confined neither to an impinging time series nor to the internal activity of the processor—by themselves they are equally meaningless. Rather, the information emerges from the iterated map that the processor applies by way of conjugating external stimuli with internal activity; it consists of the set of (stable) eigenfunctions of such a map, which for a dissipative nonlinear operator is simply the set of coexisting attractors. Biological processors of a given species, possessing identical central nervous systems, tend to generate more or less identical maps or flows, thereby creating, as it were by social consensus, universal pattern classification schemes.

– A Biological Processor Possessing Multiple Coexisting Attractors –

The processor amounts to a nonlinear dissipative operator of the form $\dot{\underline{x}} = \underline{f}(\underline{x}; \lambda)$, $\nabla \cdot \underline{f} < 0$, whose dynamical ongoings are embedded in a N-dimensional state space. Suppose there are Λ coexisting attractors (steady states, limit cycles, tori, and strange attractors) with basins divided by fractal separatrices—either fully connected or disconnected ("Mandelbrot dust").

From a biological processor we expect the formation of multiple dynamical memories (attractors) upon which we impose a number of demands if they are to emulate biological structures. A dynamical memory has to be contrasted with a static one. In the latter, all possible environmental stimuli are prestored in the processor. For large memories such a technique is impractical. Hence, instead of memorizing all possible answers to a set of environmental inputs, one stores an *algorithm* that, once triggered, generates on the spot all possible strings that may match the incoming message, assuming they share the same attractor, i.e., belong to the same basin. Such demands are input sensitivity, fault tolerance, content (context) addressability, large storage capacity, and compressibility, as well as criteria for evaluating the self-consistency of (linguistic) self-referential schemes. But before examining the above prerequisites and modeling the corresponding memories, let us note that since all linguistic schemes are by necessity discrete (in order to allow for error detection and correction), it is important to know how it is possible to deduce them in principle for continuous dissipative flows. There are two general ways:

1. Concerning a single chaotic strange attractor in an N-dimensional space, we first perform an (arbitrary) analog-to-digital conversion by way of a Poincaré mapping, which amounts to cutting the trajectory with a hyperplane, recording the isophasic crossings, and deriving a set of discrete coupled nonlinear difference equations in $N - 1$ (or fewer) dimensions.

   Next, we choose a nonoverlapping partition β on the discrete map above and derive a Markov chain of simple or multiple memory with a number of states, the partition number and transitional probabilities determined by the local slopes of the map [16]. Specifically, the conditional probability of jumping from the jth partition element to the ith in exactly ℓ iterations is given by the ratio of the value of the invariant measure $= \overline{\mu}$ of the map at the intersection of the ℓth backward iterate of the map of the element B_i with the element B_j versus the value of the invariant measure at B_j or

$$P_1(B_i|B_j) = \frac{\overline{\mu}[f^{-1}(B_i) \cap B_j]}{\overline{\mu}(B_j)}. \tag{115}$$

This formalism holds true only for cases in which the division points of the partition are mapped onto division points. A great many partitions, in fact, are non-Markovian in the sense that division points are not always mapped onto division points. This means that instead of having B_i mapped onto either B_j and/or B_k, etc., it is mapped onto part of B_j, onto part of B_k, and so on. The analysis becomes very involved [17], and a numerical program carried out by R. Feistel seems to indicate that in the case of such non-Markovian partitions, a long-range coherence in the digit time series $B_i B_j B_k B_l B_m \ldots$ may result. Namely, the appearance of a given digit B_i may influence the appearances of a considerable number of those digits that follow. At the same time, though, the overlapping of the partition intervals results in increasing entropy production in cascade iterations. In such a case, the alphabet is distorted by the flow.

2. Consider now the case of multiple coexisting attractors (see Figure 38). Following [18–20], we may write down a system of kinetic equations with respect to the probabilities P_i of occupancy of attractor i, and accept the possibility of jumping from attractor to attractor as a result of external noise, which is instrumental in shifting an initial condition from basin i (where it normally belongs) to basin j. In this formalism, the noise appears via the individual transient times τ_0 for leaving attractor i—until landing in another. Here the transition probabilities P_{ij} of jumping from attractor j to attractor i are assumed to be proportional to the area of the basin of the attractor i. The shape of the separatrices also influences the above parameters. For Λ coexisting attractors, Arecchi's equation reads

$$\dot{P_i} = -\frac{1}{\tau_{0i}} P_i + \sum_{j=1}^{\Lambda} \frac{1}{\tau_{0j}} P_{ij} Pj = \text{escape term} + \text{re-injection term.}$$

(116)

We must say here that it is physically meaningless to search for asymptotic stability in Arecchi's equations, since they describe an essentially intermittent mechanism that is structurally unstable. Further, we cannot really talk about a priori probabilities P_i. The notion of an a priori probability presumes the existence of asymptotic stability (convergence) in a trial process. Their solutions would fully determine a semi-Markov (see Appendix A) chain now having to do with a linguistic scheme among the coexisting "memories" of the system.

Thus we have a biological model processor with two distinct types of linguistic schemes:

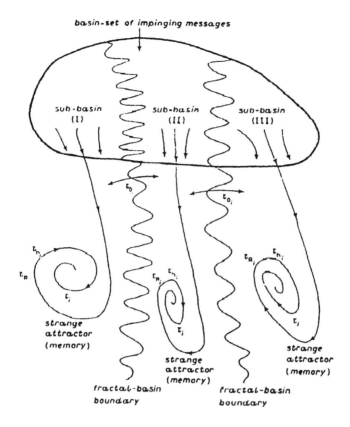

Figure 38. Sketch of a cognitive channel working after the dynamics of "generalized intermittency"; within each attractor a "micro"-intermittency may go on as well (for example, in the Lorenz system).

1. One that refers to one global memory, e.g., the memory of a basin (attractor). This scheme is instrumental in directing single messages into a particular attractor. This process promotes a slow diffusion (mixing) from one part of the attractor to another, so that the memory of the *specific* message inevitably fades.

2. Another scheme involves establishing connectivity and communication among otherwise independently existing global memories (coexisting attractors). In the case in which the noise catalyzing the transitions between attractors comes from within the processor itself, so much the better. Happily, as we already know from experimental results, neurophysiological evidence from the (human) central nervous system tends to support the view that the above "scanner" of the memories might be identified with the

thalamocortical pacemaker, whose one-dimensional signature is the recorded EEG.

Let us now return to the group of characteristics a memory should possess in order to qualify as being "biological": input sensitivity, fault tolerance, and content addressability can all be handled by accepting a chaotic attractor as a memory. Recognition of an object from a distorted or partial input—which implies a large Hamming distance of the initial stimulus from the attractor—can be achieved since the underlying dissipative dynamics will drag the initial condition onto the attractor in a transient time during which there is a progressive abstraction of the signal from the initial dimension N to the final information dimension of the attractor D_i, thereby ensuring an average compression factor $C_i = N - D_i$. The process is also context addressable, since only the signals belonging to the basin of the particular attractor will be affected by the processor.

Next, the attractor must be strange in order to ensure dynamical flexibility. Steady states and limit cycles possessing dimensions 0 and 1, respectively, are excellent devices as compressors but poor as memory banks. The issue of the best compromise is raised in [25], where we found an optimum nonzero resolution ϵ^* under which the dynamical capacity for memory storage by a strange attractor (that is, its information dimension) is maximized—without jeopardizing compressibility. Of course, for $\epsilon < \infty$, D depends on the partition and no longer expresses a topological property, as is the case when $\epsilon \to 0$. Still, the derived expression of D^* is dynamically meaningful, and determines the most parsimonious partition for the processor involved, as well as the optimum code or length of string

$$M^* = \log_2 \frac{1}{\epsilon^*}, \tag{117}$$

which may describe the attractor memory. The fact that in some examples M^* turns out to be near the "magical" number 7 ± 2, known from experiments on cognitive psychology, may not be entirely accidental.

– Chaotic Dynamics of Logical Paradoxes –

Most systems of information science (e.g., linguistics) contain multiple, hierarchical feedback loops that inevitably make them self-referential (see Figure 39). If such systems do not possess closure or logical consistency, they are termed "paradoxical" (see Appendix C). The truths— if any—revealed by such systems are associated with self-consistency,

Figure 39. The caricature of a "closed system" taken from R. Rucker, *The Fourth Dimension* (Houghton Mifflin, Boston, 1984), displays the futility (and soberness) of a self-referential scheme that is not even paradoxical, since it is self-consistent. Information from the outside is needed to dispel this tautology. In any open dissipative system, such a communication with the external source of broken symmetry is possible.

which implies the existence of stable attractors in the iterative and recursive system we call a self-referential sentence.

The search for self-consistency of a self-referential sentence then leads to the formulation of a boundary-value problem. The resulting stable eigenfunctions (if any) play the role of coexisting stable patterns (memories). It follows immediately that logic or rationality, thus far static concepts, can be dynamically implemented and replaced by evolutionary stability [45]. If an external stimulus, instead of being categorized and disposed of, is intermittently "circulating" in the processor (that is, oscillating between two or more coexisting repellers), we are talking about inconsistency or paradox. A certain self-referential linguistic scheme may be simulated as a discrete map (a digital feedback loop)—plain or hierarchical. We may classify the lack of consistency of such a scheme into two categories:

• The inability to display stable attractors ("cognitive receptors") as eigenfunctions of the above map ("This sentence is false.").

• The inability to distinguish the ongoings at two different hierarchical levels ("Chicago is a trisyllable.").

An example involving both cases is "This sentence contains two errors." At one (grammatical) level, the sentence is false. At a higher (cognitive) level it is true, but then it is false again, and so on—we get an interlevel contradiction. Let us also give two counterexamples whose consistency is associated only with specific basins of initial conditions, while for another choice of initial conditions they remain contradictory.

Take as a first example the symbolic self-referential loop: "This sentence contains X_0 0s, X_1 1s, X_2 2s, X_3 3s, X_4 4s, X_5 5s, X_6 6s, X_7 7s, X_8 8s, and X_9 9s." Let the 10-dimensional vector $\underline{X} = (X_0, X_1, X_2, \ldots, X_9)$ start from a randomly selected initial condition, the next iteration being obtained by checking the truth of the previous one. For example, if $X_a = (0, 1, 2, 3, 4, 5, 6, 7, 8, 9)$, the next iteration yields $X_b = (2, 2, 2, 2, 2, 2, 2, 2, 2, 2)$. Next, $X_c = (1, 1, 11, 1, 1, 1, 1, 1, 1, 1)$, followed by $X_d = (1, 12, 1, 1, 1, 1, 1, 1, 1, 1)$, and finally, $X_e = (1, 11, 2, 1, 1, 1, 1, 1, 1, 1)$, which is clearly the attractor corresponding to the chosen initial condition that makes the self-referential sentence above self-consistent, i.e., true. One can find another point attractor, as well, a period-2 limit cycle. By the same token, sentences like "I can't say this" oscillate between two cognitive repellers.

In biology any metabolic step can be considered as a mapping; an immunological contradiction may arise in cases, for example, where part of a given chromosome codes for proteins that act as antigens to proteins produced by other parts of the same chromosome. Or it may happen that the lack of biochemical receptors on membrane sites (e.g., LDL or insulin receptors) causes a certain substance to circulate in the blood for a long time instead of being disposed of. Or consider the dissipative multihierarchical genetic map between the levels of nucleic acids and amino acids. Here again we have a self-referential scheme. Yet, in a healthy organism the eigenfunctions of this map (different types of proteins and *specialized* cells) are *coexisting* stable attractors trapped in a "web" of fractal separatrices (basin boundaries). In cases in which the stability of those attractors is upset and the topology of fractal boundaries changes (due to a shifting of some control parameters by noise), an attractor "crisis" may develop as a "collision" across a fractal boundary, and a functional illness (e.g., a malignancy) may erupt as a result. In perception, examples of ambiguous patterns (like the Necker cube or the "vase and two opposing figures") provide cases of stimuli moving intermittently between two coexisting cognitive repellors—leaving recognition inconclusive.

– Neurophysiological Evidence About the Role of EEG and Modeling the Role of the Thalamocortical Pacemaker –

A description of the state of the art in EEG research is given elsewhere ([46], [8], and references therein). Specific thalamic nuclei are capable of self-sustained oscillatory activity. Via fibers emanating from these nuclei and projecting on various cortical areas, as well as fibers leading back from cortical areas to other nonspecific thalamic nuclei, a thalamocortical-corticothalamic pacemaker is established. A one-dimensional macroscopic manifestation of this activity is the EEG, which sums up instantaneous subthreshold dendritic potentials from any lead on the scalp. It appears that the thalamocortical pacemaker acts as a scanner of the set of cortical neurons on a time-division-multiplexing basis: In the absence of external stimulation, this non-linear oscillator takes up, on a "forced" intermittent basis, individual subsets of cortical postsynaptic-membrane dendritic oscillations and entrains or phase locks them, thereby forming semicoherent neuronal groups that constitute the hardware of the attractor involved. This is accomplished by plastic modification of the synapses. The sequence of jumps from attractor to attractor may then be simulated as a semi-Markov process.

Here the pacemaker plays two roles: On the one hand, it is responsible for stimulating the attractors one by one, and on the other hand, it is instrumental in making them commute as if under the spell of external noise. One might ask what happens to these attractor-memories when the pacemaker leaves them and grabs onto another subgroup of cortical neurons. Do they dissolve into oblivion? The answer is that the pacemaker simply awakens these memories, since by making the cortical group involved coherent it helps that group to elevate itself above the ambient neuronal noise. The *consolidation* of memory, however, may be achieved via synaptic-membrane-genome interactions [1,47], e.g., by stimulating genes in the neuronal genome in which the genes give rise to proteins that renew ("recoat") the postsynaptic-membrane sites of the population involved, thereby ensuring a long-term engram of the particular memory.

When the stimulation comes from the environment (via the peripheral nervous system), the degree of arousal of the ascending branch of the reticular formation increases, and the specific thalamic nuclei responsible for the generation of the initial oscillatory activity get polarized by amounts of time that are roughly proportional to the speed of information pumped from the peripheral nervous system or proportional to the intensity of the impinging external stimuli on the peripheral receptors. The result is that the simple semi-Markov sequence describing the intermittent processing by the thalamocortical pacemaker turns via a phase transition, as it were, to a composite semi-Markov process (see Appendices C and D). This means that during the time interval τ_{h_i} that the specific thalamic nuclei are polarized, the oscillatory activity of the pacemaker stops and the system gets stuck at the previous attractor-memory *in excess* of the usual residence time τ_{R_i} that one would expect from simple intermittency. After selecting the next attractor—but before moving to it—the pacemaker remains at the previous attractor by an amount of time equal to the interruption interval of the specific thalamonucleic activity. So now the scanner works in a different mode, namely metastable chaos.

The distribution of holding times is, of course, unknown. In some models [48] it is feasible to consider a geometrical distribution. Two questions now arise: Does the thalamocortical oscillator itself possess a low-dimensional attractor in spite of its manifestly irregular (noisy) activity and, second, how can one determine the dimension of the individual attractor-memories of the system? The answer to the second question remains open. The first question has been addressed quite recently [49,50] by collecting EEG time series during epilepsy, deep sleep, and awakening, and then trying to infer the dimension of the underlying dynamics. A basic difficulty in such an enterprise is the essentially nonstationary character of the EEG, especially during the awakening regime. Nevertheless, people came up with low dimensions, ~ 2.1 and ~ 4.1 for the epileptic and the deep-sleep regime, respectively, and with high dimensions, $a + b$, $b \sim O(a)$, and $a \sim 7 - 8$ [50,51], in the case of the awakening regime, which de facto shows the unreliability of the underlying method when the *modus operandi* of the pacemaker is the metastable chaotic (partial covering of many attractors).

Thus the general model emerging from the combination of the dynamical processes going on within each individual strange attractor-memory on the one hand, and on the other hand, between coexisting attractor-memories is as follows: For a given bioprocessor, a (small) subset of environmental stimuli is partitioned into a number Γ of coexisting attractor-memories whose formation is mediated by a nonlinear dissipative operator (map). These attractors are separated by fractal basin boundaries. The entropy of such a partition (of the messages to the attractors) and the degree of compressibility afforded by each attractor $C_i = N - D_i$ (where N is the dimension of the raw environmental messages and D_i is the dimension of the individual memory) give the two essential macroparameters characterizing the cognitive channel between environment and processor at a given hierarchical level. [This may not be accomplished in one single hierarchical step; in a second step the attractors (if numerous at a lower level) will play the role of the members of a hyperbasin towards a new hierarchy of fewer hyper-attractors, and so on.]

The attractor memories involved establish further communication via the thalamocortical pacemaker—within the processor. The entire activity has two aspects: *chaos* and *hyperchaos:*

• The *intramemory* activity (which involves rehearsal and consolidation) refers to a Markovian process within each attractor; this amounts to a slow diffusion from one part of the attractor to the other, i.e., a progressive "smearing-out" (mixing) of the specific initial stimulus. The memory of the basin as a whole, though, remains intact. In the long run, we do not memorize specific events but rather, sets of events, unless we possess a phase-coherent attractor, where regeneration of transinformation is possible, thereby ensuring persistence of memory of any individual member of the basin as well.

• The *intermemory* activity refers to establishing an intermittent connectivity *between* the individual stable memories. In the absence of external excitation this is a simple semi-Markov process; in the presence of stimulation, though, it turns into a composite semi-Markov process with total holding time distribution depending not only the intrinsic residence times, but also on the statistics of the external stimulus modulating the pacemaker itself.

We end this series of speculations with one last topic, namely, the spectrum of intermemory jumping activity. If we consider the slow process of self-diffusion within each individual attractor-memory as resulting from an autocorrelation function $\exp(-t/\tau_i)$, $i = 1, \ldots, \Lambda$, involving a single time scale τ_i, then the corresponding spectrum is simply a Lorenzian,

$$\frac{\tau_i}{1 + \omega^2 \tau_i^2}. \tag{118}$$

If we have more attractors and weight them in some way on a scale-invariant fashion (e.g., as $1/\tau_i$), we might get as the spectrum of the processing activity

$$S(\omega) \sim \int_{\tau_1}^{\tau_2} \frac{\tau}{1 + \omega^2 \tau^2} \frac{d\tau}{\tau} = \left. \frac{tg^{-1}\omega\tau}{\omega} \right|_{\tau_1}^{\tau_2}, \tag{119}$$

and if $r_2/r_1 \gg 1$, then $S(\omega) \sim 1/\omega$ over the correspondingly larger range of frequencies (see Appendices E, F, and G).

Is there any physical model that could justify the existence of the notorious "$1/f$ noise" for a biological processor involving the sequential jumping among coexisting memories? In [18–20] an interesting possibility is suggested: Whenever an event is conditioned by a sequence of previous ones, assuming that the probability per unit time of each step is independent of the others, the probability of the final event is then the product of the probabilities of the individual preliminary events:

$$P = \prod_{i=1}^{\Lambda} P_1 \quad \text{and} \quad \log_2 P = \sum_{i=1}^{\Lambda} \log_2 P_i, \Lambda \gg 1.$$

Thus, the density of $\log_2 P$ follows the normal distribution provided that the central-limit theorem holds. A variable X obeying a log-normal distribution has a density $\sim 1/X$ over a wide range. Applying this argument to the time constant $\tau \sim 1/P$ of an event conditioned by a chain of previous ones, we get a $1/\tau$ distribution under the scaling assumed above.

To apply this theory to our physical model of jumping from attractor to attractor, we consider a case [18] where there is a "leakage" from one attractor to the next (perhaps via a crisis), and the last one can be reached only via a unique chain and not via multiple paths. Also, the basin of stimuli feeds mainly the first attractor, with all other attractors having negligible basins of attraction. Alternatively, it is possible to reach a $1/f$ spectrum [20] even if the processor possesses only two attractors separated by a *fractal* boundary. Then the noise responsible for making the attractors commute will force the trajectory of any initial condition to perform a random walk. Spectra of EEG, though, under all behavioral conditions studied so far, do not exactly conform to $1/f$ noise. This *might* be the case if one could identify τ_i as $\tau_{r_i} + \tau_{h_i}$, i.e., that full memory mixing takes place before jumping to the next attractor, and that $t_{0_i} \gg \tau_i, \tau_{R_i}, \tau_{h_i}$.

Note that we have four (different) time scale distributions:

- The average *relaxation* times τ_i within each attractor, which we may operationally define within an individual attractor as the time required for the transinformation to fall, say, by $1/2$. This transinformation $I(t)$ is not uniquely defined, but depends on the map or flow (f) that we choose and the given partition β with M state subintervals B_i on this map (or quantized flow). The transinformation is given by the expression

$$I(t) = \sum_{i=1}^{M} \sum_{j=1}^{M} \bar{\mu}[f^{-t}(B_i) \cap B_j] \log_2 \frac{\bar{\mu}[f^{-t}(B_i \cap B_j]}{\bar{\mu}[f^{-t}(B_i)]\bar{\mu}(B_j)} \text{ bits.} \quad (120)$$

$I(t)$ is the information about the state of the system t time units ahead contained in an initial condition. The slope of $I(t)$ is a measure of the loss rate of initial information, that is, the rate of fading of the specific initial condition.

- The *transient* times τ_{0_i} from one attractor to another.

- The *residence* times τ_{R_i}, which depend on the mechanism of the forced intermittency before the pumping of messages from the outside. In this regime, the thalamocortical pacemaker is continuously *on*.

- The *holding* times τ_{h_i}—after the biasing of the specific thalamic nuclei and the selective interruption of their pacemaking function take place. The way of convolving τ_{R_i} and τ_{h_i} to get an overall residence time on the ith attractor is an open problem.

Unlike the EEG, "good" music seems to comply with the $1/f$ spectrum. Elucidating this disparity ("Why then do people like $1/f$ music?") appears to be one of the great intellectual challenges of chaotic dynamics as applied to biological—and especially to human—information processing.

– Summing Up and Outlook –

A long-standing goal of reliable information processing (and avoidance of catastrophes due to communication breakdown) is to find a design of cognitive devices (processors) characterized by input-sensitive, and yet fault-tolerant, content (context)-addressable memories. Digital computers cannot satisfy such a demand on any practical time scale. Analog nonlinear networks, however (mimicking to some extent biological neuronal ensembles performing parallel processing), can accomplish global cognitive tasks from imperfect inputs if their nonlinear dissipative dynamics possesses multiple (coexisting) stable attractors (steady states, limit cycles, tori, and strange attractors) separated by fractal (nowhere differentiable) basin boundaries. For a memory with the dual demands of good compressibility and large information dimension (dynamical storing capacity), only strange attractors will suffice. The attractors also need to be chaotic in order to comply with the extreme sensitivity of biological receptors to stimuli. In such cases, then, the stable attractors, which compress (e.g., abstract and store) any initial condition (stimulus) belonging to their own basin of attraction, play the role of dynamically stored-retrievable memories (patterns).

The attractors come about as the stable eigenfunctions of the nonlinear operator that the processor applies in conjugating externally (environmental) impinging stimuli with the internal (processor's) dynamical activity. These patterns are reachable though transients from any initial cue (state stimulus) belonging to their domain of attraction (basin) in the state space in which the dynamics of the processor is embedded. Such a cue may be far away in the state space (its Hamming distance from the attractor involved may be large), so it may contain just a small (and distorted) portion of the pattern under retrieval. Nevertheless, the operating dynamics, will pull down this cue together with its neighbors and make it converge and eventually land on the pertinent attractor-memory.

Clearly, such a type of retrieval is also content-addressable, since the initial cue containing incomplete information will converge only onto the attractor to which the basin belongs. Noisy perturbations affecting the control parameters of the system (e.g., the coupling constants of the operating variables) may blur the separatrices and change appreciably the topology of these boundaries beyond a certain critical shifting of the numerical values of those control parameters—thereby making the coexisting attractors "leak" onto each other. As a result, some subsets of initial conditions will jump from one attractor to another. The noise may also (through a crisis) completely delete some attractor-memories, create others, or fuse some others together. So under the action of external noise, some stimuli are transformed into perpetual transients (so that their destination remains inconclusive), while some others are "reinterpreted" or rescheduled to different stored patterns (they may elicit different memories than before).

External noise may also be instrumental in inducing transitions among the existing attractors (without involving crises), i.e., the noise causes stored memories to establish mutual connections on a sequential or a time-division-multiplexing basis. If the noise is supplied from within the processor itself, so much the better. Happily, this seems to be the case for the mammalian brain. In our model of biological processing, the thalamocortical pacemaker, whose behavior is externalized as the EEG, is the processor's noisy agent responsible for stimulating the attractors out of semicoherent groups or cortical neurons, permuting them on a time-division-multiplexing basis and provoking crises involved in the annihilation, creation, and fusion of attractors. The thalamocortical pacemaker during sleep and epileptic seizures is a strange attractor of rather low dimension (~ 4.1 or ~ 2.1, respectively). During the execution of mental tasks or REM sleep, the oscillator is really noisy—of high dimension in the state space. It may happen, though, that even in the absence of external noise some of the coexisting attractors are intrinsically unstable (repellers). In such cases, a given stimulus belonging to the basins of those repellers oscillates forever among them, as in the well-known example of the ambiguous perception of "the vase and the two opposing figures."

Lacking asymptotic stability, such ambiguous stimuli are considered logically inconclusive—or even paradoxical—and their persistence may cause a "deadlock" in the cognitive process, since such stimuli, instead of being categorized and disposed of, circulate endlessly in the processor. In such a case, the organism cannot habituate, a well-known cognitive symptom of schizophrenic subjects, for instance, which embodies *par excellence* a paradigm of communication breakdown with the environment.

A final comment is in order: If we *really* intend to understand biological information processing, we have to attribute a *nonlocal* property to any biological processor. This comes about when we realize that a mapping between environmental stimuli and the internal processor's activity presumes that the external stimuli *are* initial conditions, that is, solutions of the processor's dissipative dynamics whose eigenfunctions (if any) are the attractors, i.e., the abstract forms (or physical laws) of environmental stimuli. Otherwise, if a biological processor is just a local (in the trivial physical sense) and unsuspected evolutionary byproduct of the laws of nature, how can this byproduct compress the very laws that acting upon matter gave rise to its own information? Pragmatically speaking, we might with some trepidation speculate that the physical world has no independent existence (someone must always be around in order to "reduce" the wave function). Rather, it is embedded into a nonlocal intelligence (a nonlinear dissipative operator) that manifests itself in information processing, i.e., in mapping external stimuli onto the internal processor's dynamical activity. The stable eigenfunctions of such a map (the attractors) constitute the patterns observed.

Consequently, intelligence appears as an emergent evolutionary epiphenomenon and a prerequisite for the collapse of the physical world from a pure state of "suspended animation" into a set of distinct categories. Such an attitude is familiar, of course, for the treatment of subatomic particles, and constitutes the time-honored and recurring theme of quantum-mechanical measurement. An elementary particle, nevertheless, is a good example of a hierarchy "biting its own tail," as it were, since it also constitutes a very complex and very abstract macroscopic pattern. We may couch the argument in the language of chaotic dynamics, asserting that an electron, for example, can be considered as a basin of attraction in operator space. Depending on the initial conditions, that is, the way we prepare the experiment, there are two stable attractors: the corpuscule-like pattern and the wave-like pattern, complementary and mutually exclusive. Conversely, quite recently we have obtained theoretical predictions and experimental evidence [52,53] of macroscopic variables, like the phase difference across a Josephson junction interrupting a superconductor ring, that display quantum-mechanical behavior. Such variables belong to macroscopic systems controlled by microscopic triggers, i.e., chaotic systems whose intrinsic instability literally explains the passage from time-reversible microscopic laws to irreversible macroscopic dynamics. On a more modest level we may point to three (still unresolved) problems:

• The problem of the (static) partitioning of 2^N strings (messages) of length N into Λ coexisting attractors with dimensions $D_i \ll N$. The difficulty lies in the fractal character of the basin boundaries (connected or, even worse, unconnected). In such a case, no attractor can claim a clean basin of undisputed jurisdiction of radius ϵ, however small, since the finite probability of intrusion of a "tongue" of the fractal boundary immediately sends any initial condition within ϵ onto another attractor. Since, this happens for all attractors, the trajectory may hesitate for a long time, thereby creating very long random-walk-like transients.

• The problem of estimating the residence time mass function $f(i, j, \tau_j)$, which determines the statistics of residence times on each attractor j before the jump to attractor i takes place. This parameter depends on the intrinsic dynamics of a generalized (forced) intermittency, after which the individual attractors are permuted by the thalamocortical pacemaker in the absence of externally pumped messages.

• The problem of estimating the (composite) holding time mass function $h(i, j, \tau_j)$, where now messages pumped from the peripheral to the central nervous system interrupt the activity of the specific thalamic nuclei (by amounts of time depending on the speed of peripheral signaling), thereby causing the thalamocortical pacemaker to hold on to a given attractor an *extra* amount of time.

Here we have *two* intermittent mechanisms superimposed—one intrinsic, the other extrinsic—and we do not yet know how to synthesize them. Finally, let us conclude by mentioning once again the physical meaning of the three basic concepts involving characteristics of attractors as far as information processing is concerned: *dimensionality,* which stands for the capacity of the dynamical memory (and its compressing or "descriptive" ability); *transinformation,* which represents the amount of memory persistence (either in a single attractor or between attractors); and *basin width,* which characterizes the fault-tolerance ability of the attractor or its predictive ability.

– Electrical Activity of the Brain: Should It Be Chaotic? –

Brain-like structures must have evolved by performing signal processing, initially by minimizing tracking errors on a competitive basis. Such systems should be highly complex but at the same time disordered. The functional trace of the cerebral cortex of the (human) brain is a good example.

The electroencephalogram (EEG) appears particularly fragmented during the execution of mental tasks, as well as during the recurrent episodes of rapid eye movement (REM) sleep. A stochastically regular or a highly synchronized EEG, on the other hand, characterizes a drowsy (relaxing) or epileptic subject, respectively, and indicates—in both cases—a very incompetent information processor.

We suggest that such behavioral changeovers are produced via bifurcations that trigger the thalamocortical nonlinear pacemaking oscillator to switch from an unstable limit cycle to a strange attractor regime or, more correctly, from simple intermittency to multifractal chaos and vice versa. Our analysis aims to show that the EEG characteristics are not accidental but inevitable, even necessary, and therefore functionally significant.

An information processor (analog or digital) is a cognitive device that tracks and identifies the parameters of an unknown signal or pattern, which is usually contaminated by thermal (equilibrium) noise (white or colored, additive or multiplicative). In order to accomplish this task, the processor has to perform three distinct operations in the following sequence.

• Produce from within a wide variety of (spatio-temporal) patterns or "templates."

• Cross-correlate (i.e., compress) each of those patterns with the incoming one.

• Select or filter out on the basis of some pre-established hypothesis testing or consensus criteria the pattern that forms the greatest cross-correlation with the unknown signal or trigger. (The filtering is usually nonlinear, in order to create and enhance contrast, a fact which, by sharpening contours, makes recognition simpler. Selected groups of cerebral neurons—and the photocopy machine!—do just that.)

To track a signal's timing is of the essence. The simplest tracker in use in communication engineering practice is the phase-locked loop (PPL). This suggests that the existence of self-sustained nonlinear dissipative oscillators (i.e., elements displaying adaptive behavior) at the hardware level of the processor is a prerequisite for cognitive operation.

Functionally stable oscillators, in contrast to static (switching on-off) devices, offer a number of evolutionary advantages. Namely,

1. timekeeping,

2. dynamic information storage (dynamic memory) and, when triggered by very simple stimuli, possibly

3. an extremely broad spectrum of complex behavioral repertoires.

Finally, the oscillators must necessarily possess asymptotic stability. One cannot carry out reception and cognition tasks—which by involving phase locking or compressibility are dissipative (i.e., irreversible)—via Hamiltonian (reversible) subsystems; hence, the universality of the family of so-called Van der Pol oscillators in communication engineering.

Parsimony, which undoubtedly possesses survival value, requires that the locally generated dynamical patterns (the attractors of the processor) should not always be "on." Rather, they should emerge upon request (i.e., upon triggering from externally impinging stimuli) from a set of available dynamical elements and some basic, rather simple recursive rules (algorithms or maps) for combining those elements.

Below we present a sketch for a dynamical model of a brain processor. Individual neuronal oscillators in the cerebral cortex constitute the above-mentioned set of dynamical elements. The thalamocortical oscillator, on the other hand, is the adaptive agent that performs two distinct operations:

1. It provides pacemaking activity, resulting in the formation of internal synchronized or coherent (spatio-temporal) neuronal patterns, each of which stands for an attractor. By making such neuronal groups coherent, the pacemaker helps them elevate themselves

above the ambient thermal noise level. It also enables them to distinguish themselves from coexisting neighboring neuronal formations in brief time intervals.

2. It generates the recursive rules governing the sequential appearance of those coherent patterns—attractors on a time-division-multiplexing scheme (in a homogeneous and/or multifractal way).

– The Model –

We now regard the thalamocortical pacemaker as a high-amplitude, dissipative self-sustained nonlinear (relaxation) oscillator that in the absence of any environmental input is intermittently free running on an unstable limit cycle with a fundamental (sampling) frequency $\sim$ 10 Hz (the "α rhythm"). This means that the oscillator has an internal bias responsible for this instability.

The way such a strong oscillator can synchronize groups of smaller oscillators has been sketched in [21]. If our conjecture is correct, we should expect the corresponding EEG to exhibit some pseudo-periodicity. But it should also show some randomness, in the sense that successive amplitude segments of this activity (each segment corresponding to a given stationary state/pattern at the cortical level formed by the vector sum of the amplitude of phase-locked neuronal phasors of the pertinent cortical subset) should show no statistical correlation with each other. The optimal sampling of a random process is at random time intervals: thus, the sampled modalities whose embodiments are the sets of synchronized neurons should not overlap. This seems to be the case: the amplitude probability density function of such an EEG follows a normal distribution, as one might expect from the Central Limit Theorem.

Consider now what may happen in the regime of REM sleep or the performance of a mental task: information is pumped along the ascending branch of the reticular formation system, carrying from within or from the outside specific sensory inputs to be identified by the cortex.

Our thalamocortical oscillator is now under the spell of a fluctuating input, and beyond specific (threshold) values of some control parameter, a cascade of bifurcations may set in, as a result of which the intermittent limit cycle is deleted and the oscillator now follows part of a strange attractor. In other words, it becomes metastably chaotic; its scanning manifestations on the cortical neuronal subgroups turn into a spasmodic and nonperiodic oscillation, giving rise to a semi-Markov chain.

This hypothesis is corroborated by long-established experimental evidence [22]. Increasing arousal of the ascending reticular formation branch leads to polarization of the pacemaking-specific thalamic nuclei, thereby interrupting their sampling function for time intervals concomitant with the degree of arousal, i.e., the intensity of stimulation and therefore the rate of information transfer from the environment. So under excitation the sampling of the cortex from the pacemaking thalamic nuclei may switch from homogeneously chaotic to multifractal.

We witness the establishment of a semi-Markov chain with different corresponding probabilities U_i per stationary synchronized state i (internal pattern)—produced from a transitional probability matrix P_{ij} between successive states-patterns. So, and this is the crucial point, the amount of holding time that the cortical processor spends at the specific modality i depends on the amount of disruption time caused in the scanning pacemaker at i by the information input. This time usually—but not always—increases with the rate of the frequency of the incoming trigger.

Evidence supporting this view comes from analyses of the amplitude probability distribution of the EEG, indicating changes from Gaussian to skewed forms during the performance of mental tasks [23]. We are afraid, however, that the interpretation put forth in the above reference, that the skewness in the probability density function reflects some "increasing degree of cooperativity among the neuronal generators," is irrelevant. Changes of coupling among neuronal oscillators embodying a pattern may arise from different causes; they result only in changing (enhancing or deteriorating) the degree of coherence of the *specific pattern* to which they contribute. But, of course, this has nothing to do with the algorithm of time-division multiplexing or the policy of switching among patterns.

We interpret the observed skewness as indicating a statistical linear correlation $(P_{ij} \neq 0)$ or a Markovian characteristic between the successive amplitude EEG samples, i.e., between successive synchronized neuronal states, each coding for a single sensory modality. Under such linear interdependence, the Central Limit Theorem does not hold—as is the case for all *intermittent* regimes.

Suppose, finally, that the degree of arousal of the reticular formation ascending branch, or the average rate of information pumping, increases further. We conjecture that in such cases the scanning process of cortical subgroups is either extinguished (the oscillator becomes quenched), or the scanning rate increases concomitantly with the degree of arousal. In the limit, we may consider that all the subgroups of cortical neurons are synchronized at the same time, a symptom of epileptic convulsions. We do not possess reliable experimental data, and certainly a proper theory is still lacking. Nevertheless, in a recent computer simulation [24] hints have been seen that when the coupling between the environment and a limit cycle oscillator already in the chaotic regime becomes supercritical, one may witness a regression from the chaotic regime back to a limit cycle regime. The same regression can be achieved by external noise (see [25]).

The hint is relevant to the possible behavior of our thalamocortical pacemaker, which under strenuous external (phonetic and acoustic) triggering at its fundamental frequency or one of its harmonics may switch back from the chaotic to a strictly entrained regime with such a high amplitude of oscillation that it can synchronize large numbers of cortical neurons simultaneously.

To sum up, we have tried to understand why a cognitive system should be chaotic in order to perform effective signal processing. The answer can be set in both the time and frequency domains.

1. By creating a semi-Markovian time-division multiplexing, the system allows separation of sensory modalities-attractors and associates for each one of these modalities a time-processing interval commensurate with the rate of sensory input;

2. Turbulent chaos (nonequilibrium, low-dimension noise) as a *modus operandi* of the thalamocortical pacemaker under mild excitation contains a broad spectrum of temporal (and spatial) frequencies. Thus it can constrain "patches" of postsynaptic functional areas in the cortex, and create coherent patterns that can match a wide variety of incoming spatiotemporal patterns.

Specifically, cognition is manifested in the cortex as a result of a matching process between pairs of spatiotemporal patterns, each containing a great number of elementary units (neurons). In each pair, one pattern (the same for all pairs) is the unknown information that is embodied in incoming triggers, coded either in sequences of pulses from the peripheral nervous system or, if it comes from other areas of the central nervous system, encoded in strings of macromolecular (neurotransmitter/hormonal) releases from presynaptic endings.

The second pattern of the pair is one of the patterns-attractors created by the processor. It constitutes a prestored spatio-temporal mosaic embodied in a set of partly synchronized post-synaptic membrane receptors. The cross-correlation between the two patterns of each pair takes place dynamically via energy exchanges between equal or neighboring frequency pairs (ω_i, ω_j) shared by both spectra, and is estimated quantitatively by the coherence function

$$\text{coh}_{ij}(\omega) = \frac{\phi_{ij}(\omega)}{(\phi_i(\omega)\phi_j(\omega))^{1/2}}, \tag{121}$$

where $\phi_{ij}(\omega)$ is the cross spectrum and $\phi_i(\omega)$ and $\phi_j(\omega)$ are the autospectra of the two matched patterns.

The result of the cross-correlation in phase and amplitude determines the degree of cognition between the incoming and the preset of the unknown and the expected patterns. What mechanism in the brain decides which coherence function is predominant? It appears that the reticular formation might again be involved. Some plausible ways of carrying out the nonlinear filtering involved have been discussed in the paper [26].

In this application we have attributed the phase locking between individual neuronal oscillators (and therefore the formation of attractors in the cortex) only to the thalamocortical pacemaking activity. Undoubtedly there are other, perhaps equally important, very complicated mechanisms for extending long-range spatiotemporal coherence among cortical neurons, mechanisms which have received very little attention in the literature (exceptions granted, e.g., [27]). We only mention the possible role played in that respect by the cortical intercellular electrolyte. The intercellular fluid appears to contain extensive hydrated networks of complex membrane-bound macromolecules (mucopolysaccharides and mucoproteins). Removal of calcium allows those molecules to bind with water and is concomitant with their spreading in a loose hydrated network. Restoration of calcium reverses this process, unbinding water and contracting the molecular net up to five orders of magnitude in volume.

The existence of very weak electromagnetic fields appears to have a strong influence on Ca^{2+} movement in the intercellular electrolyte, as well as on the impedance of the fluid. Thus we may envision new ways of establishing and controlling long-range coherence in the brain, thereby influencing the formation of attractors in the cortex: random environmental (electromagnetic) fields. These fields, acting as external noise, may provide the mechanism for switching from attractor to attractor.

– An Example of a Fractal Basin Boundary (Separatrix) –

Grebogi et al. [32,33] give the following example of a two-dimensional non-invertible map:

$$\Theta_{n+1} = 2\Theta_n \ (\text{mod } 2\pi),$$
$$z_{n+1} = \lambda z_n + \cos\Theta_n, \qquad 1 < \lambda < 2, \ 0 < \Theta < 2\pi.$$

The Jacobian matrix of this map has eigenvalues 2 and λ. Since both exceed 1, there are no attractors with finite z. In fact, all initial conditions produce orbits that asymptotically go either to $z = +\infty$ or $z = -\infty$ as $n \to \infty$. So we consider $z = +\infty$ and $z = -\infty$ as two coexisting attractors, asking: What is their basin boundary? This means we seek a function $f(\Theta)$ such that an initial condition $z > f(\Theta)$ asymptotically leads the orbits to $z = +\infty$, while those $z < f(\Theta)$ lead the orbits to $-\infty$.

Starting with the above map, Grebogi et al. find the expression

$$z = f(\Theta) = -\sum_{k=1}^{\infty} \lambda^{-(k+1)} \cos(2^k\Theta). \tag{122}$$

Since $\lambda > 1$, this sum converges. Now consider $df/d\Theta$. From the above expression we obtain

$$\frac{df}{d\Theta} = \frac{1}{2}\sum_{k=1}^{\infty}\left(\frac{2}{\lambda}\right)^{k+1} \sin(2^k\Theta). \tag{123}$$

Since $\lambda < 2$, this sum diverges and so $f(\Theta)$ is *not differentiable*. In fact, Grebogi et al. find that the separatrix $z = f(\Theta)$ has an infinite length and a fractal dimension

$$d = 2 - \frac{\log\lambda}{\log 2} = 2 - \log\lambda. \tag{124}$$

A final and very important aspect of the information dimension of multiple coexisting attractors-classifiers is the problem of their invariance under smooth coordinate transformations. In other words, we are interested in knowing whether or not the degree of abstraction or compressibility of the basin of initial conditions is universal. We do not have a general answer to this question. Ott et al. [34] have recently investigated the possible invariance of a number of different types of dimensions of strange attractors and have found some specific coordinate transformations ensuring the invariance of only the information dimension. Clearly, much more work has to be done in this direction.

– Experimental Results in Visual Pattern Perception –

What are the crucial features of a visual pattern and how does the optical (human) cortex [in cooperation with the sensor (forea) and the optical muscular apparatus] deal with them? The fovea is a small area in the center of the retina, characterized by the highest concentration of photoreceptors. In a remarkable series of experiments conducted in the early 1970s, Noton and his associates performed a number of investigations which are discussed below. To begin, we look at Figure 2. The left part represents the famous bust of Queen Nefertiti. The right side shows the trajectory of the eye movement of a human observer (as recorded by A. L. Yarbus of the Institute for Problems of Information Transmission in Moscow)—as the person scanned Nefertiti's head and neck in the process of perception. The experiments were performed in an attempt to settle the controversy of gestalt (parallel, global) versus sequential (step-by-step, or iterative) pattern recognition. (An essential prerequisite for the meaningfulness of the attempted comparison is, of course, that the pattern must be sufficiently extended in space to allow for scanning by the sensor.) The results of the experiments heavily support the hypothesis of serial piecemeal perception and recognition. Specifically, two questions are answered:

• What are the features of the pattern that the optical cortex selects as the *key* items for identifying the object?

• How are such features integrated and related to one another to form the complete internal representation of the object?

First of all, we have good evidence that the optical system is hierarchical [35], and that between the successive hierarchical levels mappings take place giving rise to feedforward-feedback loops. The higher levels (cortex) receive information from the lower levels and respond by sending commands to the sensor's muscular apparatus to either move in order to "phase lock" with the feature under investigation, or to move away and assume the next algorithmic step of scanning the pattern. The experiments indicate that the above scanning algorithm is far from smooth and homogeneous. There are parts of the pattern that hold the majority of the information about the subject for the processor. The fixation of the holding time of the receptor tends to cluster around the parts characterized by sharp curves (small radii of curvature)—angles or, in general, around areas where the curve is nondifferentiable. It appears that the angles are the principal features the brain uses to store and recognize the pattern. There is neurophysiological justification for such a preference revealed through the painstaking experiments of Hubel and Wiesel (see [35]). It appears that there are angle-detecting neurons in the frog's retina and angle-detecting neurons in the visual cortex of cats and monkeys. Recordings obtained from the human visual cortex by Marg [36] indicate that this result can be extended to the human visual cortex as well. The scanning of Figure 2 bears witness to the heavy preference of the processor for areas of the face characterized by sharp curves. Such features are complex (they need many bits for their most compact description and are therefore endowed with high information content).

Let us digress for a moment here and give a proof (coming straight from Shannon information theory) for the above statement, namely that the information content of edges or parts of a contour with small radii of curvature is large. As experiments from cognitive psychology show [37], the information associated with a contour or curve is not uniformly distributed along the contour but is more or less concentrated at certain regions. It is the information associated with points of the curve where the *absolute value of the curvature passes through a local maximum* (where the curve "turns most") that is retained, and claims the lion's share in the perception, recognition, and reconstruction of objects represented by contours. The (human) visual system organizes the information associated with a contour by *eliminating* most of it and retaining a few "hot spots" or key features—to be acted upon by the higher (cognitive) processing cortical centers.

Let us then write down how much information is gained from the measurement of *position* and the measurement of *direction.* Suppose we want to measure a quantity x. The first measurement gives a rather coarse result; namely, x is bounded between x_1 and x_2. The second measurement refines the above interval to $x_3 > x_1$ and $x_4 < x_2$. The information gained is

$$I \cong \log_2 \left| \frac{x_2 - x_1}{x_3 - x_4} \right| \text{ bits.} \tag{125}$$

Likewise, if we intend to measure a certain *direction* emanating from a center 0, we perform a series of measurements that refer to angles within which the desired direction is included.

Of course, we measure the directional information gain in terms of the information gained from the measurement of the corresponding arc lengths on a circle of arbitrary radius r and center 0. If $s(a)$ and

$s(b)$ are the two successive arc lengths, then the information gained is

$$I = \log_2 \left[\frac{s(a)}{s(b)} \right] \text{ bits.} \tag{126}$$

Since, however, $s(a) = r \cdot a$ and $s(b) = r \cdot b$,

$$I = \log_2 \left[\frac{a}{b} \right] \text{ bits,} \tag{127}$$

where a and b, the (coarser) and the (finer) angles, respectively, bound the direction under investigation. While in measuring lengths *two* measurements at least are necessary to assess the information gained, for angles one measurement is enough, since we know *a priori* that this angle will be contained within the interval 0 to 2π.

So the information gained by measuring an angle a is

$$I = \log_2 \left[\frac{2\pi}{a} \right] \text{ bits,} \tag{128}$$

and, of course, two successive measurements of angles a and b, $a > b$, provide a gain of information for the direction under investigation of

$$-\log_2 \left[\frac{2\pi}{a} \right] + \log_2 \left[\frac{2\pi}{b} \right] = \log_2 \left[\frac{a}{b} \right] \text{ bits.} \tag{129}$$

Before applying the above expression to the information content of a contour, let us consider some simple examples of how this measure can be used [38]. If $a = \pi$, the angle is a straight line and the information content is

$$I = \log_2 \left[\frac{2\pi}{\pi} \right] = \log_2 2 = 1 \text{ bit.} \tag{130}$$

This measurement specifies that the observed direction lies in one or two half planes. Also,

$$I \left[\frac{\pi}{2} \right] = 2 \text{ bits,} \tag{131}$$

or, generally,

$$I \left[\frac{\pi}{2^n} \right] = n + 1 \text{ bits.} \tag{132}$$

Successive bisections then add one bit at each stage; smaller angels correspond to greater information. For example, an arrowhead used as a pointer or an information marker catches the attention because it contains more information than that contained in a large angular sector. Another example is the fact that the triangular shape, which provides the *least* information, is equilateral. Indeed if a, b, and c are the angles of the triangle, $a + b + c = \pi$, the corresponding information is

$$I = \log_2 \left[\frac{2\pi}{a}\right] - \log_2 \left[\frac{2\pi}{b}\right] + \log_2 \left[\frac{2\pi}{c}\right] = \log_2 \left[\frac{(2\pi)^3}{abc}\right]. \qquad (133)$$

The information I is a minimum for $a = b = c = \pi/3$ and

$$I_{\min} = \log_2 \left[\frac{(2\pi)^3}{(\pi/3)^3}\right] = \log_2 6^3 = 3\log_2 6 = 3(1 + \log_2 3) \text{ bits.} \quad (134)$$

More generally, the N-sided polygon providing least information is equilateral (so rectangles provide less information than, for example, parallelograms or trapezoids). Let us apply these results to the problem of locating those parts of a contour where the information is mostly concentrated.

We subdivide a contour into short segments of equal length from some point A to some final point B, as shown in Figure 40. Each point of the segment (save the two ends A and B) can be considered as the vertex of an *angle* formed with the two neighboring points. In the figure, $\Gamma O \Gamma$ is such an angle.

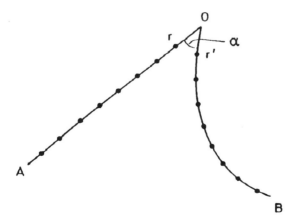

Figure 40. Linear approximation of a contour.

Associated with each angle is its measure of information as given above, and for each successive pair of angles along the path from A to

B we can calculate the gain (or loss) of information upon passing from one angle to the next. If, for example, the portion of the contour under consideration consists of two adjacent sides of a rectangle, including the right angle, we see that as we move by small steps from A to O along AO, each angle is a right angle whose corresponding information is

$$\log_2 \left[\frac{2\pi}{\pi} \right] = 1 \text{ bit.} \tag{135}$$

But the information *gain* in passing from one right angle to the next is

$$\log_2 \left[\frac{\pi}{\pi} \right] = 0 \text{ bits,} \tag{136}$$

since the angle remains *unchanged*. When the right angle with vertex 0 is finally reached, there is a *positive* gain in information equal to

$$\log_2 \left[\frac{\pi}{\frac{\pi}{2}} \right] = 1 \text{ bit.} \tag{137}$$

At the next step, passing from the right angle to the right angle again, there is an information *loss* equal to

$$\log_2 \left[\frac{\frac{\pi}{2}}{\pi} \right] = -1 \text{ bit.} \tag{138}$$

The right angle then is the only place in the contour where the contour is curved. So only there, at the point of the extreme curvature, is information concentrated. More generally, we now may associate at each point of the contour a radius of curvature

$$K = \lim_{\Delta s \to 0} \frac{\Delta a}{\Delta s} = \frac{da}{ds}, \tag{139}$$

where Δs is the elemental length of the curve. We see immediately that

$$I = \log_2 \left| \frac{\Delta a_1}{\Delta a_2} \right| = \log_2 \left[\left| \frac{k_1}{k_2} \right| \right] \text{ bits,} \tag{140}$$

which is the required result.

Dynamical analysis of the sequence of fixation of the receptor at the different states—features of the pattern—suggests a format for the interconnection of these states within the overall internal representation. The result of this experiment reveals that the sensor directed by the brain is essentially involved with two types of scanning pathways:

regular and irregular. Specifically, as we state in [36]: "The eyes usually scanned the pattern following—*intermittently* but repeatedly—a fixed path that we have termed the 'scan path . . . .' The occurrences of the scan path are separated by periods in which the fixations are ordered in a less regular manner ... scan paths usually occupied from 25 to 35% of the subject's viewing time, the rest being devoted to less regular eye movement." We may conclude that the internal representation (or the mapping) of a pattern in the memory system is an assemblage of features (states) mediated essentially by a feedback loop: a sequence of sensor-motor traces recording, abstracting, and subsequently reporting to the brain a "Markov chain" as it were of state features—whereupon the brain directs the next move of the peripheral activity.

The time intervals during which the system holds a given state obviously have to do with the excitation of the directing neuronal tissue—which in turn depends on the degree of curvature of the feature state involved [35]. This Markov chain algorithm is subsequently stored in the brain isomorphically as, say, a pattern of circulating electrical activity. When the observer subsequently encounters the same pattern, he recognizes it by matching it with the "feature ring" of the Markov chain that constitutes the internal representation in his memory, state by state. Matching or recognizing then consists in calculating the distance or the cross-correlation between two Markov chains: the one memorized and the one that runs as the observer re-examines the pattern. Clearly, the first chain directs the steps of the second through the brain-sensor-motor loop, and so learning takes place. Beyond a few reruns, the object becomes familiar as the cross-correlation above tends to 1. Under such circumstances, it would then be rather improper to try to gather the information conveyed from observing Nefertiti's bust by calculating the entropy of the pattern.

3. Concluding Remarks

In general, we may envisage a dynamical theory of preselection and preadaptation in biological information processing. Inhomogeneous dissipative chaotic flows are due to multifractality—both within an attractor and in the intermittent jumps among coexisting attractors (memories) in a biological processor responsible for drastically limiting the numbers of possible species. Natural selection then operates on the small subset of preselected attractors. *Phase transitions on multifractal sets* (supporting inhomogeneous-intermittent-chaotic flows) may be considered as the driving mechanism behind *punctuated pre-evolution* in biological systems, performing linguistic processing at both groups of levels—the *syntactic* and the *semantic*.

Indeed, a phase transition on a multifractal set amounts to an *abrupt change in the shape of the (normalizable) invariant measure of the set* and, hence, to a sudden change of the weighting factors carried by each column of the set on each symbol of the imposed (Markovian) partition. We propose then a new *selection principle*, one that acts via inhomogeneous intermittency in the phase space (e.g., via a multifractal strange chaotic attractor—the thalamocortical pacemaker). It is responsible for a highly nonlinear linguistic filtering process, one which drastically limits

- The grammatically legitimate words at the *syntactical* level;

- The "interesting" key features of a pattern at the *semantic* level.

This is an *a priori* selection principle—whereas natural selection is an *a posteriori* principle acting on *phenotypes*, e.g., on *attractors* via shifting control parameters and resulting in *generation/elimination/ fusion* ("crises") of coexisting attractor-categories. Multifractal chaotic dynamics applied to information-processing devices may thus be instrumental in:

- Drastically *simplifying* hardware without compromising the breadth of the functional repertoire (software);

- Reducing or compressing memory capacity. Instead of storing all details via pixelization of a pattern, one limits attention to very few key features (associated with small radii of curvature) coding for disproportionally high amounts of stored information. So, through an *inhomogeneous intermittent* dynamics in state space, we filter out only key words and/or key features. The principle of natural selection in linguistics involves further scrutiny of those small, *a priori* selected subsets.

Acknowledgment

The authors thank Dr. S. Nicolis for critical proofreading.

Appendix A: The Complexity of Random Time Series

If a pattern is described or coded using a string of binary symbols, the length of the shortest string can be computed as follows. Let p be the *a priori* probability of 0s and $1 - p$ that of the 1s in a fair coin-tossing realization of length N digits (bits). There are $\binom{N}{pN}$ such sequences or

$N!/\{(pN)![N(1-p)]!\}$. Hence, $\log_2\binom{N}{pN}$ bits suffice to uniquely specify such a sequence, which amounts to $\log_2\binom{N}{pN}/N$ bits per digit. Using Stirling's formula $\xi! \approx \sqrt{2\pi}\xi^\eta \exp[-\eta]$, $\eta \gg 1$, and applying it to the above relation, we obtain the expression

$$S(p) = -p\log_2 p - (1-p)\log_2(1-p)\frac{\text{bits}}{\text{per digit}}, \qquad (141)$$

for uniquely specifying an N-bit sequence of 0s and 1s with pN 0s and $(1-p)N$ 1s (as $N \to \infty$). However, this is not all: To reconstruct a *finite-length* sequence of N 0s and 1s, where p is the *a priori* probability of 0s, one needs to specify the number of 0s. For finite N, this number is not known exactly, i.e., one needs an extra amount of information of $\log_2(N+1)$ bits. Thus, we need $N+1$ digits instead of N just to make the difference between N 0s and N 1s. So the total complexity of the pattern is

$$N' \sim \log_2(N+1) + \log_2\binom{N}{pN} \sim \log_2 N + NS(p), \quad N \gg 1, \quad (142)$$

or

$$S(p)\frac{\text{bits}}{\text{per digit}} \qquad \text{as} \qquad N \to \infty. \qquad (143)$$

Chaitin [28] showed that the overwhelming majority of the above sequences are incompressible, i.e., random. Indeed, consider the number of sequences of length N that are incompressible up to K bits, i.e., $N' \le N - K$. This number is

$$\Sigma = 2^1 + 2^2 + \ldots + 2^{N-K-1} = 2^{N-K} - 2.$$

The percentage of compressible sequences is then

$$\eta = \frac{\Sigma}{2^N} \sim 2^{-K}, \qquad N \gg 1.$$

Thus for, say, $K = 10$, only one sequence in 1,000 can be specified by an algorithm 10 bits shorter than the sequence itself.

Although there are numerous attractors at each cognitive level (for a complex processor), the percentage of those that really deserve the name, that is, possess an information dimension $C = N - K(K \sim 0(N))$, is just $\eta = 2^{-K}$, which is insignificantly small.

Appendix B: Semi-Markov Chains

In a semi-Markov chain, the time between transitions follows a given probability density function (instead of being fixed at a definite interval as in a simple Markov chain). This transition time also depends on the particular transition under consideration. When the process enters a state (e.g., attractor) i, we know that this determines the next state j to which the system will move according to state j's conditional probability P_{ij}. However, after j has been selected—but before making the transition $i \rightarrow j$—the process "holds" or "gets stuck" for a time r_{ij} in state i. The holding times are positive, integer-valued random variables, each governed by a probability mass function $h_{ij}(r)$, called the *holding time density function*. If we do not know its successor state, the probability that the system will spend τ time units in state i is then

$$xi_i(\tau) = \sum_{j=1}^{\Lambda} P_{ji} j_{ji}(\tau),$$

where Λ is the total number of successor states. Thus, a waiting time is merely a holding time that is unconditional on the destination state. The mean waiting time $\langle \tau_i \rangle$ is related to the mean holding time $\langle \tau_{ji} \rangle$ as

$$\langle \tau_i \rangle = \sum_{j=1}^{\Lambda} P_{ij} \langle \tau_{ji} \rangle.$$

In this model, the process first selects its next state using the transition probabilities P_{ij}, and then selects its time of transition from a holding time mass function $h_{ij}(\tau)$, which is conditioned on the actual transition that's made. Equivalently, we can reverse the process to allow the system first to select its time of transition and then a new state conditional on the transition time. For the physical situation we have in mind—the "spasmodic" thalamocortical pacemaker—this sequence of events seems more natural. The waiting time distribution $\xi_i(\tau)$ defined above is the distribution of holding times unconditional on destination state. In the present, alternative description of the process, $\xi_i(\tau)$ would have to be specified for all states. Then we have to specify the probability of making a transition to each state, given the time interval that the system has held in its present state before making the transition.

Let $P_{ji}(\tau)$ be the probability that the process now in state i at time τ will make the transition to state j. Thus, the quantities $P_{ji}(\tau)$

are transitional probabilities conditioned on holding time. We call them conditional transitional probabilities. We have

$$P_{ji}(\tau) = \frac{P_{ji}h_{ji}(\tau)}{\xi_i(\tau)} = \frac{P_{ji}j_{ji}(\tau)}{\sum_{j=1}^{\Lambda} P_{ji}h_{ji}(\tau)},$$

which is our result.

Choosing $h_{ji}(\tau)$ supplies all the information one needs to fully determine a semi-Markov chain. In the above formalism, τ stands for the total residence time(s) $\tau_{R_i} + \tau_{h_i}$ in the intermittent dynamics [so $h_{ji}(\tau)$ may be deduced from this]. It is assumed here (in a semi-Markovian formalism) that the transient times from attractor to attractor are negligible. Unfortunately, in the real dynamics of generalized intermittency this never happens. In some cases the transients may be extremely long, especially when the boundaries are connected fractals and the trajectory that "takes off" from one attractor hesitates for a long time before deciding on whose attractor's domain it will enter. A symbolic semi-Markovian formalism in which all four time scales τ_{0_i}, τ_i, τ_{R_i}, and τ_{h_i} involved are taken into account is still not available.

Appendix C: Deductive Versus Inductive Cognitive Tasks

To better appreciate the difference between a mechanistic and a biological processor, let us note that the first has an exclusive area of intellectual activity, the field of deductive problems, whereas the second can tackle inductive problems as well. Let us classify all cognitive tasks into two broad categories:

- Exercises or deductive problems whose outcomes always exist are unique and are implicitly contained in the initial premises, provided that these premises form a complete and self-consistent set.

- Complex problems of the inductive type whose outcomes may or may not exist may not be unique (e.g., multiple outcomes) and are not necessarily nested in the initial conditions but are in the part generated by the evolution of an iterative and recursive procedure or the flow itself. Under such circumstances, we say that if there is evolutionary (asymptotic) stability in the dynamical flow in the state space, then there is a discriminable pattern or a solution.

In dealing with problems of the former category, one goes through regressive steps on a sequence that eventually converges and gives the solution. No *new* information is produced during this process, although admittedly the agent who deals with the problem may be ignorant of how to turn the implicit into the explicit, or how to choose among different solution strategies. In problems of the second type, convergence cannot be guaranteed. Such problems are dynamical and their outcome can only be understood in a historical, that is, an evolutionary context. When we say that the information required for the solution of an inductive problem is not hidden exclusively in the initial conditions, it means there is no *a priori* truth ("$\alpha - \lambda \eta \theta \epsilon \iota \alpha$") to be revealed. In other words, we mean that at certain moments the evolving scenario, that is, the trajectory of the system in state space, branches off or bifurcates into multiple asymptotically stable or unstable regimes (attractors or repellers) in ways that depend on critical values of the control parameters. These control parameters stand for the coupling among the variables, and their numerical values are shifted by ever-present (internal or external) fluctuations. When these parameters assume their critical values, the coupling between the variables concerned changes in ways that precipitate instabilities, meaning that infinitesimally small changes in the parameters lead to finite changes in the variables. Beyond these instabilities, the (new) structure(s) may not retain any trace or memory of the previous regimes, although they evolved from them via quite deterministic processes. Similar "loss of memory" takes place in systems without bifurcation, but with a sensitive dependence of their flows on initial conditions.

The (human) mind, it appears, possesses the unique capability of mapping the external (as well as part of the organism's internal) world; that is, it compresses long and complex strings of impinging environmental stimuli ("observations") and then uses these minimal-length algorithms to simulate physical phenomena, thereby revealing the "laws" of nature. This process of self-organization and category formation is implemented, we theorize, via a set of coexisting attractors in the cognizant apparatus, each of which attracts (and therefore compresses) whole subsets of initial conditions whose total sum constitutes the set of recognizable external stimuli. This set of initial conditions forms the basins of the attractors, and the process of partition and category formation in the mind of the processor involves the topology of the separatrices among the individual subsets of the basin.

Consider the case of fractal, fully connected open basin boundaries. If an initial condition (message), specified with a given (however small) error ϵ, is within that distance ϵ to a fractal separatrix with dimension $d \sim O(N)$, the attractor to which it moves cannot be predicted with certainty. Thus there is a fraction of the basins consisting of "fuzzy" messages. The topology of the fractal basin boundaries depends on the processor's parameters, so a shifting of these parameters, or an uncertainty δ in their exact values, also affects the ability to predict the attractor from the initial condition independently of the precision ϵ with which this initial condition has been chosen.

If the system parameters vary, an attractor may "inflate" so as to come to a collision with a fractal basin boundary. A crisis ensues, and the attractor may be destroyed altogether and replaced by a long transient. Moving the control parameter the other way, creation of an attractor may take place [32,34]. In the even more interesting case of fractal, fully disconnected basin boundaries ("dust"), there is no unique specification of the dimensions of these boundaries, so the fundamental problem of calculating the entropy of partition of external stimuli to coexisting attractors remains open.

Appendix D: Information Transfer Through Generalized Multihierarchical Intermittency

In Figure 38 we have shown the basin of all possible environmental stimuli (2^N strings of length N) partitioned among Γ attractors. Most of the above series are incompressible transients and the rest fall into some of the coexisting attractors. This problem of static partitioning becomes very difficult due to the fractal character of the basin boundaries. In biological information processing, though, the problem lies elsewhere. Since there exists an active internal agent providing an intermittent regime of jumps from attractor to attractor, our interest focuses on the time persistence of a given attractor-memory or the estimation of the transinformation ($I(\xi;t)$, i.e., the average amount of information contained in a prediction at time t corresponding to ξ jumps into the future. If $P^{(\xi)}(i;j,t)$ is the (joint) probability that at time t the processor is at the attractor j and on the attractor i ξ jumps later, the nonstationary transinformation is given by

$$I(\xi;t) = \frac{\sum_{i=1}^{\Lambda} \sum_{j=1}^{\Lambda} P^{(\xi)}(i;j,t) \log_2 P^{\xi}(i;j,t)}{P(i;t)P(j;t)} \text{ bits.} \tag{144}$$

All of the above quantities can be calculated in principle from the solution of Arecchi's kinetic equations. [Equation (144) shows that I is epoch-dependent.] Of course, the basic practical difficulty in solving these equations lies in the estimation of the transitional probabilities

P_{ij}, which again can be traced to the difficulty in knowing the shapes of the individual basin boundaries.

Now concerning a single attractor (resulting from a flow or map f), we introduce a stroboscopic process $f_t \to f_{t_n}$, $t_n = n(\Delta T)$, where $n = 0, 1, 2, \ldots$, and choose a partition β on the attractor of M, consisting of B_i elements. Let $\bar{\mu}$ be the invariant measure. Then the entropy of the partition β with respect to the measure $\bar{\mu}$ will be

$$S_0(\beta) = -\sum_{i=1}^{M} P_i \log_2 P_i \quad \text{where} \quad P_i = \bar{\mu}(B_i). \tag{145}$$

The conditional entropy of a partition $f^{-t}(\beta)$ with respect to the partition β will be

$$S_c(f^{-t}(\beta)/\beta) = -\sum_{i=1}^{M} \sum_{j=1}^{M} P_{j/i}(t) \log_2 P_{j/i}(t), \tag{146}$$

with

$$P_{i/j}(t) = \frac{\bar{\mu}[f^{-t}(B_j) \bigcap B_i]}{\bar{\mu}(B_i)}. \tag{147}$$

The transinformation of f, which stands for the amount of memory persistence of an initial condition on the attractor involved, is $I(t) = S_0 - S_c(t)$, where $t = 0, 1, 2, \ldots$ stands for the discrete time or the iteration number. The above analysis assumes that all transients have already died out, so that one can speak of processes whose statistics are given by the invariant measure $\bar{\mu}$. Under this regime, S_0 is also time independent. The quantity $I(t)$ may decay monotonically or not, depending on the partition chosen and the degree of coherence of the attractor involved, i.e., on its mixing properties. Thus, $I(t)$ is a measure of the memory persistence. For $t > t_n$ for which $I(t_n) = 0$, we may say that the memory of the initial condition $I(0)$ has been completely erased.

Finally, note that the regime within each attractor may also be intermittent. We have two different categories of intermittency. There is "common" intermittency, which involves a jumping process among coexisting repellers, and there is "forced" (or "generalized") intermittency, which involves a jumping process among coexisting stable attractors. Hence noise is required to provide the jumps. Thus, in general, we have the superposition of two intermittent regimes (intra- and inter-attractor). The commuting process may be semi-Markovian, even in the absence of holding times (due to the polarization of the specific thalamic nuclei). The semi-Markovian character of any intermittent process is due to the fact that after reinjection from one regime onto another, the residence time in it is not fixed but comes from a mass density function characterizing the type of intermittency involved.

All strange attractors perform phase mixing along the *transverse* directions. Along the (longitudinal) direction of the flow itself, though, mixing and loss of phase coherence vary according to the specific attractor, the values of its control parameters, and the specific partition. Regeneration of transinformation due to phase coherence or imperfect mixing is an interesting manifestation of memory persistence, one that results in an oscillatory behavior of $I(t)$ in attractors whose continuous spectra are interspersed with occasional high spikes, like Rössler attractor [54]. This phenomenon of memory persistence of an individual initial condition on a given attractor should be pursued further for the modeling or redesigning of biological information channels based on the dynamics of intermittency.

Appendix E: On the Relationship Between the Rank Distribution and $1/f$ Noise

Let us first perform a rather rough estimate. Let $L_1, L_2, \ldots, L_\Lambda$ be the lengths of the words produced by the Λ coexisting attractors with duration (relaxation times) $\tau_1, \tau_2, \ldots, \tau_\Lambda$, respectively. The probability of the occurrence of the sentence $L_1 L_2 \ldots L_\Lambda P$ can be expressed by the product

$$
P = \prod_{i=1}^{\Lambda} P(\tau_i) \prod_{j=1}^{\Lambda} P(\tau_{0_j}) \text{ or } P = p_1 p_{12} p_2 p_{23} p_3 \cdots p_\xi p_{\xi+1} \cdots .
$$

Taking into account Bayes' rule that $p_i p_{ij} = p(ij)$, $\sum_i \sum_j p_i p_{ij} = 1$, we rewrite P as follows:

$$
P = p_1 p_{12} p_2 p_{23} p_3 \ldots = p(1, 2) p_2 p_{23} p_3 \ldots
$$

$$
= p(1, 2, 3) p_2 p_3 \ldots = p(1, 2, 3, \ldots, \Lambda) \prod_{i=2}^{\Lambda-1} p_i.
$$

If $p(1, 2, 3, \ldots, \Lambda)$ is normal, P obeys a log-normal distribution

$$
P(\tau) \sim \frac{1}{\pi \sigma \tau} \exp \left[-\frac{\left(\log \frac{\tau}{\langle \tau \rangle} \right)^2}{2\sigma^2} \right],
$$

and so, over some interval of τ, $P \sim 1/\tau$ [7]. Under such circumstances, only the time scale τ_i (out of τ_i, τ_{0_i}, τ_{R_i}, and τ_{h_i}) is contributing. So only self-diffusion creates the series of words (the sentences) under the condition, of course, that $\tau_i \leq \tau_{R_i} + \tau_{h_i}$. Now let us make things more precise.

It takes τ_i sec to produce one word on a strange attractor, where t_i determines the word length. The probability of the appearance of a word in a period of time τ is $p(\tau) = p(i, j, k, l, m, \ldots)p_{M \to N}$, where $p(i, j, k, l, m, \ldots)$ is the probability of successive steps in a Markov chain produced inside the strange attractor M, and $p_{M \to N}$ is the probability of transition from the strange attractor M to the strange attractor N when we let ϵ be the precision of observation (fixation) of an initial condition in a subbasin of attraction. The probability of a "rebound" onto the strange attractor M is given as follows when the orbit comes into the *fractal* basin boundary between strange attractors M and N:

$$p_0 \sim 1 - f \sim 1 - \epsilon^{D - D_0}, \tag{148}$$

where D is the dimension of the state space, D_0 is the fractal dimension of the boundary $(D - D_0 \ll 1)$, and f stands for the probability (uncertainty) of the initial condition close to the fractal basin boundary [33] of jumping to the attractor N. The precision ϵ, on the other hand, is related to the word length $(\sim r)$ and the pertinent Lyapunov exponent λ as $\epsilon \exp(\lambda \tau) \sim 1$. So $\epsilon \sim \exp(-\lambda \tau)$. We thus obtain the expression of the probability of rebound p_0 as

$$p_0 \sim 1 - (\varepsilon^{-\lambda \tau})^{D - D_0} = 1 - e^{-\lambda \tau (D - D_0)}. \tag{149}$$

In order to obtain the dynamical expression of Zipf's law of the word ranking distribution, we must calculate the probability of the word length $p(\tau)$. To this end, we seek the probability of the prolongation of the word length by one time unit, i.e., $\tau \to \tau + 1$. We obviously have

$$p(\tau + 1) \sim p(\tau)p_0 = p(\tau)\left(1 - e^{-\lambda \tau (D - D_0)}\right). \tag{150}$$

Therefore,

$$p(\tau) \sim p(1) \prod_{l=1}^{\tau - 1} \left(1 - e^{-\lambda c(\tau - l)}\right), \tag{151}$$

where $c = D - D_0$. So the remaining task is to calculate explicitly $p(1)$.

Let p_i be the probability of coming into the ith strange attractor. Then

$$p(1) = \prod_{i=1}^{\xi} p_i. \tag{152}$$

The number ξ stands for the number of attractors visited, and since we may count the same attractor visited many times as different attractors, ξ may be taken large even if the number of coexisting attractors is as small as two. Now the crux of the matter is that the p_i may be considered *independent* of each other. This is a consequence of the *fractal* character of the basin boundaries; indeed, under such circumstances it is very uncertain where the initial condition will finally land due to the mixing (or the information loss), causing essentially the "final state sensitivity" [33]. If, on the other hand, the basin boundaries are formed by a smooth manifold, then the p_is depend on the shape (or the topology) of the stable and the unstable manifold. Taking now the logarithm of Eqs. (152) and (151), we have

$$p(\tau) = \prod_{i=1}^{\xi} p_i \prod_{l=1}^{\tau-1} \left(1 - e^{-\lambda c(\tau-l)}\right). \tag{153}$$

Then we get

$$\ln p(\tau) \sim \sum_{i=1}^{\xi} \ln p_i + \sum_{l=1}^{\tau-1} \ln \left(1 - e^{-\lambda c(\tau-l)}\right). \tag{154}$$

As to $\xi \gg 1$, the Central Limit Theorem holds, meaning that the first term on the right of Eq. (154) follows the Gaussian distribution. Let

$$\ln p_0(\tau) \equiv \sum_{i=1}^{\xi} \ln p_i \ln p_l(\tau) \equiv \sum_{l=1}^{\tau-1} \ln \left(1 - e^{-\lambda c(\tau-l)}\right).$$

The distribution of $P_0(\tau)$ is log-normal. Consider now the second term, $\ln p_l(\tau)$. If we put $z = e^{-\lambda cy}$, $dz = -\lambda c e^{-\lambda cy} \, dy$, then $dy = -(\lambda c)^{-1} dz/z$. Substituting, we get

$$\ln p_1(\tau) \simeq \int_{e^{-\lambda c}}^{e^{-\lambda c(\tau-1)}} -\ln(1-z) \frac{dz}{\lambda cz}$$

(we have replaced the sum with the integral). For highly sensitive attractors, $\lambda_c \gg 1$. Hence $z \ll 1$, which leads to

$$\frac{\ln(1-z)}{z} \sim \frac{1}{1-z}.$$

The resulting calculation is then of the standard type:

$$\int_a^b \frac{\ln(1-z)}{z} \, dz = \int_a^b -\sum_{n=0}^{\infty} z^n \, dz = -\sum_{n=0}^{\infty} \left(\frac{b^{n+1}}{n+1} - \frac{a^{n+1}}{n+1} \right)$$
$$= -\left(\ln \frac{1}{1-b} - \ln \frac{1}{1-a} \right).$$

Therefore,

$$\ln p_1(\tau) \sim \frac{1}{\lambda c} \left[\ln(1 - e^{\lambda c}) - \ln(1 - e^{-\lambda c(\tau-1)}) \right]. \tag{155}$$

The second term on the right-hand side of Eq. (153) gives the τ-dependence of $p_1(\tau)$. So the τ-dependence of $\ln p_1(\tau)$ gives the result

$$\ln p_1(\tau) \sim (\lambda c)^{-1} \exp^{-\lambda c(\tau-1)} + \text{const}$$

when τ is large. Therefore,

$$\ln p(\tau) \sim \ln p_\tau - (\lambda c)^{-1}(1 - e^{-\lambda c(\tau-1)}) + \text{const}$$

when τ is small, but greater than 1, or

$$\ln p(\tau) \sim \ln p_\tau - (\lambda c)^{-1} e^{-\lambda c \tau} + \text{const}$$

when τ is large. In any case, $\ln p_1(\tau)$ does *not* lead to $1/f$-like behavior and contributes only to some perturbations or corrections to a log-normal term. Only the first term, $\ln p_0(\tau)$, produces the log-normal distribution under the supposed (and more interesting) regime of fractalness in the basin boundaries. Specifically, the distribution of $\ln p_0(\tau)$ is

$$F(\ln p_0(\tau)) = \frac{1}{\sqrt{2\pi\sigma^2}} \exp\left(-\left[\ln \frac{p_0(\tau)}{\langle p_0(\tau) \rangle} \right]^2 / 2\sigma^2 \right), \tag{156}$$

where $\langle p_0(\tau) \rangle$ is the mean and σ^2 is the square of the dispersion. The probability that the variable $p_0(\tau)/\langle p_0(\tau) \rangle$ lies in the interval of width $d(p_0(\tau)/\langle p_0(\tau) \rangle)$ at $p_0(\tau)/\langle p_0(\tau) \rangle$ is

$$G\left(\frac{p_0(\tau)}{\langle p_0(\tau) \rangle}\right) d\left(\frac{p_0(\tau)}{\langle p_0(\tau) \rangle}\right) = \frac{1}{2\pi\sigma^2} \exp\left(-\left[\ln\frac{p_0(\tau)}{\langle p_0(\tau) \rangle}\right]^2 / 2\sigma^2\right)$$
$$\times \frac{d(p_0(\tau)/\langle p_0(\tau) \rangle)}{p_0(\tau)/\langle p_0 \rangle}. \quad (157)$$

In the appropriate interval of $p_0(\tau)/\langle p_0(\tau) \rangle$, the distribution G follows $1/f$-like behavior [7], where $p_0(\tau)/\langle p_0(\tau) \rangle = f$. By the very meaning of $p_0(\tau)$, the rank s_τ of the word with duration $\sim$ length τ is proportional to $p_0(\tau)^{-1}$. Therefore, $f = p_0(\tau)/\langle p_0(\tau) \rangle = \langle x_\tau \rangle / x_\tau$ gives the log-normal, and so the $1/f$ distribution

$$G\left(\frac{p_0(\tau)}{\langle p_0(\tau) \rangle}\right) d\left(\frac{p_0(\tau)}{\langle p_0(\tau) \rangle}\right) = \frac{1}{2\pi\sigma^2} \exp\left\{-\left[\ln\frac{x_\tau}{\langle x_\tau \rangle}\right]^2 / 2\sigma^2\right\}$$
$$\times \frac{d(x_\tau/\langle x_\tau \rangle)}{x_\tau/\langle x_\tau \rangle}.$$

So from Eq. (156) we deduce that the rank distribution follows $1/f$ in some parameter interval. Here τ stands for the correlation loss time, namely, the time at which the maximum value of the mutual information $I(0)$, or transinformation, falls by $1/e$. The other time parameters of the problem, τ_{R_i}, τ_{h_i}, τ_{0_i}, do not appear explicitly, meaning that they are embedded into the assumption of statistical independence of p_i that is the hypothesis of the fractal character of the basin boundaries.

Appendix F: Fractional Scaling Exponents Giving a Weighting Function $p(\tau) \sim 1/\tau$ in the System with Many Time Scales

How should similarity in the structure of the chaotic systems possessing coexisting strange attractors be incorporated under the condition that weighting function $p(\tau) \sim 1/\tau$? Under the assumption that the correlation decays exponentially with timescale τ_i on each strange attractor i, namely, $c_i(t) = e^{-t/\tau_i}$, our problem can also be considered as that of determining the distribution of fractal times [55,56] such that the events with τ_i occur in succession. In the system with many time scales, the correlation can be expressed as

$$c(t) = \sum_{\tau_i} e^{-t/\tau_i} = \int_0^\infty e^{-t/\tau} p(\tau)\, d\tau. \quad (158)$$

Since the transition occurs at the fractal basin boundaries, the probability of the transition from one attractor to the other with ϵ uncertainty is $f \sim \epsilon^c$, where $c = D - D_0$. The distribution of time necessary for transition via self-similar basin boundaries can give the weighting function $p(\tau)$. Since the situation is similar to that of the motion in a fractal potential, $p(\tau)$ can be expressed by the superposition of a weighted Poisson process:

$$p(\tau) = \int_0^\infty e^{-\lambda t} h(\lambda)\, d\lambda. \tag{159}$$

Assuming λ scales as $\lambda = \epsilon\beta$ and $h(\lambda)\, d\lambda = f(\epsilon)d\epsilon$, which means that the transition rate is scaled up by ϵ, we obtain

$$h(\lambda)\, d\lambda = \frac{1}{\beta} \lambda^{\frac{1+c}{\beta} - 1}\, d\lambda$$

and so

$$p(\tau) = \int_0^\infty \frac{1}{\beta} e^{-\lambda t} \lambda^{\frac{1+c}{b} - 1}\, d\lambda = \frac{1}{\beta} \Gamma\left(\frac{1+c}{\beta}\right) \tau^{-\frac{1+c}{\beta}}. \tag{160}$$

Therefore, the correlation function is given as

$$c(t) = \int_0^\infty \frac{1}{\beta} \Gamma\left(\frac{1+c}{\beta}\right) e^{-t/\tau} \tau^{-\frac{1+c}{\beta}}\, d\tau,$$

and so the spectrum $S(\omega)$ is expressed as

$$S(\omega) = Re \int_0^\infty c(t) e^{i\omega t}\, dt = A \frac{1}{\omega^{2 - \frac{1+c}{\beta}}},$$

where

$$A = \frac{1}{\beta} \Gamma\left(\frac{1+c}{\beta}\right) \frac{\pi}{2} \operatorname{cosec} \frac{\pi}{2}\left(2 - \frac{1+c}{\beta}\right). \tag{161}$$

Hence, the existence of the integral gives the condition on the exponent: $0 < 2 - (1 + c)/\beta < 2$. If the exponent of the transition rate has the same exponent as that of the fractal basin, i.e., $\beta = c$, we need more than three dimensions of phase space to get an ω-γ spectrum $(0 < \gamma < 1)$. More interestingly, we conclude that the scaling exponent of the transition rate should be $\beta = c + 1$ in order to obtain just the $1/\omega$ spectrum.

Appendix G: On the Partition of Basins of Attraction into Coexisting Attractors Under a Regime of Fractal Basin Boundaries

As has been shown numerically in [34], the process of categorization of a basin (an attracting set) of messages—initial conditions—onto Λ coexisting attractors coming from a chaotic dissipative dynamics $dx/dt = f(x; \mu)$ and $\nabla \cdot f < 0$ may be *inconclusive* in the case where the boundaries are fractals. Indeed, if ϵ is the uncertainty of a given member of the basin (a word of length D, where D is the dimension of the embedding state space), then in the case of fractal basin boundaries the "uncertainty of categorization" f, that is, the probability of having an initial condition normally (e.g., for $\epsilon \to 0$) belonging to attractor i to fall on attractor $j \neq i$ scales with ϵ as $f(\epsilon \sim \epsilon^g$, where $g = D - D_0 << 1$, with D_0 being the dimension of the fractal basin boundary.

References

[1] J. S. Nicolis, *Dynamics of Hierarchical Systems: An Evolutionary Approach.* Springer-Verlag, 1986.

[2] R. Crauss, "Language as a symbolic process in communication—A psychologic perspective." *Am. Sci.*, 56 (3) (1968), 265–278.

[3] B. L. Whorf, *Language, Thought and Reality.* MIT Press, 1956, pp. 212–213.

[4] J. S. Nicolis, "Chaotic dynamics as applied to information processing." *Rep. Prog. Phys.*, 49 (1986), 1109–1187.

[5] G. Nicolis and C. Nicolis, *Phys. Rev.*, A38 (1988), 427.

[6] G. K. Zipf, *Human Behavior and the Principle of Least Effort.* Addison-Wesley, 1949.

[7] E. W. Montroll and M. F. Shlesinger, *J. Stat. Phys.*, 32 (1983), 209.

[8] B. J. West, "An essay on the importance of being nonlinear." *Lecture Notes in Biomathematics 62.* Springer-Verlag, 1985.

[9] B. Roy Frieden, *Found. Phys.*, 16 (1986), 883.

[10] J. S. Nicolis and I. Tsuda, "On the parallel between Zipf's law and $1/f$ spectra in systems possessing multiple attractors." *Prog. Theor. Phys.*, 1989.

[11] M. Schroeder, *Fractals, Chaos and Scaling Laws.* Freeman, 1991.

[12] J. S. Nicolis, *Chaos and Information Processing—A Heuristic Outline.* World Scientific, 1991.

[13] A. Katsikas and J. S. Nicolis, *Il Nuovo Cimento,* 12 (1990), 2.

[14] I. Antoniou, A. Katsikas, S. Pahaut, and B. Saphir, "Musique du chaos." Communication in "Temps, complexité et oeuvres musicales," *Les Treilles,* Jul. 1990.

[15] I. Antoniou and A. Katsikas, preprint.

[16] J. S. Nicolis, "Chaotic dynamics as applied to information processing." *Rep. Prog. Phys.,* 49 (1986), 1109–1187.

[17] B. Pompe, J. Kruscha, and R. W. Leven, *Z. Naturforsch.,* A41 (1986), 801.

[18] F. T. Arecchi and A. Califano, *Phys. Lett.,* A101 (1984), 443.

[19] F. T. Arecchi, R. Badii, and A. Politi, *Phys. Rev.,* A32 (1985), 402.

[20] F. T. Arecchi, "Noise traping at the boundary between two attractors: A source of $1/f$ spectra in nonlinear dynamics." Preprint, 1986.

[21] J. S. Nicolis, G. Nicolis, and M. Benrubi, *Proc. 1st Int. Conf. Info. Sci. Systems,* Vol. 2, N. S. Tzannes and D. Lainiotis, eds. Hemisphere, 1977, p. 569.

[22] G. Moruzzi and H. W. Magoun, *Clin. Neurophys.,* 1 (1949), 455.

[23] M. Verzeano et al., in *The Neural Control of Behavior,* R. E. Whalen, R. F. Thoson, M. Verzeano, and S. Wienberger, eds. Academic Press, 1970, p. 523.

[24] R. M. May, *Ann. NY Acad. Sci.,* 357 (1981), 267.

[25] J. S. Nicolis and I. Tsuda, *Bull. Math. Biol.,* 47 (1985), 343.

[26] W. Kilmer and W. S. McCulloch, in *The Information Processing in the Nervous System,* J. Leibovic, ed. Springer, 1969, p. 152.

[27] W. R. Adey, *Int. J. Neurosci.* (1972).

[28] G. Chaitin, *Am. Sci.,* 256 (1975), 47.

[29] L. Glass and R. Perez, *Nature,* 256 (1973), 360.

[30] G. Nicolis and I. Prigogine, *Proc. Natl. Acad. Sci. USA,* 78 (1981), 654.

[31] R. M. Mason, *Nature,* 311 (1984), 19.

[32] C. Grebogi, S. McDonald, E. Ott, and J. Yorke, *Phys. Lett.,* A99 (1983), 415.

[33] C. Grebogi, E. Ott, and J. Yorke, *Phys. Rev. Lett.,* 50 (1983), 935.

[34] E. Ott, W. D. Witters, and J. A. Yorke, *J. Stat. Phys.,* 36 (1984), 687.

[35] D. Noton, *IEEE Trans. Syst. Man Cyber.,* 6 (1970), 439.

[36] D. Noton and L. Stark, *Sci. Am.,* 224 (1971), 34.

[37] F. Attneave, "Some informational aspects of visual perception." *Phys. Rev.,* 61 (1954), 183–193.

[38] H. Resnikoff, *The Illusion of Reality.* Springer-Verlag, 1989.

[39] G. Nicolis, S. Rao, G. Rao, and C. Nicolis, in *Coherence and Chaos in Dynamical Systems.* Manchester Univ. Press, 1988.

[40] S. Luria, *36 Lectures in Biology.* MIT Press, 1975.

[41] T. Tel, "Fractals, multifractals and thermodynamics: An introductory review." *Z. Naturforsch.,* A43 (1988), 1154–1174.

[42] J. S. Nicolis, "Should a reliable information processor be chaotic?" *Kybernetes,* 11 (1982), 269.

[43] A. Destexhe, J. A. Supelchere, and A. Babloyantz, *Phys. Lett.,* A132 (1988), 101.

[44] J. S. Nicolis, G. Mayer-Kress, and G. Haubs, "Non-uniform chaotic dynamics with application to information processing." *Z. Naturforsch.,* A38 (1983), 1157–1169.

[45] J. Maynard Smith, *Evolution and the Theory of Games.* Cambridge Univ. Press, 1982.

[46] J. S. Nicolis, *Kybernetes,* 11 (1982), 123.

[47] J. S. Nicolis and M. Benrubi, *J. Theor. Biol.,* 59 (1976), 360.

[48] J. S. Nicolis, E. N. Protonotarios, and M. Theologou, *Int. J. Man. Mach. Stud.,* 10 (1978), 343.

[49] A. Babloyantz and C. Nicolis, U. L. B. preprint, 1985.

[50] A. Babloyantz, in *Dimensions and Entropies in Chaotic Systems,* G. Mayer-Kress, ed. Springer-Verlag, 1986, p. 241.

[51] S. P. Layne, G. Mayer-Kress, and J. Holzfuss, in *Dimensions and Entropies in Chaotic Systems,* G. Mayer-Kress, ed. Springer-Verlag, 1986, p. 246.

[52] A. J. Leggett, *Contemp. Phys.,* 25 (1984), 583.

[53] M. Martins, M. Devoret, and J. Clarke, *Phys. Rev. Lett.,* 55 (1985), 1543.

[54] J. D. Farmer, J. Crutchfield, H. Froeling, N. Packard, and R. Shaw, "Power spectra and mixing properties of strange attractors," in *Nonlinear Dynamics,* R. Helleman, ed. *Ann. N.Y. Acad. Sci.,* 357 (1980), 453.

[55] B. B. Mandelbrot, *Fractals—Form, Chance, Dimension.* Freeman, 1977.

[56] M. F. Shlesinger, *Ann. N.Y. Acad. Sci.,* 504 (1987), 214.

[57] G. Nicolis, C. Nicolis, and J. Nicolis, "Chaotic dynamics of Markov partitions." *J. Stat. Phys.,* 54(3/4) (Feb. 1989), 915–924.

CHAPTER 9

Cooperation and Chimera

ROBERT ROSEN

Department of Physiology and Biophysics
Dalhousie University
Halifax, Nova Scotia B3H 4H7, Canada

1. Introduction

If you pick up any American coin, you will see somewhere on it the Latin motto *E Pluribus Unum:* Out of many, one. That one, or "unum," is what I shall call a chimera.

In mythology, Chimera was a monster built up from parts of lions, of goats, and of serpents. Variously, it was described as having the head of the first, the body of the second, and the tail of the third; or as having three heads, one of each. It had other peculiarities of its own, such as (according to Homer, among others) the breathing of fire. Some have suggested that it represented a volcano in Lydia, where serpents dwelt at the base, goats along its slopes, and lions at the crest. Accordingly, the word "chimera" has come to mean any impossible, absurd, or fanciful thing; an illusion.

In biology, aside from being the name of a strange-looking fish, chimera refers to any individual composed of diverse genetic parts; any organism containing cell populations arising from different zygotes (cf. McLaren, 1976). Indeed, the fusion of two gametes to form a zygote may itself be looked upon as a chimerization: a new individual produced from parts of others.

In general, chimera formation, in which a new individual, or a

Cooperation and Conflict in General Evolutionary Processes, edited by John L. Casti and Anders Karlqvist.
ISBN 0-471-59487-3 ©1994 John Wiley & Sons, Inc.

new *identity,* arises out of other, initially independent individuals, is a kind of inverse process to *differentiation,* in which a single initial individual spawns many diverse individuals, or in which one part of a single individual becomes different from other parts.

The role of chimera formation, as a phenomenon in itself, has received nothing like the amount of study in biology as it should. Primarily, it has been looked upon as aberrant: a curiosity or pathology when it occurs spontaneously, and mainly involving directed human intervention when it does not. Grafting, in all its manifestations, provides the most immediate examples. And yet if one looks for them, chimeras are everywhere around us. Ecosystems are chimerical; social systems are chimerical; man-machine interactions are chimerical; even chemical reactions can be thus regarded.

In the present chapter, we shall look briefly at the evolutionary correlates of chimera formation. In particular, we shall consider chimera in the sense of an adaptive response based on modes of cooperative behavior in a diverse population of otherwise independent individuals in competition with each other.

As we shall see, taking chimera seriously, in the sense of being a new *individual* with an identity (genome) and behaviors (phenotype) of its own raises some deep epistemological and system-theoretic questions. These range from the efficacy of "reduction" of a chimera to constituent parts, all the way to "sociobiology" (i.e., what is phenotype anyway?) and beyond.

One way of illustrating these questions arises from considering a true natural chimera: the hermit crab. It is part arthropod (the crab itself), part mollusk (its adopted shell), and part echinoderm (the anemones which grow on its shell). These parts together form a single functioning whole, with its own behaviors and its own identity. As such, what is its genotype? What is its phenotype? How are the two related, both to each other and to the identities and behaviors of its constituents? How did it "evolve"? What, in short, are the relations between the initial "pluribus" and the resulting "unum"?

2. Some Generalities about "Evolution"

Nowadays, the word "evolution" is used very carelessly. Thus, one talks about the "time evolution" of an arbitrary dynamical process, as if "evolution" were synonymous with "change of state" and hence with arbitrary dynamical behavior. But this usage is simply an equivocation, which permits anyone studying any kind of dynamics to call himself an "evolutionist," and to say (vacuously) that "everything evolves."

From the very beginning, however, the word "evolution" was inextricably linked to certain modes of *adaptation*. What is it that qualifies a particular dynamical behavior, a change of state, as being adaptive? I suggested long ago (cf. Rosen, 1975) that the term "adaptation" requires the prior stipulation of that *individual* whose survival, in some sense, is favored or enhanced by it. Unless we do this, the very same change of state, or behavior, can be "adaptive" for a subsystem but maladaptive for a larger system containing it—and vice versa. Thus, for instance, a neoplastic cell may well be more adaptive than its normal counterparts, but neoplasia is highly maladaptive for the organism to which the cell belongs. Conversely, restriction of adaptive capability, which is often bad for a cell as a cell, can favor the survival of the entire organism.

Likewise, what favors a species may "kill" its ecosystem. What favors a firm might "kill" the economy, as we have seen all too recently. Indeed, it is seldom true that, as a president of General Motors asserted long ago, "What's good for GM is good for the country." Garrett Hardin, has called such situations "tragedies of the commons."

The contrapositive of this attitude was stated by Dr. Pangloss in Voltaire's *Candide* (lampooning Leibnitz): "Private misfortunes make up the General Good. Therefore, the more private misfortunes there are, the greater the General Good."

The source of the "Invisible Hand" which rules over this kind of cacophony, , was called *natural selection* by Darwin in the context of biological evolution. Its seat was placed not in the adapting individuals themselves but rather, in their environments (which of course included all other individuals). In Darwin's view, this Hand was based on, and driven by, competition, the "struggle for existence." This view necessarily assigns to "environment" a number of distinct roles. First, the environment was the repository of what its resident individuals were struggling for. At the same time, it was also the repository of what each individual was struggling against. And finally, it made the decision as to who was winning and who was losing, a judgment as to fitness.

As we recall, the basic question for us will be: In what sense is a chimera, a cooperative thing, an adaptive response in an evolutionary context? How could such a thing arise, let alone prosper, in a situation presumably driven by competition?

Our point of departure will be the statement above, that to qualify as an adaptation, a behavior must enhance the survival of the individual manifesting it. There are a number of vague terms in this characterization: "individual"; "survival"; even "behavior." Before we can approach our basic question, we must make these heuristic terms more precise. To do so requires digressions into epistemology and ontology, to which we must now turn our attention.

3. Some System Theory

The *mechanism* that most biologists espouse provides definite answers to some of the questions we have just raised: it stipulates what an "individual" is and what "behavior" is; what "survival" means, however, does not have any straightforward mechanistic counterpart. Yet, as we have seen, *adaptive* behavior is linked precisely to "survival," and Darwinian evolution depends on adaptation. Biologists, in general, feel that they dare not stray from mechanistic strictures without falling into vitalism—hence the conundrums which have always plagued what passes for evolutionary theory (cf. Rosen, 1991). So let us start with the easy questions first; let us talk about Mechanistic representations of individuals and their behaviors.

A behavior, in general, is represented as a change of state. What changes a state is the assignment of a tangent vector to the state and interpreting it as the temporal derivative of state. Now a tangent vector, like any vector, has both magnitude and direction. The *direction* is determined by, or represents, the *forces* which are posited to be responsible for the behavior; the *magnitude* is determined by what I shall call *inertial properties,* or inertial parameters, which modulate between the posited forces and the system behaviors elicited by them.

In terms of Aristotelian causality, when we treat the resultant behavior as an *effect,* and ask "Why?" about it, we get three answers:

    a. Because of the initial *state* (material cause).

    b. Because the associated tangent vector points in this direction (efficient cause).

    c. Because the tangent vector has this magnitude (formal cause).

This, in fact, is the very essence of mechanism.

The *identity* or (intrinsic) nature of the system which is doing the behaving is tied to the *formal causes* of its behaviors, its inertial characteristics, and how it sees forces. It is not tied to the behaviors themselves in any direct fashion, although it is the behaviors (states and changes of state) which we actually see. It is thus natural to call these "identity-determining" aspects *genome,* and the behaviors themselves, of which genome constitutes formal cause, as *phenotypes.*

In Weissmannian terms, phenotypes constitute soma, while their formal causes are *"germ-plasm."*

Now let us turn to the forces themselves; the efficient causes of behaviors. First, let us consider *internal forces*. These are the efficient causes of behaviors of the system in isolation, the behaviors *in an empty environment*. This concept would be vacuous unless "parts" of our system could exert forces on other "parts" as well as respond to them inertially.

Parenthetically, let us note that the identity of a system, which we have tied to its inertial aspects, has a dual in what we may call its *gravitational* aspects: its capacity to exert forces. We measure inertial properties of a system by looking at how the system responds to forces; we measure gravitational properties by seeing how the system affects behavior in other systems. At some sufficiently deep level, mechanism mandates that inertial and gravitational parameters must coincide; e.g., inertial mass must equal gravitational mass.

The main thing to note is the requirement that *behavior cannot affect identity* in an isolated system. A behavior cannot act back on its formal causes. Or, stated another way, an isolated system cannot change its own identity.

Now let us suppose that the environment of our system is not empty. That is, we cannot answer "Why?" questions about system behaviors in terms of internal forces alone. We shall call an environmental force *admissible* if it, like the "internal forces," does not change the system's identity. That is, its only effect on the system is a change in *behavior*. A nonadmissible force, one which can change the system *identity*, would make that very identity *context-dependent*, make it depend on a larger system for its determination. In particular, reductionism in general rests on a presumed context independence of identity, namely, that the "parts" of a larger system are *the same*, either *in situ* or *in vacuo*.

Clearly, the change in behavior arising in a system from external, environmental sources (i.e., the discrepancy between actual behavior in a nonempty environment and the corresponding behavior in vacuo) constitutes forced behavior. In its turn, forced behavior is divided into two classes, traditionally called *autonomous* and *nonautonomous*. In the former case, the tangent vector depends on state alone in its direction; in the latter case, it also depends on time. Autonomous systems must not be confused with closed systems, nor autonomous forces with internal forces.

Nonautonomous admissible forcings are often called *inputs,* and the resultant system behaviors *outputs.* The aim in studying them is to express output as a function (the transfer function) of input. In autonomous situations, where time is not involved in characterizing the environment, the goal is to express behavior as a function of time alone.

The restriction to *admissible* forcings, which show up only in system behavior and never in system identity, is crucial to all these kinds of studies, of course. That restriction on *environment* becomes more and more severe as we deal with larger and larger *systems.* Moreover, only admissible environments, which keep identity context-independent, allow us to do traditional mechanics at all. Much of what we say henceforth, particularly our comments on "survival," pertain to *what happens in nonadmissible environments.*

4. Ontological Aspects

Epistemology is concerned with the causal underpinnings of system behavior, i.e., with unraveling why behavior changes state the way it does. Ontology, on the other hand, is concerned with existence and creation. Hence it is concerned with why a given system has come to have the identity it has, and the states it displays instead of something else.

Thus, when we describe chimera as a new *individual,* with its own new identity and its own behaviors (i.e., as a system in its own right), we are asking inherently ontological questions. In particular, we are focusing on chimera as an adaptive response of other systems, other individuals, with their own behaviors and epistemologies.

In general terms, it is simply presumed that ontological questions can be *reduced* to merely epistemological ones. Roughly, the argument is that the "creation" of a new individual is simply another way of talking about the *behaviors of some larger system;* perhaps the entire material universe. Stated another way: Any question relating to the ontology of something finds its answers precisely in how that bigger system is behaving, i.e., merely how that *bigger* system is changing state.

On the other hand, the need to go to a *larger* system to answer ontological questions, rather than smaller ones as we try to do in all analytical approaches to behavior (epistemology) in systems already given, is a little jarring, to say the least. If nothing else, it requires us to give up the view that *analysis* (as a tool for understanding a given system's behavior in terms of some of its subsystems) and *synthesis* (the assembly or reassembly of such analytic parts, to recreate the system itself) are generally inverse processes (cf. Rosen, 1988). Indeed, we must view *synthesis* as highly *context-dependent,* in total contrast to the completely context-independent ("objective") analytical units that reductionism seeks.

It is not our intention here to discuss these matters in depth (cf. Rosen, 1991). However, they arise naturally in the course of our present discussion; cooperation and competition are concepts straddling the boundary between epistemology and ontology, analysis and synthesis, behavior and creation. As such, they challenge both.

5. *"Survival"*

As we stated earlier, in order to qualify as adaptive, a particular system behavior must be referred to a specific individual whose survival is thereby favored. Thus, the same *behavior,* the same set of "objective" circumstances, can be adaptive or not, contingent on the "individual" to which it is referred. On the other hand, though "behavior" is a mechanistic concept, "survival" is not; it is uniquely biological.

Intuitively, "survival" must pertain to the *identity* of a system, which we have tied to the formal causes of its behaviors. That is, survival is tied to the system's array of inertial and gravitational parameters; to what we have called its *genome.* As we have seen above, *admissible* environments, admissible forcings, do not affect such formal causes. They show up rather as efficient causes of behaviors. Hence, survival must inherently involve *nonadmissible* environments, environments which can change system identity. More precisely, survival must involve the interplay between system *behavior* and the effect of such a nonadmissible environment on the identity of something. As we have seen, this fact alone is sufficient to remove survival from the realm of mechanism, in which only admissible forces are allowed.

What behavior can always do, however, is to have an impact on the environment itself. Indeed, we recall that even the most mechanistic system is not only inertial (i.e., not only responds to forces) but is also gravitational (i.e., it generates forces) in a way that depends directly on *genome* or identity. These gravitational effects of system on environment are difficult to discuss in conventional mechanistic settings, because typically the "environment" is assigned no states, and hence does not even "behave" in the traditional sense; it is a gravitational *source,* not an inertial sink.

Nevertheless, we must discuss the capacity of a system to change its environment as being the crux of our characterization of behavior as adaptive, and adaptivity as favoring survival. We will employ a language we have developed elsewhere (cf. Rosen, 1979) for somewhat different purposes—but which turns out to be appropriate in the present context.

Very generally, we want to express an *effect* (a behavior, or change of state, which we shall locally call $\dot{x}$) as a function of its *causes*. As we have seen, conventionally there are three such causes: an initial state x_0 (material cause), an identity or genome α (formal cause), and a total force F (efficient cause). Thus, we have

$$\dot{x} = \dot{x}(x_0, \alpha, F). \tag{1}$$

In conventional mechanics, these *local* relations are given global validity and constitute the "equations of motion" of the *system*. What we have called the total force F may or may not depend on time (i.e., the system may be autonomous or not).

If the forces F are *admissible* (i.e., do not affect identity), this is expressed by the formal condition

$$\partial \alpha / \partial F = 0. \tag{2}$$

As we have seen, this is the conventional setting for mechanistic environments. On the other hand, this condition fails in nonadmissible environments; i.e., we must have

$$\partial \alpha / \partial F \neq 0. \tag{3}$$

Additionally, we must express the impact of behavior on the environment (i.e., on F). This will be embodied in the formal expression

$$\partial F / \partial \dot{x}. \tag{4}$$

In this language, then, a behavior $\dot{x}$ is *adaptive* if its impact on the environment F lowers the impact of the environment on the genomes. The condition for adaptivity of $\dot{x}$ is thus

$$\frac{\partial}{\partial \dot{x}} \left(\frac{\partial \alpha}{\partial F} \right) < 0. \tag{5}$$

How adaptive the behavior $\dot{x}$ is (i.e., how "fit" the behavior, and by extension the genome, which is its formal cause) is clearly measured by how negative (5) is in the given situation; the "fittest" behavior is the one for which (5) is as small as possible.

Clearly, the expression (2) says that genome is environment-dependent, or context-independent; i.e., there are only admissible forces involved. If (2) fails [i.e., (3) holds], the environment F is nonadmissible; genome is itself changing or behaving, and hence there will be a nonzero temporal derivative:

$$d\alpha/dt \neq 0.$$

Thus, we may rewrite the adaptivity condition (5) for $\dot{x}$ as

$$\frac{\partial}{\partial \dot{x}} \left(\frac{d\alpha}{dt} \right) < 0. \tag{6}$$

In the language of Rosen (1979), we may say that $\dot{x}$ is adaptive to the extent that it is an *inhibitor* of genome change in a nonadmissible environment.

We will conclude this section with two brief remarks. First, the entire machinery of Rosen (1979) may be taken over into the present context of adaptation and survival. We recall that the interpretation we gave to this machinery was "informational," in the sense that our partial derivatives answered questions of the form: "If A changes, what happens to B?" We interpreted these answers, this "information," in terms of activations and inhibitions, agonisms and antagonisms, and showed that they transmute into a system of ordinary rate equations if, and only if, a corresponding web of differential forms having these partial derivatives as coefficients are all *exact*. Otherwise, there are no overarching rate equations, and the situation becomes *complex* in my sense (cf. Rosen, 1991).

The second observation is that adaptivity, in the sense that we have just described, does involve interactions between behaviors and their genomes. Not direct interactions, but indirect ones, through their gravitational effects on environments that are affecting genomes. It is these interactions that constitute the activations/inhibitions, agonisms/antagonisms, and the like, which determine adaptivity and fitness. Thus, we can see in another way that considerations of adaptivity require "more causal entailment" than purely mechanical approaches to systems allow. This "extra entailment" can be regarded, in a perfectly rigorous way, as the *final causations* that seem inherent in adaptation itself.

6. When Is Chimera Formation Adaptive?

In a certain deep sense, the combination of "system" and environment to constitute a single "universe" is already an instance of chimerization.

But for our purposes, this is not a very useful situation, since "environment" does not in itself constitute a system in any causal sense; in particular, it does not have in itself a meaningful "identity" or genome. Rather, what we shall do is to identify a second system in the environment of the first, and instead of looking at the interplay between our initial system and its "undifferentiated" environment, divide it into the interplay between the two systems and *their* environment.

Our original system, we recall, had behaviors $\dot{x}$, states x, an identity α, and an environment F. Our new system will, analogously, have behaviors $\dot{y}$, states y, an identity β, and an environment G. Each system is now part of the environment of the other. In general, we will allow our environments to be nonadmissible.

We will allow our systems to interact with each other in the broadest possible way, along the lines indicated in the preceding section. Thus, we will say that the systems *compete* when the behaviors of each of them lower the fitness or survivability of the other. We can express this as follows:

$$\frac{\partial}{\partial \dot{x}} \left(\frac{\partial \beta}{\partial G} \right) > 0; \qquad \frac{\partial}{\partial \dot{y}} \left(\frac{\partial \alpha}{\partial F} \right) > 0. \tag{7}$$

We recall that the behavior $\dot{x}$ is part of the total environment G seen by the second system; likewise, the behavior $\dot{y}$ is part of the total environment F seen by the first. Thus, it is not generally true that the most *adaptive* behavior $\dot{x}$, with respect to its environment F as a whole, will also be the most "competitive," i.e., the one which will make $(\partial / \partial \dot{x})(\partial \beta / \partial G)$ as large as possible. In other words, fitness of a behavior $\dot{x}$ does not generally mean "most competitive," measured entirely against another system.

Likewise, we will call an interaction *cooperative* if the following conditions obtain:

$$\frac{\partial}{\partial \dot{x}} \left(\frac{\partial \beta}{\partial G} \right) < 0, \qquad \frac{\partial}{\partial \dot{y}} \left(\frac{\partial \alpha}{\partial F} \right) < 0. \tag{8}$$

In this case, the behavior of each system increases the survival or fitness of the *other*. It thus *indirectly increases its own fitness* by favoring something which *inhibits* the impact of its total environment on its genome.

This kind of cooperative strategy constitutes, in the broadest sense, a *symbiosis* of our two systems. It is not yet a chimera in our sense, in that it does not yet have a real identity (genome) and behaviors of its own; it is as yet only a kind of direct product of the individual systems which comprise it.

Nevertheless, we can already see that total fitness of a system depends on behaviors $\dot{x}, \dot{y}$ satisfying a cooperative condition (8) instead of a competitive condition (7). Under such circumstances, we can say that cooperation will be favored over competition. Moreover, the *causally* independent behaviors $\dot{x}, \dot{y}$ in these situations (i.e., each of which can be causally understood as *effects* without invoking the other system at all) have become *correlated*. The statement of this correlation looks like what, in mechanics, is called a *constraint*. This observation, in fact, provides the basic clue that leads to the next adaptive step—from symbiosis to chimera.

In the situation we have been discussing, we have supposed no *direct interaction* between our two systems, only *indirect* interaction through activations and inhibitions of environmental effects. In other words, the behaviors of each are governed entirely by inertial responses of each to an undifferentiated gravitating environment, and each system's *gravitational* properties are likewise expended entirely into this environment. This indirect interaction results in our correlation of causally independent behaviors of the two systems separately.

But if the systems do interact directly, their joint behaviors are no longer *causally* independent; we can no longer answer "Why?" questions about the one without invoking the other. A pair of such systems in direct interaction constitutes a new system, with its own behaviors $\dot{z}$, its own states z, its own identity γ, and its own environment H. In terms of the original systems, the *interaction itself* takes the form of constraints, identity relations between behaviors, states, and genomes, out of which $\dot{z}, z$, and γ are built.

In any case, we can now ask about the survival of this new composite system, i.e., about

$$\frac{\partial \gamma}{\partial H}$$

and the adaptivity of its own behavior; i.e., about

$$\frac{\partial}{\partial \dot{z}} \left(\frac{\partial \gamma}{\partial H} \right).$$

Finally, these quantities must be compared with the corresponding quantities for the original free systems. It is easy to write down, in these terms, the conditions under which a pair of interacting systems, in which behaviors are constrained rather than correlated, survives better (i.e., preserves more of *its* genome in a nonadmissible environment) than either system can by itself.

7. *"The Survival of the Fittest"*

So far, we have couched our discussion entirely in terms of an *individual,* with its own identity (genome), its own behaviors, and its own (generally nonadmissible) environment. We have identified "survival" with persistence of genome, and measured "fitness" of a behavior $\dot{x}$ in terms of inhibition of rate of change of genome. Even in these terms, "survival of the fittest" is already highly nontautological.

However, Darwinian evolution does not yet pertain to this view. To say that a behavior $\dot{x}$, manifested in an individual *now,* has *evolved* is to raise an ontological question about $\dot{x}$, i.e., about behaviors of behaviors. In a traditional mechanistic sense, we can claim to understand a behavior $\dot{x}$ completely in traditional Aristotelian causal terms without involving any "larger" system. But when we answer the question "Why $\dot{x}$?" with a putative answer "Because $\ddot{x}$," i.e., because of some dynamics on some *space* of behaviors or phenotypes, we are in a different realm. In fact, we are dealing with an entirely new cast of individuals (the ones with $\ddot{x}$ as phenotypes), with their own identities and environments.

Many of the deep puzzles and conundrums of evolutionary "theory" arise from this opposition. A major factor here is that biologists, as a group, seek simplistic mechanistic explanations for behaviors $\dot{x}$, but do not want corresponding explanations for $\ddot{x}$. We have touched on this peculiar situation elsewhere (cf. Rosen, 1991).

We have so far omitted from our discussion the primary connection between the individuals that are adapting and the (entirely different) "individuals" that are evolving. That connection rests on a uniquely biological *behavior,* manifested by our adapting individuals, which we may for simplicity call "reproduction." In the most elementary terms, this behavior is manifested in the reorganization of environmental materials, along with its own, to generate new individuals with their own identities and behaviors.

Intuitively, the result of this reorganization of the environment, arising from a (gravitational) behavior of our original individual, is to increase the "ploidy" of its genome and of the behaviors in the new individual of which identity is formal cause. As with any behavior, we can ask "How adaptive is it?" But as we have seen, before we can answer such a question, we must specify the *individual* whose survival is favored by it.

When we begin to address such questions, we find ourselves suddenly transported to quite a different universe, one whose "states" are now specified by these "ploidies" and with how they are changing over a quite different time scale. In this universe, the original adapting individuals are *turning over* (in fact, very rapidly in the new time scale). But "behavior" in this new universe is just rates of change of ploidies. And this is the universe that must be related to the "behavior of behavior," the ontological universe mentioned above.

In this new universe, the basic concept of "survival" needs to be completely redefined, to refer to *ploidies* of genomes rather than to the genomes themselves. Symbolically, it now refers to $d\|\alpha\|/dt$, rather than to $d\alpha/dt$ or $\partial\alpha/\partial F$. Accordingly, adaptiveness and fitness must be *redefined*, according to whether or not they activate $d\|\alpha\|/dt$, i.e., according to how they affect ploidy, rather than, as was the case before, how much they protect an individual genome from a nonadmissible environment F. Without going into too many details, we may just state that there is no reason that these two notions of adaptiveness should coincide; they refer to entirely different "individuals."

The long-time behaviors of these "ploidies" are the province of Darwinian evolution. In particular, "survival" means keeping a ploidy positive. From this viewpoint, the "fittest" behavior of an *individual* is the one for which

$$\partial(d\|\alpha\|/dt)/\partial\dot{x}$$

is as great as possible, not the one for which

$$-\partial(\partial\alpha/\partial F)/\partial\dot{x}$$

is largest. From an evolutionary point of view, these are the "fittest" that do the surviving in the new situation.

As always, the nonadmissible environment is regarded as a source for "new genomes" (i.e., an environment that makes a "ploidy" go from zero to something positive). And with the new genomes come new behaviors. Thus, any dynamics on a space of "ploidies" is reflected in a corresponding dynamics on a phenotypic space of *behaviors*. That dynamics is supposed to generate precisely the $\ddot{x}$, i.e., provide the ontogeny of behaviors manifested by original individual *phenotypes*.

Thus, it is traditional in this context to measure the "fitness" of a *genome* α by its effect on its own ploidy, i.e., by

$$\partial\|\alpha\|/\partial\alpha.$$

Competition between two genomes α, β is expressed by the conditions

$$\frac{\partial \|\alpha\|}{\partial \beta} < 0, \qquad \frac{\partial \|\beta\|}{\partial \alpha} < 0$$

analogous to (7) above. The hope is that these quantities suffice to drive an autonomous dynamics expressed locally by the differential relations

$$d\|\dot{\alpha}_i\| = \sum_j \frac{\partial \|\alpha_i\|}{\partial \alpha_j} d\|\alpha_j\|,$$

perhaps supplemented by an *admissible* forcing term derived from the originally nonadmissible environment F; and further, that such a dynamics can be projected onto a corresponding "behavior" space, or phenotype space, where it will assume the form of an action principle. But we shall not go into these matters here.

8. When Is Chimera the Fittest?

As we said at the outset, adaptation is a meaningless concept unless it is tied to an individual whose survival is enhanced by it. Otherwise, adaptation just disappears into dynamics. If we choose our "individuals" differently (and change our idea of what survival means correspondingly), our notions of what is adaptive will generally change as well.

From the outset, we have tied our identification of an individual to its "genome," and hence with the formal causes of its behaviors. A behavior itself could be adaptive or not, in this context, depending on its effect on genome preservation. In this context, then, the "fittest" behaviors are those that minimize change of genome (identity) in the face of environments that can change it. We have given some very general conditions for the "fitness" of chimera formation (i.e., cooperation) in this situation.

In general, one of the basic requirements for identifying the kinds of "individuals" to which such arguments apply is that they themselves must not, by virtue of their own behaviors, be able to change their own genomes. That is, we must not allow their behaviors to act back on their own formal causes. This is true quite generally; in biology, it is the content of the "Central Dogma."

However, behaviors do affect the "ploidy" of genome, seen in the context of a population. Such a population may itself be regarded as a new "individual," whose own behaviors can be regarded as adaptive or not. This is the context of *evolutionary* adaptation. In this new context, adaptations are measured by the growth of ploidies.

At heart, the Darwinian concept of "survival of the fittest" rests on an identification between the two entirely different kinds of "fitness" of individual behaviors we have just sketched, pertaining to two entirely different kinds of individuals, and hence two entirely different measures of survival and adaptation. Conceptual difficulties with evolution have always grown from the fact that the two need not coincide.

In particular, there is no reason why behaviors which maximize the survival of an individual should also maximize its fecundity, or indeed, vice versa. To the extent that these entirely different measures of "fitness" diverge, to speak of fitness as an abstract property of an individual behavior simply creates an equivocation. Especially so since "fecundity" arises from individual behaviors which may be very far from adaptive in the sense of individual survival.

Equivocations of this kind spawn apparent paradoxes. A simple example is the so-called Galileo Paradox, which involved the "size" (i.e., cardinality) of sets. Galileo asked which is "bigger," the set of all integers or the set of even integers. Judged simply by inclusion, and according to the Euclidean axiom that a whole (all the integers) is bigger than any proper part (the even integers), we conclude that the former is clearly "bigger." But, as measured by enumeration, we must conclude that the two are the "same size." Two entirely different measures of "size" are involved here, measures which coincide for finite things but which can disagree for infinite ones. Indeed, in the hands of Georg Cantor, the Galileo Paradox ended up as the diagnostic property separating that which is finite from that which is not.

So it is too with concepts like "fitness," and especially so when we attempt to compare fitnesses. In the present situation, we treat a chimera as a new individual and ask about its "fitness" as such. Moreover, just as in the Galileo paradox, we try to compare its "fitness" to that of other individuals, especially against those of its constituents, in terms of the advantages of cooperation against noncooperation.

What I have tried to do in the above discussion is rather more modest. I have tried to argue that chimera formation culminates in the generation of new kinds of individuals; it causes new identities, new genomes, and new behaviors to emerge which could never be generated in any other way, certainly not by processes of differentiation alone. I have tried to give conditions under which chimera formation is adaptive, in the sense that it favors the survival of its constituents, more so than the survivals they could achieve otherwise. But I certainly cannot say that if fitness is measured in terms of ploidies, as it is when we speak of Darwinian evolution, such considerations have any significance at all. The two issues are clearly not the same.

Indeed an establishment of relations between these two distinct playgrounds for adaptiveness, the evolutionary and the physiological, would in itself be an instance of chimera formation.

References

McLaren, A. (1976). *Mammalian Chimeras.* Cambridge: Cambridge University Press.

Rosen, R. (1975). "Biological Systems as Paradigms for Adaptation," in *Adaptive Economic Models,* R. H. Day and T. Groves, eds., New York: Academic Press, pp. 39–72.

Rosen, R. (1979). "Some Comments on Activation and Inhibition," *Bull. Math. Biol.,* 41, 427–445.

Rosen, R. (1988). "How Universal Is a Universal Unfolding?" *Appl. Math. Lett.,* 1, 105–107.

Rosen, R. (1991). *Life Itself.* New York: Columbia University Press.

CHAPTER 10

Minimal Properties for Evolutionary Optimization

PETER SCHUSTER

Institute for Molecular Biology
Beutenbergstraße 11
D–07708 Jena, Germany

1. Evolution in an RNA World

Until the 1980s it was a commonly held view in biochemistry that
biopolymers are either carriers of biological information like nucleic
acids (and thus have coding functions) or specific catalysts and structure building molecules like proteins. Exceptions to the rule were
known: for example, transfer-RNA molecules, which represent media-
tors of the genetic code in the translation from nucleic acids to protein,
and ribosomal-RNA molecules, which were then understood as a sort
of skeleton holding together the ribosomal proteins. The discoveries of
Thomas Cech and Sidney Altman (Cech, 1986; Guerrier-Takada et al.,
1983; Guerrier-Takada and Altman, 1984; Symons, 1992) have shown
that in contrast to previous belief, RNA molecules can catalyze several classes of reactions with high efficiency and specificity similar to
those of protein enzymes. The notion of **ribozymes** was created for
such RNA-based catalysts. The catalytic activities of RNA molecules
are all related to cleavage or the formation of nucleotide or peptide
bonds. For RNA substrates, ligation and cleavage are highly sequence
specific. The capability of RNA to catalyze peptide bond cleavage and
formation (Noller, 1991; Noller et al., 1992; Piccirilli et al., 1992) is of
particular interest since it can be interpreted as a hint that primordial

Cooperation and Conflict in General Evolutionary Processes, edited by John L.
Casti and Anders Karlqvist.
ISBN 0–471–59487–3 ©1994 John Wiley & Sons, Inc.

ribosomes might have worked without proteins. Although the catalytic repertoire of ribozymes is rather poor compared to that of protein enzymes, it comprises the key reactions that are necessary to produce polynucleotides from oligomers and monomers.

Enzyme-free, template-induced RNA synthesis has been studied extensively by Leslie Orgel (1992). Template-induced reproduction of RNA molecules occurs regularly in cells that were infected by RNA viruses or viroids (for viruses, see e.g., Weissmann, 1974; for viroids see, e.g., Riesner and Gross, 1985). Replication of RNA molecules under nonequilibrium conditions *in vivo* has been studied extensively since the pioneering works by Sol Spiegelman in the mid-1960s (Spiegelman, 1971; Biebricher and Eigen, 1988). It provides the basis for evolutionary adaptation in the sense of Darwin's mechanism, as shown in the next two sections. Catalytic properties of RNA molecules certainly have an enormous consequence for the Darwinian scenario. Isn't it a great advantage to defeat competitors not only by outgrowing them in future generations, but also by cutting them into pieces *hic et numc* through ribozyme reactions ?

RNA molecules are thus unique in chemistry and biology since they can act not only as templates for their reproduction but are also catalysts for several reactions involving other RNA molecules or oligopeptides. It is generally believed nowadays that there was a period in the origin of life on Earth when RNA served as both genetic material and specific catalyst (see, e.g., Gilbert, 1986; Orgel, 1986; Joyce, 1991). In other words, RNA has the capability to be genotype and phenotype at the same time . A world of RNA molecules driven by a steady supply of suitable energy-rich compounds also represents a kind of biological toy universe which allows us to study the evolutionary process in its simplest form. Populations in such an **RNA world** are capable of Darwinian evolution since they adapt to changes in the environment by replication, mutation, and selection. In biology these populations form the simplest currently conceivable Darwinian scenarios and thus are the best candidates for adaptive systems of minimal complexity.

Is there room for an RNA world in the current view of chemical evolution on the primordial Earth ? In Figure 1 we show a sketch of a plausible sequence of scenarios leading to the onset of RNA-based Darwinian evolution (for more details, see, e.g., Schuster, 1981; Mason, 1991). The major problem with RNA in chemical evolution is to be seen in its highly elaborate molecular structure: for RNA template-induced reactions, compounds of high stereochemical purity are required. Single isomers with proper configurations have to be available at a high

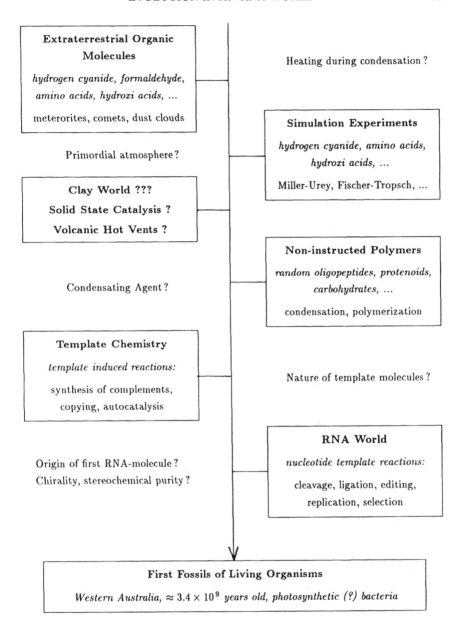

Figure 1. Some facts and open questions about the origin of life.

degree of purity, and chiral compounds such as ribose have to be present as pure single antipods. Higher amounts of other pentose or hexose molecules are prohibitive for template-induced replication, and racemic mixtures of D- and L-ribose (containing 50 % L-ribose) are real poisons for template-induced RNA synthesis on templates built from D-ribose. It was suggested therefore that less complicated molecules which could nevertheless act as templates preceded the RNA world (Orgel, 1986; Joyce, 1989). Recent progress in **template chemistry** (Orgel, 1992) makes this suggestion more and more plausible. RNA molecules, on the other hand, are the first template molecules which share most of their chemical properties with DNA, the predominant genetic information carrier at present. Information transfer from DNA to RNA, and vice versa, known as transcription and reverse transcription, is routine in present-day biochemistry. We can speculate therefore that a phylogenetic extrapolation of present-day genetic information may go back as far as to the first efficiently replicating species in a primordial RNA world. In short, RNA molecules seem to appear late in chemical evolution and very early in biological evolution.

According to the analysis of microfossils carried out by William Schopf (Schopf and Packer, 1992), the first sure remnants of microorganisms are dated 3.4×10^9 years ago. These microorganisms were most likely photosynthetic and related to present-day cyanobacteria. Needless to say, there was a long way with many intermediate steps to go from a primitive Darwinian system like the RNA world to a fully developed photosynthetic organism (Eigen and Schuster, 1985). Plausible speculations would foresee several scenarios, among them those built from

- RNA molecules and instructed proteins related by a genetic code,
- RNA molecules, proteins, and a metabolism allowing for a high degree of autonomy, or
- RNA molecules, proteins, metabolism, and membranes, including a mechanism for replication controlled cell division, etc.

The flowchart sketched in Figure 1 does not refer explicitly to these other **worlds** which inevitably lay between the RNA world and the (earliest) organisms which gave rise to the fossils reported and analyzed by William Schopf.

2. Molecular Genotypes and Phenotypes

Evolution experiments with pure RNA would face severe problems: template-induced RNA synthesis is slow and does not work for most of the templates under the conditions applied so far, and the accuracy of base incorporation is low (on the order of one error in 100 nucleotides). Another problem concerns separation of double strands, which does not occur readily under the conditions of template-induced synthesis. In order to explore the capabilities of a minimal complexity adaptive system, we are, however, not restricted to the pure RNA world. Why not use protein catalysts as highly specific and efficient environmental factors? Addition of enzymes leads to a new toy universe for studies on evolution that might be characterized as a **protein-assisted RNA world**.

It should be repeated that the enzymes are environmental factors in such a scenario and thus cannot be changed by the evolutionary process under consideration. Sol Spiegelman (1971) has indeed shown that such a universe can be created in the test tube, and in fact, it shows all features of the Darwinian scenario. RNA-dependent RNA polymerases, commonly called RNA replicases, occur in nature. Spiegelman used an enzyme isolated from bacterial cells of *E. coli* which were infected by the RNA bacteriophage $Q\beta$. This enzyme replicates RNA fast and with much higher accuracy than is achieved without a protein catalyst: many thousands of generations can be studied in experiments lasting less than a day, and the accuracy of replication is about one error in 3,000 nucleotides.

The dichotomy between genotypes and phenotypes is basic to biological evolution. Molecular biology has shown that the genetic information for the unfolding of organisms (which represent the phenotypes) is contained in the DNA (which is the genotype). The Darwinian principle of evolution is based on heredity, variation, and selection. According to our present-day knowledge, all inheritable variations are caused by changes of the genotype (i.e., by changes in the sequence of nucleotide bases of the DNA), and selection acts through the differential reproductive success of phenotypes. RNA molecules in cell-free evolution experiments are both genotypes and phenotypes. As we see in Figure 2, the genotype is again a sequence of nucleotide bases, this time in the form of an RNA molecule. Following Sol Spiegelman, we consider the spatial structure of the RNA molecule as its phenotype. Indeed, this structure is evaluated by the selection process. Systematic kinetic studies on RNA replication and test tube evolution carried out by Christof Biebricher in Manfred Eigen's laboratory (Biebricher et al., 1983, 1984, 1985; Biebricher and Eigen, 1988) revealed the mechanism of RNA replication. The rate and equilibrium constants which determine the outcome of selection under different conditions are known now. They can be determined independently from selection experiments by direct measurements. In addition, minimum requirements for efficient replication were discovered. They consist in sequence reg-

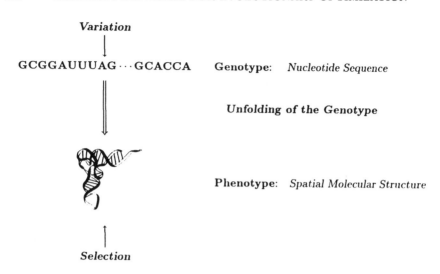

Variation

GCGGAUUUAG···GCACCA   **Genotype:**   *Nucleotide Sequence*

Unfolding of the Genotype

Phenotype: *Spatial Molecular Structure*

Selection

Figure 2. Molecular genotypes and phenotypes in RNA evolution experiments.

ularities and sufficient secondary structure to facilitate double-strand separation by the enzyme.

Selection experiments may be carried out by discontinuous renewal of the material consumed as in the serial transfer technique (Spiegelman, 1971), or under the continuous constraint maintained in elaborate evolution reactors (Husimi et al., 1982; Husimi and Keweloh, 1987). A new selection technique was used by John McCaskill in his molecular evolution experiments in capillaries (Bauer et al., 1989a): the capillaries contain a medium suitable for replication, RNA is injected, and a wave front spreads through the medium. The front velocity of the traveling wave increases with the replication rate, and hence, faster-replicating mutants are selected by the wave propagation mechanism.

The capability of RNA molecules to evolve by natural selection is used now in RNA-based applied molecular evolution to produce biomolecules with new properties. This novel evolutionary technology was originally suggested by Eigen and Gardiner (1984). Somewhat later Kauffman (1986) suggested the use of random sequences in selection experiments (for a recent review, see Kauffman, 1992). One approach to the problem is suitable for **batch experiments** (Figure 3). The essential trick of this technique is to encode the desired functions into the selection constraint. As indicated in Figure 3, sufficient variation is introduced into populations either by artificially increased mutation rates

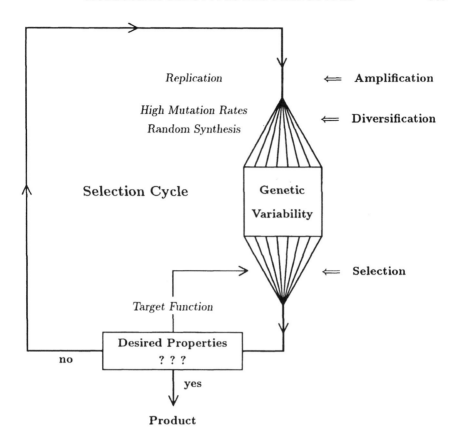

Replication

⟸ **Amplification**

High Mutation Rates
Random Synthesis

⟸ **Diversification**

Selection Cycle

Genetic

Variability

⟸ **Selection**

Target Function

Desired Properties
no   **? ? ?**

yes

Product

Figure 3. A selection technique of applied molecular evolution based on encoding of the function to be developed into the selection constraint.

or by partial randomization of sequences. The synthesis of oligonucleotides with random sequences is routine nowadays. Selected RNA molecules are amplified either by replication or after transcription into DNA by conventional cloning techniques. Several examples of successful applications of molecular selection techniques to biochemical problems are found in the current literature (Horwitz et al., 1989; Tuerk and Gold, 1990; Ellington and Szostak, 1990; Beaudry and Joyce, 1992). In reality it will often be impossible to encode the desired function directly into the selection constraint. Then spatial separation of individuals and massively parallel screening provide a solution (Figure 4). This technique, however, requires highly sophisticated equipment which is currently under development (Bauer, 1990; Bauer et al., 1989b).

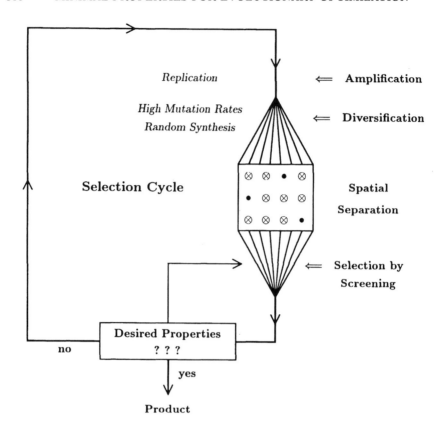

Figure 4. A selection technique of applied molecular evolution based on massively parallel screening.

3. Adaptation and Stability of Populations

The most frequently occurring genotypes (commonly called wild-type **master sequences**) can be isolated from populations and analyzed during the course of molecular evolution experiments. These investigations show how the structure of RNA molecules is adjusted in order to cope with changes in the environment. Such environmental changes are caused by adding suitable substances to the replication medium which deteriorate the conditions for reproduction. For example, heterocyclic dyes (ethidium bromide, acridinium orange, etc.), which intercalate between Watson-Crick base pairs and thus interfere with replication, can be used, or ribonucleases (RNA cleaving enzymes) can be applied (Strunk, 1992). Mutants which compensate for the change in the environment by having fewer binding or cleavage sites than the wild type appear and are enriched in the populations.

Replication errors lead to new molecular species whose replication efficiency is evaluated by the selection mechanism. The higher the error rate, the more mutations occur and the more viable mutants appear in the population. The stationary mutant distribution is characterized as **quasispecies** (Eigen and Schuster, 1977; Eigen et al., 1988, 1989) because it represents the genetic reservoir of asexually replicating populations. An increase in the error rate leads to a broader spectrum of mutants and thus makes evolutionary optimization faster and more efficient in the sense that populations are less likely to be caught in local fitness optima. There is, however, a critical error threshold (Eigen, 1971; Eigen and Schuster, 1979; Swetina and Schuster, 1982): if the error rate exceeds the critical limit, heredity breaks down, populations are drifting in the sense that new RNA sequences are formed steadily, old ones disappear, and no evolutionary optimization according to Darwin's principle is possible (Figure 5). Variable environments require sufficiently fast adaptation, and populations with tunable error rates will adjust their quasispecies to meet the environmental challenge. In constant environments, on the other hand, such species will tune their error rates to the smallest possible values in order to maximize fitness.

Viruses are confronted with extremely fast changing environments because their hosts developed a variety of defense mechanisms, ranging from the restriction enzymes of bacteria to the immune system of mammals and man. RNA viruses have been studied extensively. Their multiplication is determined by enzymes that do not allow large-scale variations of the replication accuracies. They vary the cumulative error rate, however, by changing the length of their genomes. It is adjusted to optimal values which often correspond to maximal chain lengths (Eigen, 1971; Eigen and Schuster, 1977):

$$\nu_{\max} = \frac{\ln \sigma}{p} . \tag{1}$$

Here $\nu_{\max}$ is the maximal chain length that a master sequence can adopt and still allow for stable replication over many generations, $\sigma \geq 1$ is the superiority of this master sequence in the stationary population, and p is the error rate per (newly incorporated) base and replication event. The superiority expresses differential fitness between the master sequence and the average of the remaining population. In the limit $\sigma \to 1$, we are dealing with **neutral evolution** (Kimura, 1983). Experimental analysis of several RNA virus populations has shown that almost all chain lengths are adjusted to yield error rates close to the

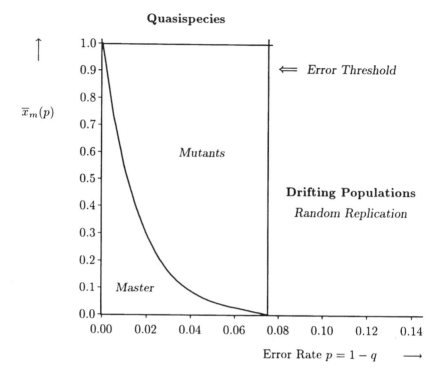

Figure 5. Evolution at the error threshold of replication. The fraction of the most frequent species in the population, called the **master sequence**, is denoted by $x_m(p)$. It becomes very small at the error threshold. Accordingly, the total fraction of all mutants, $1 - x_m(p)$, approaches 1 at the critical mutation rate.

threshold value. Thus, RNA viruses appear to adapt to their environments by driving optimization efficiency toward the maximum.

4. Evolutionary Stability of RNA Structures

Stability against mutation also has a second, less strict meaning: assume that we have a change in the sequence that does not alter the structure and the properties of the RNA molecule. We would not be able to detect such a **neutral mutation** unless we compare the sequences. Given a certain mutation rate, we may therefore ask what are the differences in stability of RNA structures against mutations in the corresponding sequences. How likely does a change in the sequence result in an actual change in the structure? In other words, we try to estimate the fraction of neutral mutants in the neighborhood of a typical sequence (**neutral** is commonly used for sequences which have properties that are indistinguishable for the selection process; we shall use the term here in the narrower sense of identical structures). This question is of statistical nature and can be answered only by proper application of statistical techniques.

First, the RNA sequences have to be ordered in a natural way. In case point mutations (single base exchanges) are predominant, the Hamming distance (d_h, counting the number of positions in which two aligned sequences differ) is an appropriate measure of the relationship between two sequences, since it counts the minimum number of point mutations required to convert one sequence into another. Moreover, the Hamming distance d_h induces a metric on the sequence space. The space of binary sequences of chain length $\nu = 4$ is shown in Figure 6 as an example.

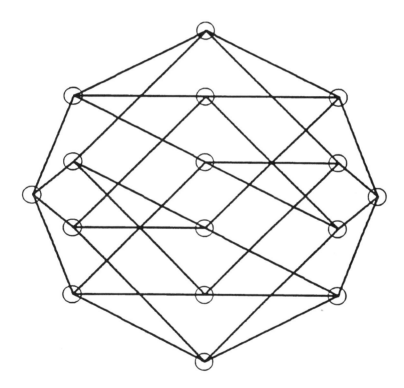

Figure 6. The sequence space of binary (**AU** or **GC**) sequences of chain length $\nu = 4$. Every circle represents a single sequence of four letters. All pairs of sequences with Hamming distance $d_h = 1$ (these are pairs of sequences that differ in one position) are connected by a straight line. The geometric object obtained is a hypercube in four-dimensional space, and hence all positions (and all sequences) are topologically equivalent.

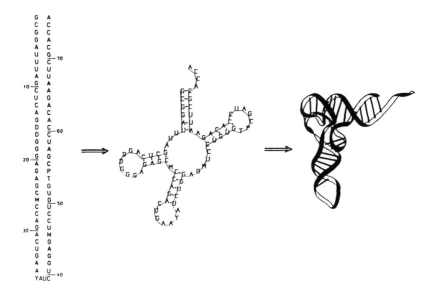

Figure 7. Folding of an RNA sequence into its spatial structure. The process is partitioned into two phases: in the first phase only the Watson-Crick-type base pairs are formed (which constitute the major fraction of the free energy), and in the second phase the actual spatial structure is built by folding the planar graph into a three-dimensional object. The example shown here is phenylalanyl-transfer-RNA (t-RNA^{phe}), whose spatial structure is known from X-ray crystallography.

RNA secondary structures are first approximations to the spatial structures of RNA molecules. They are understood as a listing of the Watson-Crick-type base pairs in the actual structure and may be represented as planar graphs (Figure 7). We consider RNA secondary structures as elements of an abstract **shape space**. Again a measure of the relationship of RNA structures which induces a metric on the shape space can be found (Fontana et al., 1991, 1993a,b). We derived this distance measure from trees which are equivalent to the structure graphs, and accordingly, it is called the **tree distance**, d_t. RNA folding thus can be understood as a mapping from one metric space into another, in particular from sequence space into shape space. A path in sequence space corresponds uniquely to a path in shape space. (The inversion of this statement is not true, as we shall mention in the next section.)

The whole machinery of mathematical statistics and time series analysis can now be applied to RNA folding. In particular, an autocorrelation function of structures based on tree distances (d_t) is computed from the equation

$$\varrho_t(h) = 1 - \frac{< d_t^2(h) >}{< d_t^2 >} . \tag{2}$$

Mean square averages are taken over sequences in sequence space $(< d_t^2 >)$ or over sequences in the mutant class h of the reference sequence $[< d_t^2(h) >$, i.e., over all sequences at Hamming distance h from the reference]. The autocorrelation functions can be approximated by exponential functions, and correlation lengths (ℓ_t) are estimated from the relation $\ln(\varrho_t(\ell_t)) = -1$.

The correlation length is a statistical measure of the hardness of optimization problems (see, e.g., Eigen et al., 1989). The shorter the correlation length, the more likely it is that a structural change is occurring as a consequence of mutation. The correlation length thus measures stability against mutation. In Figure 8 correlation lengths of RNA structures are plotted against the chain length. An almost linear increase is observed. Substantial differences are found in the correlation lengths derived from different base-pairing alphabets. In particular, the structures of natural (**AUGC**) sequences are much more stable against mutation than are those of pure **GC**-sequences or pure **AU**-sequences. This observation is in agreement with structural data obtained for ribosomal RNA (Wakeman and Maden, 1989). It also provides a plausible explanation for the use of two base pairs in nature: optimization in an RNA world with only one base pair would be very hard, and the base-pairing probability in sequences with three base pairs is rather low, and hence most random sequences of short chain lengths ($\nu < 50$) do not form thermodynamically stable structures. The choice of two base pairs thus appears to be a compromise between stability against mutation and thermodynamic stability.

5. Mapping RNA Sequences into Structures

The sequence space is a bizarre object: it is of very high dimension [since every nucleotide can be mutated independently, its dimension coincides with the chain length of RNA: $25 < \nu < 500$ for RNA molecules in test tube experiments, $250 < \nu < 400$ for viroids, and $3,500 < \nu < 20,000$ for (most) RNA viruses], but there are only a few points on each coordinate axis (κ points; κ is the number of digits in the alphabet: $\kappa = 2$ for **AU** and **GC**, $\kappa = 4$ for **AUGC**). The number of secondary structures which are acceptable as minimum free energy structures of RNA molecules is much smaller than the number of different sequences

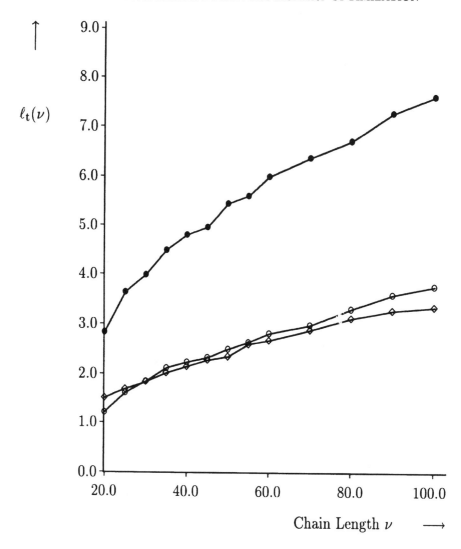

Figure 8. Correlation lengths of structures (ℓ_t) of RNA molecules in their most stable secondary structures as functions of the chain length ν. Values are shown for binary pure **GC**-sequences ($\diamond$), for binary pure **AU**-sequences ($\circ$), and for natural **AUGC**-sequences ($\bullet$). The correlation lengths are computed from plots in the ($\ln \varrho_t(h), h$)-plane by means of a least root-mean-square deviation fit.

and can be estimated by means of proper combinatorics (Schuster et al., 1993): in case of natural (**AUGC**) molecules we have about $1.485 \times \nu^{-3/2}(1.849)^{\nu}$ structures for 4^{ν} sequences. The mapping from sequence space into shape space thus is not invertible: many sequences fold into the same secondary structure. We cannot expect therefore that our intuition which is well trained with mostly invertible maps in three-dimensional space will guide us well through sequence and shape spaces. In order to get a feeling for the problem, search algorithms for the optimization of RNA structures and properties were conceived (Fontana and Schuster, 1987; Wang, 1987) and computer simulations were carried out on realistic landscapes based on RNA folding (Fontana and Schuster, 1987; Fontana et al., 1989). Here we shall adopt another strategy to obtain information on sequences and structures and use proper statistical techniques for the analysis of such an abstract object as the RNA shape space.

The information contained in the mapping from sequence space into shape space is condensed into a two-dimensional, conditional probability density surface

$$S(t, h) \;=\; \text{Prob}\left(d_t = t \mid d_h = h\right). \tag{3}$$

This structure density surface (SDS) expresses the probability that the secondary structures of two randomly chosen sequences have a structure distance t provided that their Hamming distance is h. An example of a structure density surface for natural sequences of chain length $\nu = 100$ is shown in Figure 9. We recognize an overall shape that corresponds to one half of a horseshoe with rugged details superimposed upon it. The contour plot illustrates an important property of the structure density surface: at short Hamming distances ($1 \leq \nu < 16$) the probability density changes strongly with increasing Hamming distance, but farther away from the reference sequence ($16 < \nu < 100$) this probability density is essentially independent of the Hamming distance h. The first part reflects the local features of sequence-structure relations. Up to a Hamming distance of $h = 16$, there is still some memory of the reference sequence. Then at larger Hamming distances the structure density surface contains exclusively global information independent of the reference.

In order to gain more information on the relation between sequences and structures an inverse folding algorithm which determines the sequences that share the same minimum free energy secondary structure was conceived and applied to a variety of different structures (Schuster et al., 1993). The frequency distribution of structures has a very sharp peak: relatively few structures are very common, many structures are rare and play no statistically significant role. The results obtained show in addition that sequences folding into the same

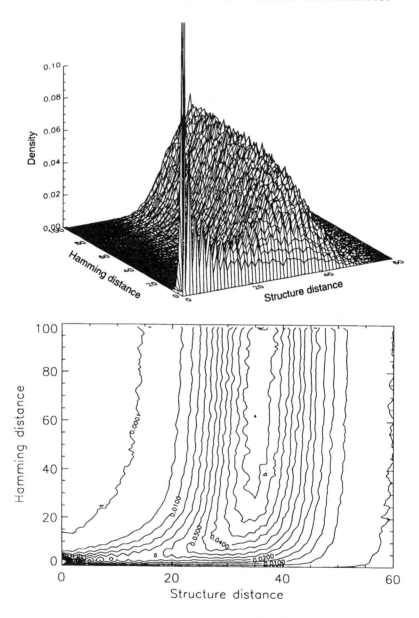

Figure 9. The structure density surface $S(t, h)$ of naturally occurring **AUGC**-sequences of chain length $\nu = 100$. The density surface (upper part) is shown together with a contour plot (lower part). In order to eliminate confusing details, the contour lines were smoothed. In this computation a sample of 1,000 reference sequences was used, which amounts to a total sample size of 10^6 individual RNA foldings.

secondary structure are, in essence, randomly distributed in sequence space. For natural sequences of chain length $\nu = 100$, a sphere of radius $h \approx 20$ (in Hamming distance) is sufficient to yield the global distribution of structure distances.

We conjecture therefore that all common structures are found in these relatively small patches of sequence space. This conjecture was proven by a suitable computer experiment: we choose a test and a target sequence at random, both having a defined structure. Then we determine the shortest Hamming distance between the two structures by approaching the target sequence with the test sequence following a path through sequence space along which the test sequence changes but where its structure remains constant (as shown in the next paragraph, such a path is **neutral**). The result for the case considered here yields an average minimum distance of two arbitrary structures of a Hamming distance of approximately 20. In order to find a given common structure of an RNA molecule of chain length $\nu = 100$, one has to search at maximum a patch of radius 20, which contains about 2×10^{30} sequences. This number is certainly not small, but it is negligible compared to the total number of sequences of this chain length, which is 1.6×10^{60}!

In order to complement this illustration of the RNA shape space, a second computer experiment was carried out which allows an estimate of the degree of selective neutrality (two sequences are considered neutral here if the fold into the same secondary structure). As indicated in Figure 10, we search for **neutral paths** through sequence space. The Hamming distance from the reference increases monotonously along such a neutral path but the structure remains unchanged. A neutral path ends when no further neutral sequence is found in the neighborhood of the last sequence. The length ℓ of a path is the Hamming distance between the reference sequence and the last sequence. Clearly, a neutral path cannot be longer than the chain length ν $(\ell \leq \nu)$. The length distribution of the neutral path in the sequence space of natural RNA molecules of chain length $\nu = 100$ is shown in Figure 11. It is remarkable that about 20 % of the neutral paths have the maximum length, and thus lead through the whole sequence space to one of the sequences, which differ in all positions from the reference but have the same structure.

A combination of information derived from Figures 10 and 11 provides insight into the structure of the shape space of RNA secondary structures, which is basic to optimization of RNA molecules already in an RNA world. Our results can be cast into four statements:

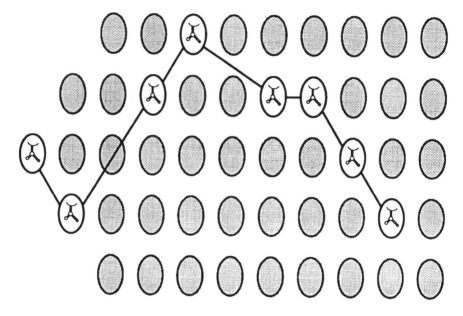

Figure 10. Percolation of sequence space by neutral networks. A neutral path connects sequences of Hamming distance $h=1$ (single base exchange) or $h=2$ (base pair exchange) which fold into identical minimum free energy structures. The sketch shows a neutral path of length $h=9$. The path ends since no to identical structure was found with $h=10$ and $h=11$ from the reference.

(1) sequences folding into one and the same structure are distributed randomly in sequence space,

(2) the frequency distribution of structures is sharply peaked (there are comparatively few common structures and many rare ones),

(3) sequences folding into all common structures are found within (relatively) small neighborhoods of any random sequence, and

(4) the shape space contains extended **neutral networks** joining sequences with identical structures [a large fraction of neutral path leads from the initial sequence through the entire sequence space to a final sequence on the opposite side—there are $(\kappa-1)^{\nu}$ sequences which differ in all positions from an initial sequence].

Combining the two statements (1) and (3) we may visualize the mapping from sequences into structures as illustrated by the sketch shown in Figure 12.

These results suggest straightforward strategies in the search for new RNA structures. It provides little advantage to start from nat-

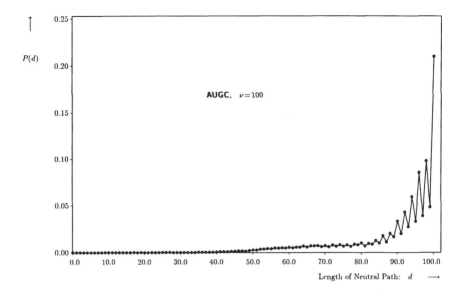

Figure 11. Length distribution of neutral paths starting from random **AUGC**-sequences of chain length $\nu = 100$. A neutral path connects pairs of sequences with identical structures and Hamming distance $d_h = 1$ (single base exchange) or $d_h = 2$ (base-pair exchange). The Hamming distance to the reference sequence is monotonously increasing along the path.

ural or other preselected sequences since any random sequence would do equally well as the starting molecule for the selection cycles shown in Figures 3 and 4. Any common secondary structure with optimal functions is accessible in a few selection cycles. The secondary structure of RNA is understood as a crude first-order approximation to the actual spatial structure. Fine tuning of properties by choosing from a variety of molecules sharing the same secondary structure will often be necessary. In order to achieve this goal it is of advantage to adopt alternations of selection cycles with low and high error rates. At low error rates the population performs a search in the vicinity of the current master sequence (the most common sequence, which is usually also the fittest sequence). If no RNA molecule with satisfactory properties is found, a change to high error rate is adequate. Then the population spreads along the neutral network to other regions in sequence space which can be explored in detail after tuning the error rate low again.

The structure of shape space is highly relevant for evolutionary optimization in nature, too. Since long neutral paths are common, populations drift readily through sequence space whenever selection constraints are absent. This is precisely what is predicted for higher or-

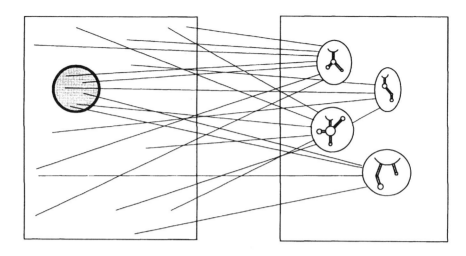

Sequence Space   Shape Space

Figure 12. A sketch of the mapping from sequences into RNA secondary structures as derived here. Any random sequence is surrounded by a ball in sequence space which contains sequences folding into (almost) all common structures. The radius of this ball is much smaller than the dimension of sequence space.

ganisms by the neutral theory of evolution (Kimura, 1983), and what is observed in molecular phylogeny by sequence comparisons of different species. The structure of shape space also provides a rigorous answer to the old probability argument against the possibility of successful adaptive evolution (Wigner, 1961). How should nature find a given biopolymer by trial and error when the chance to guess it is as low as $1/\kappa^\nu$? Previously given answers (Eigen, 1971; Eigen and Schuster, 1979) can be supported and extended by precise data on the RNA shape space. The numbers of sequences that have to be searched in order to find adequate solutions are many orders of magnitude smaller than those guessed on naive statistical grounds. If one of the common structures has a property which increases the fitness of the corresponding sequence, it can hardly be missed in a suitably conducted evolutionary search.

Acknowledgments

The statistical analysis of the RNA shape space presented here is joint work with Drs. Walter Fontana and Peter F. Stadler and will be described elsewhere in detail. Financial support by the Austrian *Fonds zur Förderung der wissenschaftlichen Forschung* (Projects S 5305-PHY and P 8526-MOB) and by the Commission of the European Communities (Contract PSS∗0396) is gratefully acknowledged.

References

Bauer, G. J., 1990: Biochemische Verwirklichung und Analyse von kontrollierten Evolutionsexperimenten mit RNA-Quasispezies *in vitro.* Doctoral Thesis, Universität Braunschweig, BRD.

Bauer, G. J., McCaskill, J. S., and Otten, H., 1989a: "Traveling waves of *in vitro* evolving RNA." *Proc. Nat'l. Acad. Sci. USA,* 86, 7937–7941.

Bauer, G. J., McCaskill, J. S., and Schwienhorst, A., 1989b: "Evolution im Laboratorium." *Nachr. Chem. Tech. Lab.,* 37, 484–488.

Beaudry, A. A., and Joyce, G. F., 1992: "Directed evolution of an RNA enzyme." *Science,* 257, 635–641.

Biebricher, C. K., and Eigen, M., 1988: "Kinetics of RNA replication by Qβ replicase," in: Domingo, E., Holland, J. J., and Ahlquist, P., eds. *RNA genetics.* Vol. I: *RNA-directed virus replication,* pp. 1–21. CRC Press, Boca Raton, FL.

Biebricher, C. K., Eigen, M., and Gardiner, W. C., jr., 1983: "Kinetics of RNA replication." *Biochemistry,* 22, 2544–2559.

Biebricher, C. K., Eigen, M., and Gardiner, W. C., jr., 1984: "Kinetics of RNA replication: Plus-minus asymmetry and double-strand formation." *Biochemistry,* 23, 3186–3194.

Biebricher, C. K., Eigen, M., and Gardiner, W. C., jr., 1985: "Kinetics of RNA replication: Competition and selection among self-replicating RNA species." *Biochemistry,* 24, 6550–6560.

Cech, T. R., 1986: "RNA as an enzyme." *Sci. Am.,* 255/5, 76–84.

Domingo, E., 1990: "Virus quasispecies: impact for disease control." *Futura,* 3/90, 6–8.

Eigen, M., 1971: "Selforganization of matter and the evolution of biological macromolecules." *Naturwissenschaften,* 58, 465–523.

Eigen, M., and Gardiner, W., 1984: "Evolutionary molecular engineering based on RNA replication." *Pure Appl.Chem.,* 56, 967–978.

Eigen, M., and Schuster, P., 1977: "The hypercycle. A principle of natural self-organization. Part A: Emergence of the hypercycle." *Naturwissenschaften,* 64, 541–565.

Eigen, M., and Schuster, P., 1979: *The hypercycle—A principle of natural self-organization.* Springer-Verlag, Berlin.

Eigen, M., and Schuster, P., 1985: "Stages of emerging life—Five principles of early organization." *J. Mol. Evol.,* 19, 47–61.

Eigen, M., McCaskill, J., and Schuster, P., 1988: "The molecular quasispecies – An abridged account." *J. Phys. Chem.,* 92, 6881–6891.

Eigen, M., McCaskill, J., and Schuster, P., 1989: "The molecular quasispecies." *Adv. Chem. Phys.,* 75, 149–263.

Ellington, A. D., and Szostak, J. W., 1990: "*In vitro* selection of RNA molecules that bind to specific ligands." *Nature,* 346, 818–822.

Fontana, W., and Schuster, P., 1987: "A computer model of evolutionary optimization." *Biophys. Chem.,* 26, 123–147.

Fontana, W., Schnabl, W., and Schuster, P., 1989: "Physical aspects of evolutionary optimization and adaptation." *Phys. Rev. A,* 40, 3301–3321.

Fontana, W., Griesmacher, T., Schnabl, W., Stadler, P. F., and Schuster, P., 1991: "Statistics of landscapes based on free energies, replication and degradation rate constants of RNA secondary structures." *Mh. Chem.,* 122, 795–819.

Fontana, W., Stadler, P. F., Bornberg-Bauer, E., Griesmacher, T., Hofacker, I. L., Tacker, M., Tarazona, P., Weinberger, E. D., and Schuster, P., 1993a: "RNA folding and combinatory landscapes." *Phys. Rev. E,* 47, 2083–2099.

Fontana, W., Konings, D. A. M, Stadler, P. F., and Schuster, P., 1993b: "Statistics of RNA secondary structures." Preprint No. 90–02–008, Santa Fe Institute, Santa Fe, NM. *Biopolymers,* in press.

Gilbert, W., 1986: "The RNA world." *Nature,* 319, 618.

Guerrier-Takada, C., and Altman, S., 1984: "Catalytic activity of an RNA molecule prepared by transcription *in vitro*." *Science,* 223, 285–286.

Guerrier-Takada, C., Gardiner, K., Marsh, T., Pace, N. and Altman, S., 1983: "The RNA moiety of ribonuclease P is the catalytic subunit of the enzyme." *Cell,* 35, 849–957.

Horwitz, M. S. Z., Dube, D. K., and Loeb, L. A., 1989: "Selection of new biological activities from random nucleotide sequences: Evolutionary and practical considerations." *Genome,* 31, 112–117.

Husimi, Y., and Keweloh, H.-C., 1987: "Continuous culture of bacteriophage Qβ using a cellstat with a bubble wall-growth scraper." *Rev. Sci. Instrum.*, 58, 1109–1111.

Husimi, Y., Nishgaki, K., Kinoshita, Y., and Tanaka, T., 1982: "Cellstat—A continuous culture system of a bacteriophage for the study of the mutation rate and the selection process at the DNA level." *Rev. Sci. Instrum.*, 53, 517–522.

Joyce, G. F., 1989: "RNA evolution and the origins of life." *Nature*, 338, 217–224.

Joyce, G. F., 1991: "The rise and fall of the RNA world." *New Biol.*, 3, 399–407.

Kauffman, S. A., 1986: "Autocatalytic sets of proteins." *J. Theor. Biol.*, 119, 1–24.

Kauffman, S. A., 1992: "Applied molecular evolution." *J. Theor. Biol.*, 157, 1–7.

Kimura, M., 1983: *The neutral theory of molecular evolution.* Cambridge University Press: Cambridge, UK.

Mason, S. F., 1991: *Chemical evolution. Origin of the elements, molecules and living systems.* Clarendon Press, Oxford, UK.

Noller, H. F., 1991: "Ribosomal RNA and translation." *Annu. Rev. Biochem.*, 60, 191–227.

Noller, H. F., Hoffarth, V., and Zimniak, L., 1992: "Unusual resistance of peptidyl transferase to protein extraction procedures." *Science*, 256, 1146–1419.

Orgel, L. E., 1986: "RNA catalysis and the origins of life." *J. Theor. Biol.*, 123, 127–149.

Orgel, L. E., 1992: "Molecular replication." *Nature*, 358, 203–209.

Piccirilli, J. A., McConnell, T. S., Zaug, A. J., Noller, H. F., and Cech, T. R., 1992: "Aminoacyl esterase activity of the *tetrahymena* ribozyme." *Science*, 256, 1420–1424.

Riesner, D., and Gross, H. J., 1985: "Viroids." *Annu. Rev. Biochem.*, 54, 531–564.

Schopf, J. W., and Packer, B. M., 1992: "Early archean (3.3-billion to 3.5-billion-year-old) microfossils from Warrawoona group, Australia." *Science*, 237, 70–73.

Schuster, P., 1981: "Prebiotic evolution," in: Gutfreund, H., ed. *Biochemical evolution,* pp. 15–87. Cambridge University Press, Cambridge, UK.

Schuster, P., Fontana, W., Stadler, P. F., and Hofacker, I. L., 1993: "From sequences to shapes and back. A case study on RNA secondary structures." Preprint.

Spiegelman, S., 1971: "An approach to the experimental analysis of precellular evolution." *Quart. Rev. Biophys.*, 4, 213–253.

Strunk, G., 1992: "Automatisierte Evolutionsexperimente *in vitro* und natürliche Selektion unter kontrollierten Bedingungen mit Hilfe der Serial-Transfer-Technik." Doctoral Thesis, Universität Braunschweig, BRD.

Swetina, J., and Schuster, P., 1982: "Self-replication with errors - A model for polynucleotide replication." *Biophys. Chem.*, 16, 329–345.

Symons, R. H., 1992: "Small catalytic RNAs." *Annu. Rev. Biochem.*, 61, 641–671.

Tuerk, C., and Gold, L., 1990: "Systematic evolution of ligands by exponential enrichment: RNA ligands to bacteriophage T4 DNA polymerase." *Science,* 249, 505–510.

Wakeman, J. A., and Maden, B. E. H., 1989: "28 S Ribosomal RNA in vertebrates. Location of large-scale features revealed by electron microscopy in relation to other features of the sequences." *Biochem. J.,* 258, 49–56.

Wang, Q., 1987: "Optimization by simulating molecular evolution." *Biol. Cybern.,* 57, 95–101.

Weissmann, C., 1974: "The making of a phage." *FEBS Lett.,* 40, Suppl.10–18.

Wigner, E., 1961: "The logic of personal knowledge" in: Shils, E., ed. *Essays presented to Michael Polanyi on his seventieth birthday 11th March 1961.* Free Press, Glencoe, IL.

CHAPTER 11

A Perception Machine
Built of Many Cooperating Agents

ERIK SKARMAN

Saab Missiles, AB
S–581 88 Linköping, Sweden

1. Introduction

It is generally accepted nowadays that the process called perception within psychology, and state estimation within control engineering, should preferably be based on models. Models offer the possibility of giving meaning to sensory data. Sensory data can be interpreted, grouped together, and assigned to objects having names within the model. This gives at least a tentative explanation as to why humans can see anything at all in the chaos of lines and dots and shades and surfaces that our eyes provide us. To be of any value, models of the world are necessarily complex, since they have to cope with the complexity of the world. This means that models have a complex structure.

Now if we have a model with a complex structure and we want to build a "perception machine" around that model, what structure does that perception machine have? There are two preferred answers to this question (and several other less popular ones). One is that the perception machine has the same structure as the model. The other is that the perception machine has its structure of its own and regards the model only as a set of data that it uses. Most likely, the two systems described by the two answers are the same in the sense that they can be mapped onto one another.

Cooperation and Conflict in General Evolutionary Processes, edited by John L. Casti and Anders Karlqvist.
ISBN 0–471–59487–3 ©1994 John Wiley & Sons, Inc.

If somebody came up with the answer "The perception machine has a simpler structure than the model," it would seem likely that simplicity had been gained at a cost of information destruction.

The reply: "The machine has a different structure than the model, but of equal complexity" would be a bad answer, because it would force us to pay wages to an engineer who builds this different structure, and who then rebuilds it every time the world changes, necessitating the model of it to change.

The answer: "The machine has a more complex structure" is even worse, because the more complex structure is not only different, it is also more complex.

So the best answer that we can actually hope for is the answer "The same structure." In what follows, I shall try to display such a perception machine having the same structure as the model it's based on.

Clearly there is a difference between a perception machine and a model. There is no action in a model; a model is a passive description of something. The models in this paper have nodes that contain values of properties of objects constituting the model.

The key idea is that the perception machine has the same structure as the model, thus the same nodes. And every node has an agent who actively tries to compute the node value. Each of these agents does a reasonably simple job, and all the agents acting together in a cooperative way serve as a perception machine.

2. The Relational Network

Our model is what we will call a relational network, which can be thought of as an outgrowth of a so-called semantic network. The semantic network consists of nodes and arcs, where the arcs describe relations between the nodes and the nodes correspond to objects and properties of these objects. Typical relations are "is a," as in "a tiger is an animal," "is the son of," or "has the property."

The overall network has a tree-like structure, with the outermost nodes in the structure accordingly called leaves. In our view of the semantic network, the leaf nodes represent values, namely the values of the properties of the objects.

At the leaf level, there may be a separate web of relations. These relations are mathematical relations between the values. Some of these relations are associated with the tree structure of the underlying semantic network. This web of mathematical relations is our relational network.

If Joseph "is a" tiger, then the colors that Joseph can have is a one-element subset of the colors that tigers can have. In this case tigers can only have one color (they are striped), so this is a degenerate case. But if Joseph "is either a" horse "or a" donkey, then the color of Joseph is a subset of the union of the sets of possible colors for horses and for donkeys.

Other mathematical relations may express something that is not immediately derivable from the tree structure of the semantic network. Two values may be known to be equal, even though this may not be a consequence of their structural relationships. Or one value may be known to be the sum of two other values.

3. Relations and Assignments

The difference between the activeness of the perception machine and the passiveness of the model has a parallel in the difference between a relation and an assignment. A relation may be

$$x = y. \tag{1}$$

An assignment of similar appearance is

$$x := y. \tag{2}$$

Equation (1) describes a relation that exists between x and y, while Eq. (2) expresses an action.

Basically, assignments are the only things computers can do, because a computer is a machine and machines only know how to act. If you write a relational expression like (1) in a computer program, that only hides an assignment, namely

$$L := x = y,$$

which assigns the truth value of the relation to the Boolean variable L.

One property of the relation (1) is its symmetry. This has nothing to do with whether or not the relation "=" is a symmetric relation or not. It just means that if the relation (1) is to hold, then it is equally the "responsibility" for both x and y.

By way of contrast, the assignment (2) is asymmetric. The quantity x has to "suffer" for what y is, and y is left unaffected. On the other hand, if the relation (1) between x and y is broken, then the assignment (2) serves to restore it, so that (2) is a relation-restoring assignment corresponding to (1).

But the symmetry property of relations then tells us that there should be one more relation-restoring assignment. This is clearly

$$y := x.$$

What all this teaches us is that where the model has a relation, the perception machine should have two assignments, allowing data to flow in both directions.

If there are ternary relations like

$$x = y + z,$$

then there should be three relation-restoring assignments,

$$x := y + z,$$
$$y := x - z,$$
$$z := x - y,$$

allowing a free data flow between all three nodes.

– The Concept of Unification –

If x is a node and there is a relation-restoring assignment like

$$x := y,$$

then the right member of this assignment is an input to x. This input gives an information about x from another node y that is somehow known to be related to x. In the general case, a node x will have more than one input. This means that there is more than one information source claiming something about the value of x.

The task of the agent working at node x can now be described as a task of unification. The agent unifies the different inputs, or mediates between them, and lets them come to a common agreement. Of course, for such a mediation to be meaningful, each of the inputs must be willing to negotiate about the value of x. If one information source is insisting that x be equal to 7 and nothing but 7, then there is no need for any perception machine at all. So every value in every node must be thought of as having some level of uncertainty.

Thus, in the perception machine x is not a number but some data structure defining the uncertainty of x. And the right-hand member of an assignment

$$x := y + z$$

is not an addition of numbers, but a mapping that maps the uncertainties of y and z onto the uncertainty of their sum.

So far the perception machine has no dynamical properties and no state. Principally, the values of the nodes are held constant by the constancy of the sensory inputs form the outside, and by possible constant parameters in the relations. The life of the agents may be somewhat easier, however, if we allow every node to have a state that the agent can maintain. Then if we label the result of the unification between two claims x_1 and x_2 about x as $u(x_1, x_2)$, we let the agent do the following:

$$x := u(x, x_i),$$

where x is the state of the node, and x_i is one of its inputs.

When this has been done for all i, x contains what we would call

$$u(x_1, \ldots, x_i, \ldots, x_n).$$

This requires some properties (e.g., associativity) of the function u to be established once the function u has been chosen. It also requires that x have a neutral initial value, called unknown or U, with the defining property

$$u(U, x_i) = x_i.$$

4. Representation of Uncertainty

– Probabilistic Representation –

In the probabilistic case, a node contains a probability density function $p_X(x)$ representing the probability that the stochastic variable X falls in a unit volume environment of x. The quantity X is then the variable that is modeled in the node.

Of course, a function $p_X(x)$ is not easily represented in a computer. Once we have decided to represent the uncertainty with $p_X(x)$, we must again decide how to represent this function.

One way to do it is in a parametric form

$$p_X(x) = f(x, \theta_1, \theta_2, \ldots, \theta_n).$$

The parameters may be, for instance, momenta of different degrees. In that case they are computable from $p_X(x)$,

$$(\theta_1 \theta_2 \cdots \theta_n) = g(p_X(x)).$$

But f may not be invertible. If it is, we can say, somewhat loosely, that $p_X(x)$ has a closed form.

Now it is well known that nonlinearities in a model will ensure that $p_X(x)$ has no closed form. A linear model, however, preserves the Gaussian distribution, which is a closed-form structure having the mean and covariance as parameters:

$$p_X(x) = N(\hat{x}, P).$$

This is the idea behind the Kalman filter. And if our model satisfies the necessary requirements, the Kalman filter will fall out as a special case of a relation network filter.

If there is no closed form available, we may choose to represent the function $p_X(x)$ by its values in a representative set of points. Such a representation is always approximative, and heavy for the computer. Now to the meaning of unification in this case.

We may suppose that the node that describes X by $p_X(x)$ divides the network into two parts, one to the left and one to the right of the node. Each part communicates with the world, of which the network is a model, through some sensors. Each of these sensors observes an event E_l and E_r, respectively, that happens in the world. The probabilities in the network $p_X(x)$ are then to be regarded as conditional probabilities $p_X(x|E)$: that is, probabilities assuming that the events that we observe have actually taken place.

The left input to the node may tell us something about $p_X(x|E_l)$, while the right input says something about $p_X(x|E_r)$. The task of the unification is to tell us something about $p_X(x|E_r, E_l)$, i.e., the probability given that both events have occurred simultaneously. The process of unification then takes us from a probability given nothing— that is, the probability $p_X(x)$—to $p_X(x|E_r, E_l)$.

According to Bayes' rule, we have

$$p(E_l, E_r) = \frac{p_X(E_l, E_r|x)p_X(x)}{p_X(x|E_l, E_r)}.$$

If E_l and E_r are independent, which may well be the case since they happen in different ends of the system as described by the network, then we have

$$p(E_l)p(E_r) = \frac{p_X(E_l|x)p_X(E_r|x)p_X(x)}{p_X(x|E_l, E_r)}.$$

Essentially, we multiply what we had in the node as an a priori information, first by information that comes in from the right-hand part
and then by information from the left-hand part. Technically, then,
the unification process is a process of multiplication.

One needs to be careful not to do these multiplication more than
once. Put another way, one has to keep track of the independent parts.
Once we have computed $p_X(x|E_l, E_r)$, we have expressed something
about the simultaneous event E_l, E_r, which, of course, is not independent of E_l. The agent in the node should not only do the unification
work, but also keep track of that work.

– The Set Representation –

The set representation is conceptually simpler than the probabilistic
representation. If the uncertainty about the value of some node variable x is represented by a set X, this means that x falls within X, i.e.,
$x \in X$.

There is no graduation of any probability inside X. Using fuzzy
logic, and thus introducing a varying membership over the interior of X,
appears to contribute little, if anything, to the analysis.

Unification can now be easily defined. If two neighboring nodes
of x say that x is in X_1 and in X_2, respectively, then x is in both, i.e.,
$x \in X_1 \cap X_2$. So $X = X_1 \cap X_2$ or

$$X_1 \text{ unified with } X_2 = X_1 \cap X_2.$$

The sets can have any kinds of elements: enumerations of discrete
elements, unions of intervals on the real numbers, arbitrary subsets
of the plane, and so forth. In all these cases, the data have a lattice
structure with relations $\subset$ and $\supset$ and operations $\cap$ and $\cup$. Such a
lattice has a top element and a bottom element. The top element is an
appropriate universe U. It represents the unknown; that is, we know
nothing more about x than we may possibly have known already before
we started the network. In the implementations it seems practical to
work with U as a symbol for "the universal unknown," having the
property

$$\text{any } x \text{ unified with } U = x.$$

The bottom element is, of course, the empty set $\emptyset$. If U signifies that
we know very little, then $\emptyset$ indicates that we know "too much"; that
is, there is a contradiction. The contradiction is really between the
relational and semantic networks rather than within the relational network. The semantic network says that some object has a property,

but the relational network says that property has no value. Again, in the implementations it is practical to work with a symbol for $\emptyset$ or the contradiction. This symbol is usually called K and has the property

$$\text{any } X \text{ unified with } K = K.$$

Still on the matter of implementation, once it has been decided that the uncertainty about x can be represented with sets, a representation of these sets in the computer has to be found, together with algorithms to perform operations on these representations.

On the mathematical side, we have to find what an assignment like

$$y := f(x) \tag{3}$$

should correspond to for the sets.

As mathematics gives a meaning to the symbol $f(X)$, where X is a set, this is quite easy. We simply take

$$Y := f(X). \tag{4}$$

That is, Y is the set of all points that can be reached by the function $f(x)$ for some $x \in X$.

The function f in (3) may not have an inverse, but f in (4) always has. It may map some Y on a noncontiguous set or onto the empty set, but it does exist. So if our model contains a relation $y = f(x)$, then we can always implement both the relation-restoring assignments

$$Y := f(X)$$

and

$$X := f^{-1}(Y).$$

The same things hold for functions with more than one variable. For instance, it's meaningful to write $Z := Y + X$, where X, Y, and Z are sets. The equality relation $x = y$ corresponds then to the assignments

$$X := Y$$

and

$$Y := X.$$

However there may also be relations in our model that are taken directly from set theory, and which may have no real meaning for single elements.

What relation-restoring assignments should we propose for the relation

$$Z = Y U X \tag{5}$$

(corresponding to some is either $a \ldots$ or $a \ldots$ in the semantic network)? Obviously

$$Z := Y U X \tag{6}$$

is a candidate. But what about others? A candidate for the second assignment is $Y := Z - X$, where $Z - X$ means "all elements in Z that are not in X." This may be too much, however. In fact, the elements in X hide the same elements in Y, so that we don't know if they are there only by knowing (5), Z, and Y. So we ought to be more conservative and take

$$Y := Z, \tag{7}$$

$$X := Z. \tag{8}$$

The cartesian product

$$Z = Y \times X \tag{9}$$

is also interesting. First of all, the relation cannot really be restored. $Y \times X$ is always a "rectangular set" and Z is arbitrary. Then again there are different alternatives for two of the assignments, but the context in the filter with the unifications allows us to do the simplest assignments:

$$Z := Y \times X \tag{10}$$

$$Y := \text{the set of all left elements in the pairs of } Z \tag{11}$$

$$X := \text{the set of all right elements in the pairs of } Z. \tag{12}$$

Finally, we will study the relation

$$X \subseteq Y. \tag{13}$$

One of the assignments is

$$X := Y, \tag{14}$$

which would of course restore (13). The other assignment is void. A subset doesn't say much about its bigger brother. Of course, if X contains some element not in Y, then Y could be augmented with these elements. But the unification process doesn't allow that to happen. Instead, these element in X will be extinct after the unification after (14).

5. Convergence

The filter converges in a finite number of steps to a final state, which persists indefinitely if no new input signals are received. The convergence process is monotonic, in that the unification is monotonic; it can only generate smaller and smaller sets. Hence, oscillations are impossible.

A gradual, say, exponential decay is also excluded. There is a minimum finite "program" of unifications in

$$\sum_{i=1}^{n} m_i$$

steps, where n is the number of nodes and m_i is the number of inputs to the ith node.

We have described the system as a network of independent agents that do not follow any program. It's then important that the final state be independent of the unification order.

The network is principally loop-free. Rather, there are loops everywhere. If A and B are two nodes along a loop, then one can go around the loop. But one can also go from A to B along two different paths, one along the loop and the other in the opposite direction. These two paths define through composition two relations among A and B:

$$Ar_1 B \quad \text{and} \quad Ar_2 B$$

But that defines a new relation

$$A(r_1 \text{ and } r_2)B$$

between A and B. In this way every loop collapses into a single path between any two nodes along the loop.

The yet unproven claim is now that the final state to which the filter converges is optimal, in the sense that all conclusions that can be drawn from the model and from the input data actually are drawn.

6. Models of Dynamical Systems

The principal property of a (discrete-time) dynamical system is that the state changes its value from one time instant to another according to an equation

$$x(t + T) = f(x(t), u). \tag{15}$$

The relation $y = f(x, u)$ can easily be handled within the network, but there is also a change of "scope" following the change of time; the state $x(t + T)$ is not the same thing as $x(t)$.

First of all, the process of unification can only produce smaller sets. But the uncertainty about $x(t + T)$ is generally not smaller than the uncertainty about $x(t)$, just as $f(X)$ is not generally smaller than X.

Second, after a time shift from t to $t + T$, all the conclusions that could be drawn from $x(t)$ are worthless and have to be redrawn from $x(t + T)$. We then model the dynamics of a dynamical system in an ordinary "two-stroke engine" fashion as follows.

There is a new node x^+ associated with x, and related with x through

$$x^+ = f(x, u).$$

Its value is computed from x and u as for any other node.

At sampling, the following things occur:

1. All nodes like x, that have "+ nodes," like x^+, take over the values directly without unification from the + nodes.

2. All other nodes, including the "+ nodes," get the value U (universal unknown).

3. The nodes in connection with sensors receive new data from the sensors.

This is one of the strokes of the two-stroke engine. The other is the unification that now follows, which lets the filter again converge to a final state.

7. The Concept of Reservation

The term "reservation" is intended to be a term in psychology. It has some relation to such terms in psychology as self-deception and the unconscious. Most functions in a relational network map the empty set on the empty set, that is, $K \to K$. An exception is the union, but not the cartesian product, which sends

$$X \times K \to K$$

(a most useful equation).

The unification also lets K pass through, since K unified with any $X = K$. The consequence of this is that K spreads rapidly across a network once it has appeared. In practice, this is a disaster.

The set K appears in a network as a result of unification, so there is a possible protection mechanism to avoid the disaster: "Inhibit the unification at a node as soon as the unification will result in the value K."

The agent has the sole task to make unifications. So he would enter a reservation against the result he is displaying by saying, "The value of any node is not K, but it is not properly computed." The agent knows, of course, what the properly computed value is. It is K.

For the perception machine as a whole, the situation is more delicate. The properly computed percept is "K everywhere," which is clearly not reasonable. The correct percept is not available, so the machine is forced to believe in what it computes with the inhibition mechanism in action. If it knows that the inhibition mechanism is in action, then it is forced to believe in something that it knows is not true.

This is a perfect parallel to the phenomenon of self-deception, where we force ourselves to believe in something that we know isn't true. It is also plausible that a state of self-deception is always the result of an effort to avoid contradiction—if nothing else, a contradiction against the basic self-respect that humans always try to maintain.

The reservation is at best only a temporary response to the threat of a node becoming K. The long-term action should be to revise the model so that it can meet the world without being infected by the value K. This is an activity opposite to perception. Sets must be made bigger, while perception means that sets get smaller. It is a very difficult task. It is ubiquitous in human thought, but rare in artificial intelligence. It probably needs to have access to a trace of the foregoing perception work, in order to be able to find out where something went wrong.

If this process opposite to perception is called cognition, then we could say that cognition is driven by the appearance of contradictions in the perception. If the system avoids this cognitive process and remains in a state of reservation, there are a number of disadvantages.

First, the agent likes to unify and does not want to be inhibited.

Second, the appearance of K at unification is only a symptom. The value K results because sets are too small, and they probably became too small elsewhere. As long as a single node is kept inhibited, not only does that node display faulty results, but also many others.

Third, nodes are not only unification places; they are also communication links. But they don't work as communication links if the unification is inhibited. In that situation, submodels are isolated from one another.

Finally, if submodels are isolated from one another and contain faulty values, it's likely that tensions will build up between them. If they communicate via other nodes, these nodes will probably also be inhibited in due time. The whole model then splits into two halves.

8. Some Examples

– Distinguishing a Zebra from a Tiger –

Figure 1 shows a network for distinguishing a zebra from a tiger.

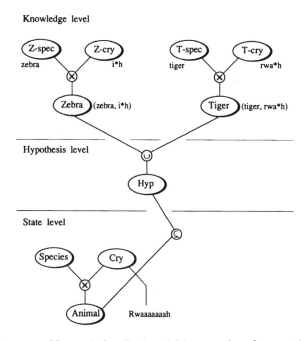

Figure 1. Network for distinguishing a zebra from a tiger.

At the beginning, we know that some animal is either a zebra or a tiger because we have seen that it is striped. Now we want to determine which it is by using its cry, that we can hear. The cry comes into the "cry" node in order to unify with it.

There is an animal node, a tiger node and a zebra node. These contain the cartesian product of the only two properties of the animals that we are interested in, their species and their cries.

The network has a state level that contains a node for the actual animal that's on the stage. It has a knowledge level containing what we know about zebras and tigers. Between them we have to build a hypothesis level that models our hypothesis that the animal is either a tiger or a zebra.

The latter has a single node containing the whole space of combinations of properties that the animal can actually have. The properties of the real animal cannot be bigger than this space; but it can be smaller, as when we know everything about the animal. In this case it is a point set.

When the agents start to work, data will flow down so that the animal node finally contains

$$(\text{zebra, } i^*h):(\text{tiger, } rwa^*h),$$

where ":" denotes union and rwa^*h is a set, namely

$$rwa^*h = rwah:rwaah:rwaaah \ldots .$$

The "species" node contains

$$\text{zebra:tiger.}$$

The "cry" node contains

$$rwa^*h:i^*h.$$

When the actual cry is heard, unification takes place in the cry node and the result is the single element

$$rwaaaaaaah.$$

The animal node will now see at its input

$$((\text{zebra:tiger}), rwaaaaaaah)$$
$$= (\text{zebra, } rwaaaaaaah):(\text{tiger, } rwaaaaaaah).$$

The first of these terms unifies to K with the contents of the animal node, but the second term matches to give the single term

$$(\text{tiger, } rwaaaaaaah).$$

Now the "species" node unifies to

$$\text{tiger.}$$

So the animal is a tiger.

The knowledge and hypothesis levels are protected from these changes by the subset relation. But the cry node was not protected. What happens if the tiger says

rwah?

That doesn't match "rwaaaaaaaah," so now, in fact, the tiger disappears. This is obviously wrong.

But the new, shorter, cry belongs to a new time. This is in fact a dynamical system. What is the state of this dynamical system? Our biological intuition tells us it is the species, which is preserved according to

$$\text{species } (t + T) = \text{species } (t).$$

If we introduce this state, the cry node will become U. But as the species node still only contains "tiger," the cry node will be refilled with the general tiger cry rwa*h.

Now what if the animal were a zebra, and the zebra disappeared behind a bush and came out on the other side? Could we be sure it is still the same zebra? Or a zebra at all? Is the species still a state?

The clear distinction between the knowledge, hypothesis and state levels, and the hesitation as to what is the state, makes this example a little more interesting than more purely technical ones.

– Navigating in a Workshop –

Figure 2 shows a mobile robot in a workshop. It has a laser that measures the distance to the nearest object in some bearing. The angle between the laser beam and the robot is known, and in some way the heading ψ of the robot is also known. In addition, the robot has access to a drawing of the workshop.

Figure 3 shows a relational network for the situation. The open ends in the diagram represent input from sensors. The drawing consists of thin sets around the locations of the wall surfaces. Laser reflections take place at the surface of the walls and not inside the concrete.

In Figure 2 we also see an example of a "thick" set. That's the "chaos area." It says: "There are lots of machines and things in there. That explains why reflexes may come from there and not from the wall behind, but these reflexes are not very useful for navigation. They allow a precision down to the size of the chaos area." A separate interpretation for the path planning system of the robot would say, "Don't go in there."

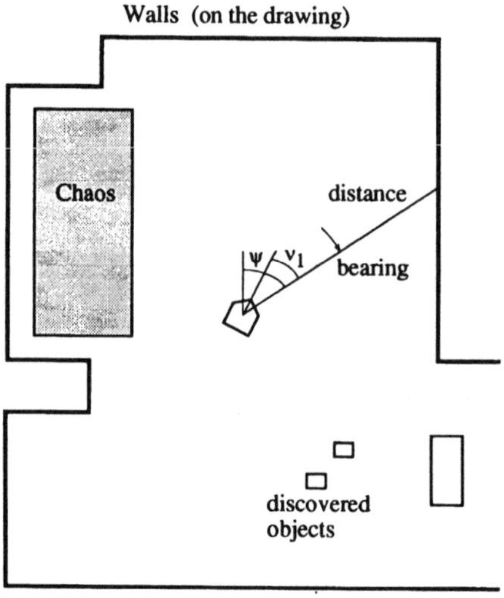

Walls (on the drawing)
Chaos
distance
ψ
v_1
bearing
discovered
objects

Figure 2. Mobile robot in a workshop.

eψ
dψ
ψ+
ψ
v_1
distance
bearing
ex, ey
dx, dy
(x, y)+
Polar
Cartesian
(x, y)
(x, y)ray
drawing

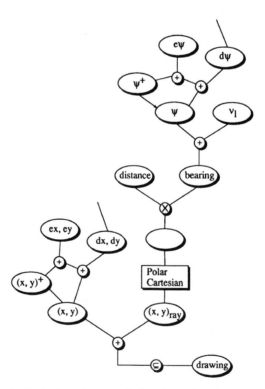

Figure 3. Relational network for the robot in a workshop.

Figure 2 also shows some discovered objects. These objects are not shown on the drawing. A laser reflex on them would result in the value K in the network. A way to handle this situation is to augment the drawing with a new object at the place where the reflection came from. In this way, the robot can partially build its own drawing.

Figure 4 shows a way to handle this without really leaving the network. It contains an A-function, or Alternative function. It is much like a union, but it "prefers" one of its input branches.

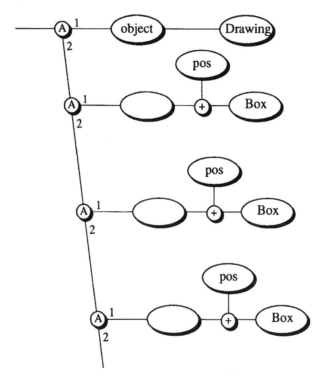

Figure 4. Augmented network for robot.

If the relation is $Z = YAX$, then we have the following assignments:

$$Z := \begin{cases} Y, & \text{if } Y \neq K, \\ YUX, & \text{if } Y = K. \end{cases}$$

$$Y := Z.$$

$$X := \begin{cases} U, & \text{if } Y \neq K, \\ Z, & \text{if } Y = K. \end{cases}$$

So A prefers to believe that $Y = Z$ as long as this is not contradictory, and in this case isolates X from Z.

Then there are a number of "boxes" available to represent new objects. They all start with unknown positions, and they are switched in whenever neither the drawing, nor any of the old boxes, can explain a laser echo.

– Machine Identification –

Figure 5 shows a sequential machine. It has a state and a next state, which are in a functional relationship. But we can also describe this so that the pairs of present state and next state belong to a transition set, which is a subset of $Q \times Q$, where Q is the state space.

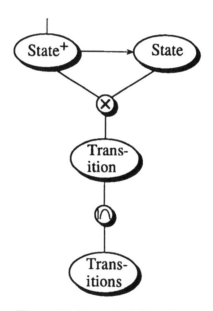

Figure 5. A sequential machine.

This is a model of a sequential machine. Once we populate the model with agents and let them start to work, the model will become the machine it models.

There is a class of machines called discrete-event machines, where each state is not mapped into a unique state, but into a set of states, which represent events that can now happen. Once one of these events occurs, the transition takes place, and new possible events are presented.

This is modeled by the same model. Events observed by sensors are unified with the contents in the State + node. Unification selects one of the possible events to go into the State node. If the sensor is decoupled from the network, then the contents of the State and State+ nodes will start to branch out to growing sets of values.

Now if the model is not a correct model of the external machine, the unification will sooner or later give the value K. Figure 6 shows how this can be used to identify a machine. It contains two tentative models of the machine. Each alternative has a name, and the bottom node contains the union of these names.

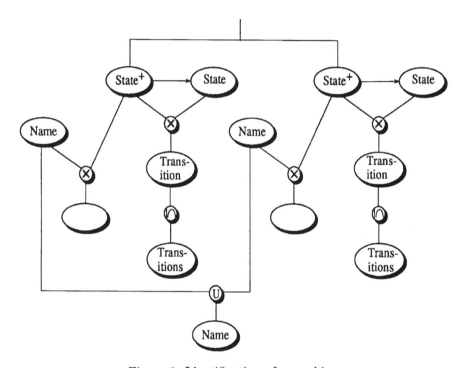

Figure 6. Identification of a machine.

Once one of the models fails to predict what happens with the real machine, the following will take place: The State+ node will contain a K. Then the State $\times$ Name node will contain the pair $(Name, K)$, which is not a pair but the empty pair (K, K). Name will then also become K, so that Name will disappear from the bottom node.

Figure 7 shows a different kind of identifier, which simply builds up a transition set by adding observed transitions to a transition set (which should be started with K unusually enough). The lower part

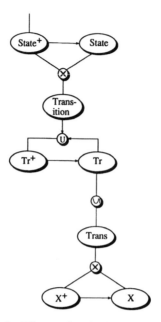

Figure 7. A different kind of machine identifier.

shows how we can connect an extra machine to the same transition set node, and let it run in advance to make predictions.

9. Post-symposium Reflections

This paper is linked to the theme of the Abisko symposium through the word *cooperation*. The perception machine contains cooperating agents. Do these agents also compete? I don't think it would be appropriate to say so. Agents who compete do so over some limited resource, such as energy. But the system here contains only information.

How about evolution? This whole concept grew out of a dissatisfaction with traditional schemes like the Kalman filter for perception. These schemes can't evolve. Once a new object in the world is discovered and found to require a new model, the whole Kalman filter has to be redesigned from the ground up. In the Kalman filter, a model is something that's closed. The relational network filter, on the other hand, is "organic" and can assimilate a new model of something very easily.

Kalman is right if he claims that this ease has its price. If we connect models to one another loosely, we lose the optimal behavior of the Kalman filter. If we connect models more tightly and make full use of all correlations between the models, we end up with precisely a

Kalman filter—both with its rigidity and optimality.

The relational network filter offers us the possibility of going from optimality to evolvability. Any living organism has to go this way. The world is so complex that there is no hope that it can be modeled with a single, closed, fully connected model, fixed once and for all.

One of the things that the artificial intelligence research effort has taught us is to admire more the human's "mechanical" skills than "pure intelligence" skills. A man can stand on two feet; he can do so even if he is wearing a heavy overcoat or is standing in a small rowboat. He can ride a bicycle, and he can lift a cup of coffee and move it to his lips.

Our experiences with control theory tell us that none of these activities is easy, and that it requires advanced forms of adaptive control. Such advanced forms usually involve Kalman filters. But a Kalman filter is a very far-fetched kind of structure, which leads me to make the following personal statement: "If anybody, through surgery or otherwise, finds a Kalman filter in the human brain, I will stop believing in Darwin." It seems to me now that the relational network filter offers a less farfetched way to move from simple perception systems to a Kalman filter. The concept of a relational network filter would help evolution with its job a little bit.

Finally, let me acknowledge a discussion with N. Katherine Hayles, who informed me that zebras are only in Africa and tigers in India, so the zebra-tiger identification system is then applicable only at the zoo.

CHAPTER 12

Language, Evolution, and the Theory of Games

KARL WÄRNERYD

Stockholm School of Economics
Box 6501
S–113 83 Stockholm, Sweden

1. Introduction

The purpose of this chapter is to sketch a formal evolutionary model of certain aspects of meaning in natural language. Natural language must surely be a most prominent candidate for an evolutionary explanation, since we cannot think of it as the result of explicit agreement among humans.

Our basic framework will be the game-theoretical model of language developed by the philosopher David Lewis in his book *Convention* (Lewis [9]). Lewis argues that Gricean, or non-natural, meaning in a language can be seen as an equilibrium property of signals in a co-ordination game between a sender and a receiver. Not all equilibria (in the game-theoretical sense) of this model have this interesting property, however, and Lewis does not address the question of why meaningful languages have actually emerged from uncoordinated noise making in the real world.

Lewis wrote before the advent of evolutionary game theory, the invention of biologist John Maynard Smith. Interestingly enough, it turns out that Maynard Smith's concept of an evolutionarily stable strategy (ESS) may be applied in a straightforward manner to Lewis's model. While the ESS notion is not explicitly dynamical, it is a reasonable

Cooperation and Conflict in General Evolutionary Processes, edited by John L. Casti and Anders Karlqvist.
ISBN 0–471–59487–3 ©1994 John Wiley & Sons, Inc.

criterion for stability in any evolutionary system. The results are very strong. It turns out that only meaningful languages are evolutionarily stable.

This exercise has at least two benefits. First, it may be seen as closing one end of Lewis's theory. And second, it suggests a solution to the game-theoretical problem of how statements in a language can have an effect in a game.

The theory developed in this paper may seem to owe more to biology than to economics and game theory. After all, the framework is that of evolutionary game theory. But it is worth pointing out that evolutionary theorizing, understood in the more general sense of explanations that show how the *explanandum* could have emerged as the result of an impersonal selection process that operates on competing candidates, has a much longer history in the social sciences than in biology. Many readers will know that Charles Darwin explicitly acknowledges the population theory of Thomas Malthus as the direct inspiration for his theory of natural selection. It is clear that he had also studied *The Theory of Moral Sentiments*, generally considered the masterpiece of Adam Smith, who is also known as the "founding father" of economics.

The most important earlier contribution to the line of research discussed in this paper is probably that of David Hume [6]. Hume saw language, along with many other social institutions and moral rules, as a spontaneously evolved convention. Language was thus one of the first subjects of modern evolutionary explanations.

This chapter is organized in the following manner. In Sections 2 and 3 I try to justify why philosophers and economists, respectively, should be interested in an evolutionary game theory of meaning in language. Section 4 then introduces the apparatus of evolutionary game theory and gives a first example of an application to communication. Lewis's coordination problem and similar models in economic theory are discussed in Section 5. In Section 6, the concept of evolutionary stability of a strategy is applied to Lewis's model, and I show in an example that only well-behaved semantic conventions are stable. Finally, Section 7 discusses some problems with the approach as a model of the origin of natural language.

2. Why the Philosophy of Language Needs Some Economics

If economic theory is a general theory of human goal-directed action and interaction, then the human use of language is part of the proper domain of economics. Human beings imagine different possible future states of the world, desire some more than others, and act so as to bring about the latter. Economics may be seen as dealing with such actions, of which speech acts are an integral part.

It is easy to find paradoxes that arise from the treatment of meaning in a language as disassociated from human speakers and particular contexts. Nevertheless, there is a tradition in linguistics that treats statements in language in this idealized fashion. Dennett [3] notes that most of linguistics can be seen as concerned with developing a theory of *hearing* rather than speaking. But hearing is only half of language.

Existing models of speech production of this type present other problems. In artificial intelligence research, the development of speaking programs is a well-established field. Such programs typically put together natural-language sentences that describe happenings in some model domain. The purpose of this is at least partly to gain some understanding of how human beings use language. The classic example of such a program is Winograd's SHRDLU (Winograd [16]). SHRDLU moves imaginary three-dimensional colored blocks around and then comments on its own activities. Another program, called COMMENTATOR, developed by Jan Fornell (see Fornell and Sigurd [5]), which has a more appealing domain of discourse, will serve as an example for us here.

COMMENTATOR's domain is a simple model of the Garden of Eden, which is displayed to the user on the screen. The Garden has a gate, a tree, a snake, and Adam and Eve. The locations of the snake and humans can be determined by the user. The program then comments on the current situation in Eden in simple language. A typical observation might be something like: "Adam is close to the tree. Eve is close to him. The snake is also there."

It struck its creator that, while having other useful features, this and similar programs lack something as models of human speech. The program's speech is *unmotivated*. The program has no purpose in uttering its sentences. It achieves nothing by pleasing its programmer and users, apart from the ensurance of its own survival. Many programs that do not speak but, for instance, show pretty pictures, also achieve that trivial goal.

Behavior like that of COMMENTATOR or SHRDLU is unusual in human beings. If we encounter people walking around uttering arbitrary true statements about the state of the world for no particular reason other than telling the truth, we will probably think of them as being insane.

It would seem that a first step toward an understanding of language would necessarily involve recognition of the fact that humans speak for particular reasons. That is, that speech is a means toward attaining economic ends in the broadest sense of that term.

3. Why Economics Needs Some Philosophy of Language

Adam Smith noted that animals are never observed to trade and that they lack language, suggesting that these two facts are connected. There is, of course, no logical implication that language is what makes exchange possible among humans. But, along with some other evidence, it is highly suggestive. It also turns out that it is hard to understand exchange theoretically without assuming some language-like mechanism.

Economic theory has avoided this problem for a long time. In the neoclassical, i.e., mainstream, theory of general equilibrium, when supply equals demand simultaneously in the markets for all goods, exchange is implicitly assumed to follow directly from the assumption that all individuals in the economy have well-defined preferences. This means essentially that they each have an internally consistent ordering of the set of possible baskets of consumption goods. Levy [8] points out that we know from laboratory experiments with rats (see Kagel, Rachlin, Green, Battalio, Basmann, and Klemm [7]) that they behave as rational consumers in purely decision-theoretic situations, i.e., that they have consistent preferences. However, Levy also reports (personal communication) that the same experimenters in a later experiment could not get pairs of rats to trade in situations with potential mutual gains from exchange. So well-defined preferences and unexploited gains from trade are not enough to explain exchange. Preferences do not imply trade in the sense implicitly assumed by neoclassical economic theory.

Game theorists have been aware of the problem for some time. Consider the following hypothetical situation. Two persons wish to meet for dinner. There are two restaurants, A and B, and each person has to go independently to one of them. Also assume that restaurant A would be the preferred alternative for both if they were both to wind up there.

We can model this as the game with payoff matrix shown below. The left-hand number in a cell of this matrix is the *payoff* to Player 1 when the players' choices are given by the cell's row and column. The right-hand number is the corresponding payoff to Player 2. These payoffs should only be thought of as the players' rankings of the various states, not as absolute quantities of something.

| | | **Player 2** | |
		Go to A	Go to B
Player 1	Go to A	2, 2	0, 0
	Go to B	0, 0	1, 1

The basic solution concept for noncooperative games, under the condition that strategy choices are made independently, centers on a situation in which no player can change his choice unilaterally and receive a better payoff from making such a change. That is, a solution in this sense is any situation such that everybody is doing the best they can given the choices of all others. A combination of strategy choices fulfilling this requirement is called a *Nash equilibrium,* or simply an *equilibrium.*

Assuming that the players cannot randomize over their actions and that no communication takes place before the choices are made, it is easy to check that this game has two equilibria: (Go to A, Go to A) and (Go to B, Go to B). These equilibria are also *strict* in the sense that a player will receive a strictly lower payoff if he deviates. Furthermore, the (Go to A, Go to A) equilibrium is *efficient* in the sense that if some other strategy combination was played, at least one player would receive a lower payoff.

Classical game theory would essentially stop right here. The prediction for behavior in an actual situation of this type would be that one of the two equilibria will be played, but we cannot say which one. But, of course, this prediction is based on the assumption that communication between the players cannot take place. But surely game theory must predict that the players would both go to restaurant A if they could communicate. Or does it?

Clearly, if the players of this simple game have the possibility of communicating before going anywhere, intuition tells us that: (1) they will agree on a restaurant and both will then go to that restaurant, and furthermore (2) they will agree to go to restaurant A.

Achieving such coordination is crucial to the idea of having a language. If the players do not manage to rule out going to restaurant B, even when they can communicate, we would be reluctant to say that they understand each other. Yet this intuition has proved notoriously difficult to capture in formal game-theoretic terms.

The reason is that verbal utterances are most reasonably modeled as being essentially cost-free. Adding the possibility of such actions to a game only nominally affects the set of equilibria. Assume the agents in the example verbally agree to go to restaurant A. If they then both go to B, that is still a Nash equilibrium. In fact, since the messages do not affect the ultimate payoffs, any combination of messages followed by an equilibrium of the original game is an equilibrium of the game extended with communication. From the standpoint of orthodox game theory, any attempt to label the inefficient equilibria as being implausible has to be based on intuition; the formal logic of game-theoretic equilibria does not exclude them. Talk is cheap and only actions that directly affect payoffs matter.

Of course, this state of affairs is unsatisfactory. Economics is to a large degree the study of coordination problems in settings where communication is not only possible but likely. It is therefore important to have a formal theory of communication. It seems the least such a theory should do is confirm our intuition about the coordination functions of language.

The area known as cooperative game theory is based on the assumption of enforceable agreements and an implicit idea of communication. It has been felt for a long time that any available enforcement mechanisms should be explicitly incorporated into models. The explicit modeling of communication mechanisms is perhaps even more urgent, since it is far from obvious why the intuition would be correct.

4. The Evolutionary Approach

To solve this problem, we must leave the logic of standard game theory behind and embed games in the larger setting of a society whose members repeatedly encounter similar situations. This reflects the notion that languages are necessarily social conventions, not mechanisms made up specifically for some particular game situation at hand.

To use the term coined by Dawkins [2], we shall consider languages as *memes,* the analogs of genes in the world of ideas and behaviors. Instead of being concerned with the choices of rational players in the sense of standard game theory, we now look directly at the survival and stability properties of various language memes themselves in a setting where they have to compete with each other for the attention of agents.

Formally, we shall still be concerned with two-player games of the following type. There is a strategy set $S = \{s_1, s_2, \ldots, s_n\}$ that contains the actions available to each player. That is, we assume that both players share the same strategy set. There is also a payoff function $P: S \times S \to R$ that maps pairs of strategy choices into payoffs measured by real numbers.

However, we now think of these games as being played by pairs of agents from a large population who randomly meet each other. Although our interpretation of selection shall be more general than genetic selection, this is the same formal framework for which biological evolutionary game theory was first developed.

Maynard Smith [11] defines an evolutionarily stable strategy as follows:

DEFINITION. *(Evolutionary Stability). A strategy $s^\star \in S$ is said to be an evolutionarily stable strategy (ESS) if, for all $s \neq s^\star$,*

$$P(s^\star, s^\star) > P(s, s^\star), \tag{1}$$

or

$$P(s^\star, s^\star) = P(s, s^\star) \quad \text{and} \quad P(s^\star, s) > P(s, s). \tag{2}$$

That is, an ESS is a symmetric equilibrium strategy that fulfills an additional requirement when the equilibrium is not strict.

Biological evolutionary game theory models agents as genetically programmed for a particular strategy. The stability properties studied are therefore those of different behavior genes. But we can just as well apply this framework to memes that spread through the learning or imitation of agents.

The idea that a solution should be stable against a small invasion of "mutants" committed to another strategy is what motivates the ESS criterion. Assume that s is a mutant strategy followed by a small proportion ϵ of the population. Then an agent playing $s^\star$ has a higher expected payoff if

$$(1 - \epsilon)P(s^\star, s^\star) + \epsilon P(s^\star, s) > (1 - \epsilon)P(s, s^\star) + \epsilon P(s, s).$$

If ϵ is close to zero, this holds if (1) holds. If $P(s^\star, s^\star) = P(s, s^\star)$, the inequality holds if (2) holds.

While the ESS concept as defined above is static, Taylor and Jonker [13] show that every ESS is an asymptotically stable fixed point of the so-called replicator dynamic, which models asexual genetic reproduction. No such explicitly dynamic justification is currently available for the memetic interpretation used here. We have few formal models of exactly how the mechanisms of learning and imitation work in detail. But clearly the requirement that in order to be called stable a strategy should do better than any competitors that might come along is reasonable also for memetic transmission. If anything, it is too strong a requirement.

It is now easy to see that in the restaurant coordination game given above, both strategies are ESSs when no communication possibilities are present. That is, evolutionary stability in itself is still not enough to rule out the players going to the less preferred restaurant B. Robson [12] suggests that the inefficient solution of games such as this one is unstable from an evolutionary point of view if communication opportunities are available. This is because if communication possibilities are not utilized, a mutant that communicates with players of its own type in order to coordinate on a better outcome could arise and eventually reproduce to supplant the original population.

As an example, let there be a mutant M_1 that can recognize members of its own type. Let M_1 play "Go to B" against nonmutants and "Go to A" against other mutants. Assuming that the nonmutant members of the population do not recognize the mutant signal, we then get a new game as follows.

		Player 2		
		Go to A	Go to B	M_1
	Go to A	2, 2	0, 0	0, 0
Player 1	Go to B	0, 0	1, 1	1, 1
	M_1	0, 0	1, 1	2, 2

Now "Go to B" is no longer an ESS. "Go to A" is still an ESS, and the mutant is a new ESS. So as a result of the possibility of communication, every evolutionarily stable configuration of this new game induces the efficient outcome of the underlying game. Robson goes on to show that in every game that has inefficient ESSs in the absence of communication, it is possible to construct a mutant that would take the population away from inefficiency.

One problem with this is that Robson does not consider the entire set of possible mutants. Once you admit the idea of a signaling mutant, what is there to stop evolution from introducing mutants that use the signal in other ways? For the analysis to be complete, we should study the stability of efficiency-inducing mutants in a situation where any kind of mutant could arise.

Consider, for example, a mutant M_2 that sends the same signal as M_1, but always plays "Go to A." The new payoff matrix is shown at the top of the next page. The game now no longer has any ESS. What it does have are three strategies ("Go to A," M_1, and M_2) that can invade each other, but which all induce the same efficient payoff. So if we cared only about stable *payoffs*, we could say that only the efficient payoff is stable.

		Player 2			
		Go to A	Go to B	M_1	M_2
	Go to A	2, 2	0, 0	0, 0	2, 2
Player 1	Go to B	0, 0	1, 1	1, 1	0, 0
	M_1	0, 0	1, 1	2, 2	2, 2
	M_2	2, 2	0, 0	2, 2	2, 2

The strategies that induce the efficient payoff satisfy the weakening of the ESS criterion that replaces the strict inequality in (2) with a weak inequality. We call such strategies *neutrally* stable.

In Wärneryd [15], I have shown that in games having two strict Nash equilibria, where one yields both players a higher payoff than the other, only communication strategies that induce the efficient payoff are neutrally stable. This result holds only for the case of 2×2 games, however.

5. The Economics of Meaning

While the kind of models discussed in the preceding section show how simple communication mechanisms allow coordination to take place, the stable states of such systems do not seem to involve anything like a language in the normal sense of the word. The models we have looked at thus far require only recognition of a certain single signal. Natural language, on the other hand, can be thought of as an equilibrium mapping from states of the world to signals. We now turn to models that, while still highly simplified, better approximate this activity.

An early example of such a model is due to Lewis [9], who uses it to develop a theory of meaning in natural language. Lewis studies a pure coordination game, in which one player has private information about the state of the world and can send a cost-free signal. Another player observes only the signal, then taking an action that, given the state, determines the payoff for both players.

At the outset, the signals do not have meanings. One could think of them as similar to lights of different colors. Lewis argues that part of the function of natural language is to act as what he calls a *signaling system*. A signaling system is a Nash equilibrium for which the sender sends a different signal for each state and the receiver responds with the correct action. In such an equilibrium, different signals may be said to *mean* different states of the world. There are also equilibria that are not signaling systems. Lewis's theory of language is thus open-ended. It explains the equilibrium nature of well-defined languages, but not why they, rather than other equilibria, should arise.

An interesting feature of Lewis's study is that it is part of a bigger project to define a *convention* as an equilibrium of a coordination game. The theory of language as having conventional aspects is resurrected as part of this project. The notion that language can be seen as a convention had been rejected earlier because there appeared to be no notion of convention that did not involve explicit agreement in a language. And such an agreement, of course, cannot be the origin of language itself.

The interest of economists in this kind of game originates with Crawford and Sobel [1], who study a more general version of Lewis's finite communication game. A particular game will have equilibria for many sizes of partition of the set of states or types of sender. There are completely nonrevealing equilibria where all sender types send the same signal. At the other extreme are the equilibria, relevant in pure coordination games, where each sender transmits a unique signal. The multiplicity of equilibria is of course particularly troubling when the agents' interests coincide exactly. We would like to be able to argue that communication possibilities allow the agents to coordinate when there are no incentives to deviate.

The game-theoretic literature on cost-free (e.g., verbal) communication now splits into two divergent strands. One, which could be called the *semantic* approach (exemplified by Farrell [4]), starts by assuming the prior existence of a language shared by the players. Equilibrium concepts are then defined that typically require the players to stick to the established meanings of messages if they do not have a positive incentive to deviate. This, of course, "solves" the coordination problem. But it begs the question of where those meanings come from. This approach provides no clue as to why our intuitive notion of the coordinating effects of verbal communication should hold, since it assumes at the outset that it does hold.

Crawford and Sobel offer two informal solutions to the multiplicity problem. They argue that a perfectly informative equilibrium would be focal in pure coordination games. This is essentially the same argument as in the semantic approach. It is more problematic here, however, since the semantic approach implicitly assumes the prior existence of a convention where signals have meanings. There are many equivalent equilibria of this type, the only difference between them being the irrelevant assignment of signals to types. Another hypothesis is that a particular perfectly informative convention would evolve from repeated interactions as individuals adjust their strategies over time.

The latter intuition can be obtained as a formal result for a class of pure coordination games closely related to both Lewis's and Crawford and Sobel's signaling games, using the apparatus of evolutionary game theory that we have already developed. The game becomes what is called an asymmetric contest in evolutionary game theory if the players are unsure of which role, sender or receiver, they play in the game. Now it turns out that only signaling systems are evolutionarily stable in the sense of Maynard Smith. This is because evolutionary strategies that are not signaling systems, even if they are equilibrium strategies, always have best replies that are signaling systems. This means that a perfectly coordinating convention could evolve from a process of trial and error, or memetic evolution. The following section gives more formal substance to this idea.

6. The Evolution of Language

Consider the following simple game, based directly on the situation discussed by Lewis. (For a more general discussion, see Wärneryd [14].) There are two players, one called S (for Sender) and one called R (for Receiver). Let the set of states of the world be $T = \{t_1, t_2, t_3\}$. The three states are equally likely. Only player S will be informed of which state of the world obtains. In each state, there is an action that both players would like player R to take. There is a different preferred action for each state. Specifically, the payoffs are as follows.

		Actions		
		a_1	a_2	a_3
	t_1	1, 1	0, 0	0, 0
States	t_2	0, 0	1, 1	0, 0
	t_3	0, 0	0, 0	1, 1

There is no conflict of interest here. The only problem is to coordinate on the correct action. We shall assume that S can do an action σ observable to R that does not affect payoffs, but which could be used to indicate the state. The action σ is a cost-free signal, and can be thought of as a verbal utterance.

We now consider the game from a position before the state of the world has been revealed to player S. A pure strategy for S specifies a signal to send for each state he may be informed of. It is thus a function ϕ_S from the set of states to the set of signals.

Assume that the set of signals is $\{\sigma_1, \sigma_2, \sigma_3\}$. Here is an example of a strategy:

$$\phi_S(t) = \begin{cases} \sigma_1 & \text{if } t \in \{t_1, t_2\}, \\ \sigma_2 & \text{if } t = t_3. \end{cases}$$

A pure strategy for player R is a rule specifying an action to be taken for each signal that he may receive. It is a function ϕ_R from the set of signals to the set of actions $\{a_1, a_2, a_3\}$. The following is an example of such a strategy:

$$\phi_R(\sigma) = \begin{cases} a_1 & \text{if } \sigma = \sigma_1, \\ a_3 & \text{if } \sigma \in \{\sigma_2, \sigma_3\}. \end{cases}$$

An interesting economic interpretation of the game is as a noncooperative version of the generic *team problem* of Marschak and Radner [10]. A team is a group of agents whose interests coincide. The team has a leader who knows the state of the world. The team problem is to convey this information to the team, so that they may take the preferred action(s). Marschak and Radner think of the team problem as an abstract model of organizations, e.g., firms. The organizational form has two parts. The information rule the leader uses to inform the team of the state of the world corresponds to a strategy for player S. The decision function the team uses to respond to signals corresponds to a strategy for R. Marschak and Radner discuss the problem from a central planning perspective, where a single agent designs the organizational structure. The central planning solution would have S send a different signal for each type and R respond with the preferred action in each case, which of course corresponds to Lewis's signaling system notion. The present example may be seen as a further step toward realism from the central planning perspective. The leader (player S) and the team (player R) have to choose their strategies independently.

An equilibrium is a strategy choice for S that maximizes his expected payoff given R's strategy, and a strategy choice for R that maximizes his expected payoff given S's strategy.

The expected payoff of any strategy profile is easy to compute. For each t, we check whether the responses of S and R will lead to R taking the preferred action. Each such "hit" adds 1/3 to the expected payoff of each player. Consider the example strategies given above. If

$t = t_1$, S will send signal σ_1 and R will respond to the signal with action a_1, which is the preferred response of both players. If $t = t_2$, S will also send σ_1, and R will again respond with a_1, which is not the preferred response. Finally, if $t = t_3$, S will send σ_2 and R will respond with a_3, which is the preferred response. Thus there are two states out of three for which the correct action would be taken, and each agent has an expected payoff of 2/3.

The example strategy profile is an equilibrium. No strategy can give more than two hits for player S, since the range of player R's strategy only includes two actions. So it is a best reply. As for player R, note that S partitions the states into two subgroups. The best R can do is to choose an action for each subgroup that maximizes his expected payoff given that the true state belongs to that subgroup. Having observed the signal σ_1, R knows that the true state is t_1 or t_2, and that they are equally likely. Actions a_1 and a_2 are then equally good responses. If R observes the signal σ_2, he knows that the true state is t_3 with certainty, so the best response is a_3. The signal σ_3 will never be received, so a_3 is as good as any other response. Since there are alternative best replies, this is not a *strict* equilibrium.

There are several equilibria of this type in which different signals denote the cells of the partition. There is now both good and bad news. The bad news is that there are equilibria where S sends the same signal regardless of the state. The good news is that the perfectly informative, or signaling system, solution is also a noncooperative equilibrium. Following the same logic as above, it is easy to verify that, for example, the S-strategy

$$\phi_S'(t) = \begin{cases} \sigma_1 & \text{if } t = t_1 \\ \sigma_2 & \text{if } t = t_3 \\ \sigma_3 & \text{if } t = t_2 \end{cases}$$

and the R-strategy

$$\phi_R'(\sigma) = \begin{cases} a_1 & \text{if } \sigma = \sigma_1 \\ a_2 & \text{if } \sigma = \sigma_3 \\ a_3 & \text{if } \sigma = \sigma_2 \end{cases}$$

make up an equilibrium strategy profile. Here S distinguishes perfectly between states, and in each state R responds with the action preferred by both. Note that a signaling system is a strict equilibrium. For any player to alter his strategy while the other player sticks to his strategy results in a lower expected payoff. With its expected payoff of 1 for each player, a signaling system also payoff-dominates any strategy profile that is not a signaling system.

There is no reason to believe that the players would necessarily hit on a signaling system in a one-shot interaction. The equilibrium concept captures this natural indeterminacy. Strategy combinations that are not signaling systems could also work out fine, depending on which state is realized.

For every S-strategy that is imperfectly informative, there is a best reply that assigns every action to some signal. Such an R-strategy could be called perfectly responsive. For every imperfectly responsive R-strategy there is a best reply that is perfectly informative. One could therefore argue that a rational player should play a perfectly informative or responsive strategy—just in case. For each strategy that is part of a signaling system, there are strategies to which it is not a best reply, however. Signaling system strategies do not dominate other strategies.

Now think of the game as being repeated many times, and played by many different players who are paired randomly from a large population. In light of our previous discussion of evolutionary stability, it seems intuitively likely that eventually a signaling convention would evolve out of trial and error.

One way of capturing this evolutionary idea is to alter the model slightly to make it an evolutionary *asymmetric contest*. Assume there is a population of many players, who meet randomly in pairs to play the above game. Each player is equally likely to have the role of Sender or Receiver in a particular match. We may now apply the concept of evolutionary stability of a strategy that we defined earlier.

A strategy of the evolutionary game is a choice of what to do for each possible role. It has a Sender part and a Receiver part. The expected payoff of a strategy $\phi_S^1\phi_R^1$ against a strategy $\phi_S^2\phi_R^2$ equals one-half times the expected payoff of ϕ_S^1 against ϕ_R^2 plus one-half times ϕ_R^1 against ϕ_S^2. The expected payoff of the signaling system strategy $\phi_S'\phi_R'$ above against itself is thus equal to 1. No other strategy is a best reply to $\phi_S'\phi_R'$. For another strategy to be a best reply, either its S part would have to do at least as well against ϕ_R' as ϕ_S', or its R part as least as well against ϕ_S' as ϕ_R', or both. This is not possible. Therefore $\phi_S'\phi_R'$ is a strict symmetric equilibrium strategy and fulfills the ESS criterion. If the population of players has adopted it, it cannot be invaded by another strategy.

In Wärneryd [14] I prove in a more general setting that only signaling systems are stable. It is easy to see that the first equilibrium of the example does not make for an ESS of the evolutionary game. Note that ϕ'_S is an alternative best reply to ϕ_R, and ϕ'_R an alternative best reply to ϕ_S. The ESS criterion now requires that $\phi_S\phi_R$ should do better against $\phi'_S\phi'_R$ than the latter against itself. But $\phi'_S\phi'_R$ against itself payoff-dominates all strategy profiles that are not signaling systems. So $\phi_S\phi_R$ is not an ESS. A population playing the suboptimal strategy can be invaded by $\phi'_S\phi'_R$. In general, given enough time we would expect some players to hit on a signaling system that is also a best reply against a suboptimal strategy that dominates the population. The signaling system would then evolve to displace the suboptimal strategy, thus becoming the ruling convention. That each symmetric equilibrium that is not a signaling system has an alternative best reply that is a signaling system provides the basis for the more general proof.

7. Concluding Remarks

In the framework of a very simple game-theoretical model, I have argued that certain conventional aspects of meaning in natural language can be seen as evolutionarily stable ways of sending and responding to signals.

We may say therefore that we understand how languages can arise that permit coordination to happen in situations where there is no conflict of interest. But of course language is also used in situations where there *is* conflict of interest. The possibility of *lying* is an important part of natural language use. In the models discussed in this paper, lying is an undefined concept and therefore meaningless.

It is easy to see why the framework adopted here cannot deal with lying. The activity of lying is only defined relative to some established semantic convention. To lie is to make manipulative use of the fact that people expect certain things to be true upon hearing certain messages. But at the starting point of our analysis, messages have no meanings. So perhaps we must adopt what was called the semantic approach above after all, assuming from the beginning that messages already have some meaning. But then for lying to take place, there must be an incentive to do it. The problem is that if the situations for which the semantic convention have evolved are such that there is an incentive to misuse the convention, then we cannot explain how the convention could have emerged in the first place. So it seems pretty dubious to assume its existence. In situations of pure conflict of interest, where there is an incentive to lie, no particular system of associating meanings with messages could be stable. This is because if such a convention was in place at some point in time, it would be exploited and society would immediately drift away from the convention.

Of course, this has to do with messages being specific to a particular game in our setup. In the real world, the same basic building-blocks of language can be used in all sorts of different interaction situations. So it would appear that we need a theory that allows the same messages to be used in different games. Lying is possible if messages that have attained meaning in a coordination game context can later be carried over to games of conflict. It is not immediately obvious how this could be modeled in a convincing manner, however.

Acknowledgments

I thank Jan Fornell, Ronald Heiner, David Levy, John Maynard Smith, and Brian Skyrms for inspiring discussions on the issues of this paper. The research reported here was sponsored by the Swedish Council for Research in the Humanities and Social Sciences.

References

[1] Crawford, V. and J. Sobel. "Strategic Information Transmission." *Econometrica,* 50(6):1431–1451, 1982.

[2] Dawkins, R. *The Selfish Gene.* Oxford: Oxford University Press, 1976.

[3] Dennett D. *Consciousness Explained.* Boston: Little, Brown, 1991.

[4] Farrell, J. "Communication, Coordination and Nash Equilibrium." *Economics Letters,* 27:209–214, 1988.

[5] Fornell, J. and B. Sigurd. *COMMENTATOR.* Working paper, University of Lund, Department of Linguistics, 1983.

[6] Hume, D. *A Treatise of Human Nature.* Oxford: Clarendon Press, 1967 (original: 1740).

[7] Kagel, J., H. Rachlin, L. Green, R. Battalio, R. Basmann, and W. Klemm. "Experimental Studies of Consumer Demand Behavior Using Laboratory Animals." *Economic Inquiry,* 13, 1975.

[8] Levy, D. *The Economic Ideas of Ordinary People: From Preferences to Trade.* London: Routledge, 1992.

[9] Lewis, D. *Convention. A Philosophical Study.* Oxford: Basil Blackwell, 1969.

[10] Marschak, J. and R. Radner. *Economic Theory of Teams.* New Haven and London: Yale University Press, 1972.

[11] Maynard Smith, J. *Evolution and the Theory of Games.* Cambridge: Cambridge University Press, 1982.

[12] Robson, A. "Efficiency in Evolutionary Games: Darwin, Nash and the Secret Handshake." *Journal of Theoretical Biology,* 144:379–396, 1990.

[13] Taylor, P. and L. Jonker. "Evolutionarily Stable Strategies and Game Dynamics." *Mathematical Biosciences,* 40:145–156, 1978.

[14] Wärneryd, K. "Cheap Talk, Coordination, and Evolutionary Stability." 1993, forthcoming in *Games and Economic Behavior.*

[15] Wärneryd, K. "Evolutionary Stability in Unanimity Games with Cheap Talk." *Economics Letters,* 36:375–378, 1991.

[16] Winograd, T. *Understanding Natural Language.* New York: Academic Press, 1972.

Index

Abstraction:
in biological information processing,
260, 297, 302
and free will, 135
in mechanistic information
processing, 294
in symbolic languages, 229
Adaptation:
and behavior, 314, 349–351, 356–357
and chimera formation, 344, 348,
351–354, 357, 358
and cooperation vs. competition,
352–353
and evolution, 345–346, 354–356, 357,
358
preadaptation, 134–135
and RNA populations, 366–368
and shape space, 378
and survival, 345, 346, 352, 353, 354,
356, 357
Adaptive behavior, 314, 349–351,
356–357
Adaptors, 37–40
accuracy of, 38–40, 43, 47, 48
defined, 21–22
and error catastrophe, 21–22
Agglomeration, 2, 6–8
competition in, 11
cooperation in, 11
and evolutionary systems, 11–12
and probability, 6–8, 9, 10
properties of, 11–12
Altruism:
and group selection, 64
and kin selection, 64, 65
in Nature, 64–66
and parental manipulation, 64–65
reciprocal, 65–66
Amino acids, 21–22, 37–40
Analogies, 118–120
Anthropomorphism, 118, 122
Aphasia, 147–150
Arecchi's equations, 300, 331
Artificial languages, *see also* Symbolic
languages
Markovian order, 279–284, 286
and Zipf's Law, 269, 275
Artificial Life, 125
Artificial life, 126–128

Assignments, relational network:
defined, 385, 386
relation-restoring, 385, 386–387,
390–392
Asynchronous updating, 51, 52, 56
Attractors, *see also* Basins of attraction;
Dynamical memories; Strange attractors,
chaotic
chaotic vs. periodic, 170, 193–194, 196
characteristics for information
processing, 291, 309–310, 311, 313
coupling, 170, 193–194, 197, 316, 317
diffusively coupled, 193–194, 197
and fault-tolerance ability, 313
and information compression, 238,
239, 299, 302, 306
and information storage, 238–239,
302
leakage among, 308
in linguistic model, 235, 236–247
as memories, 299–302
multifractal, 239–247, 260, 285,
325–326
multiple, 238
periodic vs. quasiperiodic, 179–180,
195
and quantum mechanics, 311–312
stable, 256, 303, 304
steady state, 237
and symbolic languages, 230, 232
and time scale, 308–309
Australopithecine, speech capabilities of,
153, 155
Autocorrelation function, 370–371
Axelrod, Robert, 73, 74–76

Bacterial populations:
bacteria-virus coevolution, 171, 197,
199, 200
cell division in, 172, 208–219
chaotic dynamics in, 169–172
and complexity, 181–192
and diffusive coupling, 192–197
evolution in, 171, 197–208
immune systems in, 167, 168, 173–174
models of, 170–219
modified vs. unmodified, 199–208
and rate of dilution, 174, 176–181
replication control in, 172, 208–219